DATE DUE

DEMCO 38-296

Metal Forming Handbook

Springer

Berlin
Heidelberg
New York
Barcelona
Budapest
Hong Kong
London
Milan
Paris
Santa Clara
Singapore
Tokyo

R

METAL FORMING
HANDBOOK

Springer

Consulting editor: Professor Taylan Altan
Director, Engineering Research Center for Net Shape Manufacturing
The Ohio State University, USA

Cataloging-in-Publication Data applied for

Die Deutsche Bibliothek – CIP-Einheitsaufnahme

Metal forming handbook / Schuler. – Berlin ; Heidelberg ; New York ; Barcelona ;
Budapest ; Hong Kong ; London ; Milan ; Paris ; Santa Clara ; Singapore ; Tokyo :
Springer, 1998
  Dt. Ausg. u. d. T.: Handbuch der Umformtechnik
  ISBN 3-540-61185-1

ISBN 3-540-61185-1 Springer-Verlag Berlin Heidelberg New York

© Springer-Verlag Berlin Heidelberg 1998
Printed in Germany

Cover design by MEDIO, Berlin
Layout design and data conversion by K.-D. Grotz, MEDIO, Berlin
Printing and binding by Konrad Triltsch Druck- und Verlagsanstalt, Würzburg
SPIN: 10514857   3020/ 62/ 5 4 3 2 1 0 – Printed on acid-free paper.

Preface

Following the long tradition of the Schuler Company, the Metal Forming Handbook presents the scientific fundamentals of metal forming technology in a way which is both compact and easily understood. Thus, this book makes the theory and practice of this field accessible to teaching and practical implementation.

The first Schuler "Metal Forming Handbook" was published in 1930. The last edition of 1966, already revised four times, was translated into a number of languages, and met with resounding approval around the globe.

Over the last 30 years, the field of forming technology has been radically changed by a number of innovations. New forming techniques and extended product design possibilities have been developed and introduced. This Metal Forming Handbook has been fundamentally revised to take account of these technological changes. It is both a textbook and a reference work whose initial chapters are concerned to provide a survey of the fundamental processes of forming technology and press design. The book then goes on to provide an in-depth study of the major fields of sheet metal forming, cutting, hydroforming and solid forming. A large number of relevant calculations offers state of the art solutions in the field of metal forming technology. In presenting technical explanations, particular emphasis was placed on easily understandable graphic visualization. All illustrations and diagrams were compiled using a standardized system of functionally oriented color codes with a view to aiding the reader's understanding.

It is sincerely hoped that this Handbook helps not only disseminate specialized knowledge but also provides an impetus for dialogue between the fields of production engineering, production line construction, teaching and research.

This Handbook is the product of dedicated commitment and the wide range of specialized knowledge contributed by many employees of the SCHULER Group in close cooperation with Prof. Dr.-Ing. H. Hoffmann and Dipl.-Ing. M. Kasparbauer of the *utg*, Institute for Metal Forming and Casting at the Technical University of Munich. In close cooperation with the SCHULER team, they have created a solid foundation for the practical and scientific competence presented in this Handbook. We wish to offer our sincere thanks and appreciation to all those involved.

Goeppingen, March 1998

Schuler GmbH
Board of Management

Contributors

ADAM, K., Dipl.-Ing. (FH), SMG Süddeutsche Maschinenbau GmbH & Co

BAREIS, A., Dipl.-Ing. (FH), Schuler Pressen GmbH & Co

BIRZER, F., Prof. Dipl.-Ing., Feintool AG

BLASIG, N., Dipl.-Ing. (FH), Schleicher Automation GmbH & Co

BRANDSTETTER, R., Dipl.-Ing. (FH), Schuler Pressen GmbH & Co

BREUER, W., Dipl.-Ing., Schuler Pressen GmbH & Co

FRONTZEK, H., Dr.-Ing., Schuler GmbH

HOFFMANN, H., Prof. Dr.-Ing., Lehrstuhl für Umformtechnik und Gieße-
reiwesen, Technische Universität München

JAROSCH, B., Dipl.-Ing. (FH), Schuler Pressen GmbH & Co

KÄSMACHER, H., SMG Engineering für Industrieanlagen GmbH

KASPARBAUER, M., Dipl.-Ing., Lehrstuhl für Umformtechnik und Gießerei-
wesen, Technische Universität München

KELLENBENZ, R., Dipl.-Ing. (FH), Schuler Pressen GmbH & Co

KIEFER, A., Dipl.-Ing. (BA), GMG Automation GmbH & Co

KLEIN, P., Gräbener Pressensysteme GmbH & Co. KG

KLEMM, P., Dr.-Ing., Schuler Pressen GmbH & Co

KNAUß, V., Dipl.-Ing. (FH), Schuler Werkzeuge GmbH & Co

KOHLER, H., Dipl.-Ing., Schuler Guß GmbH & Co

KÖRNER, E., Dr.-Ing., Schuler Pressen GmbH & Co

KUTSCHER, H.-W., Dipl.-Ing. (FH), Gräbener Pressensysteme GmbH & Co. KG

LEITLOFF, F.-U., Dr.-Ing., Schäfer Hydroforming GmbH & Co

MERKLE, D., Schuler Pressen GmbH & Co

OSEN, W., Dr.- Ing., SMG Süddeutsche Maschinenbau GmbH & Co

PFEIFLE, P., Dipl.-Ing. (FH), Schuler Pressen GmbH & Co

REITMEIER, C., Dipl.-Ing., Schäfer Hydroforming GmbH & Co

REMPPIS, M., Ing. grad., Schuler Pressen GmbH & Co

ROSENAUER, K., Dipl.-Ing. (FH), Schuler Werkzeuge GmbH & Co

SCHÄFER, A.W., Schäfer Hydroforming GmbH & Co

SCHMID, W., Dipl.-Ing. (FH), Schuler Werkzeuge GmbH & Co

SCHMITT, K. P., Schuler Pressen GmbH & Co

SCHNEIDER, F., Dipl.-Ing. (FH), Schuler Pressen GmbH & Co

SIMON, H., Dr.-Ing., Schuler Werkzeuge GmbH & Co

STEINMETZ, M., Dipl.-Wirt.-Ing., SMG Engineering für Industrieanlagen GmbH

STROMMER, K., Dipl.-Ing. (FH), Schuler Pressen GmbH & Co

VOGEL, N., Dipl.-Ing., Schleicher Automation GmbH & Co

WEGENER, K., Dr.-Ing., Schuler Pressen GmbH & Co

Contents

Index of formula symbols

α	rib angle, bending angle,	°
	clearance angle,	°
	die opening angle,	°
	corner angle for blanking	°
α_1	required bending angle	°
α_2	desired bending angle	°
β	draw ratio,	
	corner angle when bending	°
β_{tot}	total draw ratio	
β_{max}	maximum draw ratio	
ε	elongation, starting measurement	
ε_A	relative cross section change	%
η	efficiency	
η_A	degree of utilization of the sheet metal,	
	utilization force	
η_F	forming efficiency factor	
μ	coefficient of friction	
$\dot{V}$	volumetric flow	1/s
σ	stress	N/mm^2
σ_m	mean stress	N/mm^2
σ_{max}	largest stress	N/mm^2
σ_{md}	mean comparative stress	N/mm^2
σ_{min}	smallest stress	N/mm^2
σ_N	normal contact stress	N/mm^2
σ_r	radial stress	N/mm^2

σ_t	tangential stress	N/mm^2
σ_v	comparative stress, effective stress	N/mm^2
σ_z	critical buckling stress	N/mm^2
σ_1	greatest principle stress	N/mm^2
σ_2	mean principle stress	N/mm^2
σ_3	smallest principle stress	N/mm^2
τ_R	frictional shear stress	N/mm^2
φ	degree of deformation, strain, logarithmic/true strain	
$\dot{\varphi}$	strain rate, deformation rate, deformation speed	
φ_B	fracture strain	
φ_g	principle deformation	
$\varphi_1, \varphi_2, \varphi_3$	deformation in main directions	
A	surface	mm^2
a	blanking plate measurement, rim width, leg length during bending, slot width	mm mm
A_0	initial surface, surface of blank cross section	mm^2
a_1	blanking punch dimension	mm
A_1	surface of blank cross section, end surface	mm^2
A_5, A_{80}	ultimate elongation	%
A_G	ejector surface, surface area under pressure by the ejector	mm^2
a_R	space between the rows	mm
A_S	sheared surface	mm^2
A_{St}	cross section of the punch, surface area of hole punch	mm^2 mm^2
A_Z	blank surface, area of the blank	mm^2
b	web width, leg length during bending, strip width, section width	mm mm
B	deflection	mm
b_A	shell-shaped tear width	mm
b_E	die roll width	mm
b_S	strip width	mm
c	material coefficient	
D	blank diameter, plate diameter, mandrel diameter	mm mm

d	inner diameter, hole diameter,	mm
	(perforating) punch diameter	mm
d′	inside diameter of bottom die	mm
d_0	blank diameter, initial billet diameter	mm
d_1	diameter of the draw punch in the first drawing	mm
	operation, punch diameter, end diameter	mm
d_2	upper cup diameter, outside diameter	mm
d_3	outside flange diameter	mm
e	off-center position of force application	mm
E	elasticity module	N/mm^2
F	force	kN
f_1, f_2, f_3	offset factors	
F_A	ejection force	kN
F_B	blank holder force	kN
F_b	bending force	kN
F_G	counterforce	kN
F_{Ga}	ejection force	kN
F_{Ges}	total machine force	kN
F_N	normal force	kN
F_{N0}	rated press force, nominal load	kN
F_R	radial tension force, friction force, vee-ring force	kN
F_{Ra}	stripping force	kN
F_{Re}	reaction force	kN
F_S	blanking force for punch with flat ground	kN
	work surface, shearing force	kN
F_{ST}	slide force	kN
F_t	tangential compression force	kN
F_U	pressing force, forming force,	kN
	maximum drawing force	kN
g	gravitational acceleration	m/s^2
h	forming path, drawing stroke, distance, height,	mm
	punch displacement;	mm
	lubrication gap	μm
H	plate thickness	mm
h_0	initial billet height, height of blank	mm

h_1	final height of a body after compression	mm
h_1'	intermediate height, height of the truncated cone	mm
h_2	cup height	mm
h_E	die roll height	mm
h_G	flash height	mm
h_R, H_R	height of vee-ring	mm
h_{S1}	smooth cut section in case of fracture	%
h_{S2}	minimal smooth cut section in case of shell-shaped fracture	%
i	side cutter scrap	mm
k	correction factor	
$k_{2\alpha}$	correction coefficient (angle)	
k_f	flow stress	N/mm^2
k_{f0}	flow stress at the start of the forming process	N/mm^2
k_{f1}	flow stress towards the end of the forming process	N/mm^2
k_{fm}	mean stability factor	N/mm^2
k_h	correction coefficient (height)	
k_R	springback factor	
k_S	shearing resistance, shearing strength, relative blanking force	N/mm^2 N/mm^2
k_w	deformation resistance	N/mm^2
k_{wm}	mean deformation resistance	N/mm^2
l	rib length	mm
L	strip length, mandrel length	mm
l_a	rim length	mm
l_e	web length, strip length	mm
l_R	length of vee-ring	mm
l_S	length of sheared contour cut	mm
m	mass, module of a gear	kg
M_x	eccentric moment of load around the x axis	kNm
M_y	eccentric moment of load around the y axis	kNm
P	performance, drive power	W, kW
p	pressure	N/mm^2
p_B	specific blank holder pressure	N/mm^2

p_G	average compressive stress on the counterpunch	N/mm^2
p_i	internal pressure	N/mm^2
p_j	compressive stress at the wall of the bottom die	N/mm^2
p_m	mean (hydraustatic) pressure	N/mm^2
p_{St}	average compressive stress on the punch, average forming pressure	N/mm^2 N/mm^2
q_G	specific counterforce, counterpressure	N/mm^2
r	radius	mm
R	corner radius	mm
r_a	external radius of an inside contour	mm
R_a	external radius of an outside contour	mm
R_{eL}	lower yield strength	N/mm^2
$R_{p0,2}$	compression limit	N/mm^2
r_i	inside bending radius, internal radius of an inside contour	mm mm
R_i	internal radius of an outside contour	mm
r_{i1}	inside radius at the die	mm
r_{i2}	inside radius at the workpiece	mm
R_m	tensile strength of the material	N/mm^2
R_t	surface roughness	μm
R_w	roller radius	mm
R_z	surface roughness	μm
s	sheet metal thickness, wall thickness, blank thickness	mm mm
s_R	position of the center of force (x_s- und y_s: coordinates of the force), center of gravity	mm
t	pitch	mm
t_w	roller pitch	mm
u	blanking clearance	mm
U	speed/stroking speed, cut contour circumferences, punch perimeter	1/min mm
v	counterbalance value during bending, compensation factor	mm mm
V	feed step, volume	mm mm^3

V_0	starting volume, overall volume, part volume	mm^3
V_1	intermediate volume, compensation value	mm^3
V_1'	intermediate volume, compensation value	mm^3
V_2	intermediate volume, compensation value	mm^3
V_d	volume displaced during deformation	mm^3
W	deformation/forming work	Nm, kNm J, kJ
w	die width	mm
W_b	bending work	Nm
W_d	drawing work on double-action presses, draw energy of a double-action press	Nm, kNm Nm, kNm
W_e	drawing work on single-action presses, draw energy of a single-action press	Nm, kNm Nm, kNm
w_{id}	referenced deformation work, specific forming work	Nmm/mm^3
W_N	nominal work for continuous stroking	Nm, kNm
W_S	blanking work, blanking energy, shearing work	Nm, kNm
x	correction factor	
x_s	location of the resulting blanking force in the x direction	mm
y_s	location of the resulting blanking force in the y direction	mm
z	no. of teeth of a gear, no. of workpieces	
z_w	roller feed value	mm

1 Introduction

Technology has exerted a far greater influence on the development of our past than most history books give credit for. As late as the 19th century, craftmanship and technology were practically synonymous. It is only with the advent of mechanisation – through the use of machines – that the term technology took on a new meaning of its own.

Today, technology is one of the bastions of our modern lifestyle and the basis for our prosperity, in which metal forming technology plays a central role. Alongside the manufacture of semi-finished products through rolling, wire drawing and extrusion, the production of discrete components using sheet metal and solid forming techniques is of major significance. Its fields of application range from automotive engineering, production line and container construction through to the building construction, household appliance and packaging industries.

The machine tool, with its capacity to precisely guide and drive one or more tools for the machining of metal, has become a symbol of economic metalworking. In the past, the work processes typically seen in metal forming technology used to be executed in a series of individual operations on manually operated machine tools. Today, however, automatic production cells and interlinked individual machines through to the compact production line with integrated feed, transport, monitoring and finished part stacking systems are the state of the art. Developments in this field created the technological basis to allow the benefits of formed workpieces, such as a more favorable flow line, optimum strength characteristics and low material and energy input, to be combined with higher production output, dimensional control and surface quality.

As a reputed German manufacturer of machine tools, the company SCHULER has played a determining role in this development over a period of more than 150 years: From the manually operated sheet metal shear to the fully automatic transfer press for complete car body side panels.

Over the millenniums, the handworking of metal by forming reached what may still today be considered a remarkable degree of skill, resulting in the creation of magnificent works in gold, silver, bronze, copper and brass. It was only in around 1800 that iron sheet produced in rolling plants began to find its way into the craftsmen's workshops, requiring completely new processing techniques: In contrast to non-ferrous metals, the much harder and more brittle new material could be more economically worked with the aid of machines.

In 1839, master locksmith *Louis Schuler* founded a modest workshop comprising primarily a tinsmith's shop, as well as a blacksmith's forge and a smithy. Driven by his Swabian business sense, he considered the possibilities opened up by the newly available, cheaper iron sheet. He was quick to realize that the increased input required in terms of physical strength and working time, and thus the manufacturing costs involved in producing the finished article were far too high to benefit from the favorable price of the iron sheet itself. Step by step, *Louis Schuler* accordingly began to replace manual work processes by mechanical fixtures and devices. He began to mechanise his workshop with sheet shears, bending machines and press breaks, which were considerable innovations in those days.

Inspired by the World Exhibition in London in 1851, *Louis Schuler* decided to concentrate his activities entirely on producing machines for sheet metal working. His production range was continuously extended to include sheet metal straightening machines, metal spinning and levelling benches, eccentric presses, spindle presses, turret, crank and drawing presses, both mechanically and hydraulically powered, notching presses as well as cutting and forming tools and dies. As early as 1859, he exported his first sheet metal forming machines.

At the end of the 1870s, *Schuler* registered his first patent for "Innovations in punching dies, shears and similar". In 1895, he patented "Hydraulic drawing presses with two pistons fitted into each other", and in the same year was also awarded first prize at the Sheet Metal Industry Trade Exhibition in Leipzig. With expansion of the production program, the workforce as well as the company premises had undergone continuous growth *(Fig. 1.1)*. The Schuler machine tool company

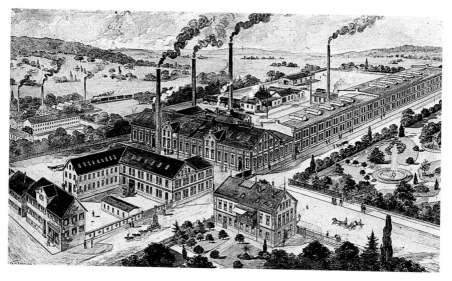

▲ **Fig. 1.1** L. Schuler, Machine tool factory and foundry, Goeppingen, around 1900

was one of the foresighted enterprises of the day to pioneer the process of differentiation taking place in the field of machine tool engineering. As a supplier of machines and production lines for industrial manufacture – in particular series production – the company's reputation increased rapidly.

The increasing export volume and a consistent process of diversification in the field of forming technology led to an early process of globalisation and to the development of the international SCHULER Group of Companies.

The SCHULER Group's process of globalisation got under way at the beginning of the sixties with the founding of foreign subsidiaries. Today, SCHULER runs not only eight manufacturing plants in Germany but also additional five production facilities in France, the US, Brazil and China. Alongside its world-wide network of sales agencies, SCHULER has also set up its own sales and service centers in Spain, India, Malaysia and Thailand.

An internationally-based network of production facilities coordinated from the parent plant in Goeppingen permits rapid response to the changes taking place in the targeted markets. Production in overseas locations brings about not only a reduction in costs but also creates major strategic benefits by increasing "local content" and so ensuring an improved market position.

The North and South American markets are supplied locally. The NAFTA area is coordinated by Schuler Inc. in Ohio, while South America's common market, the Mercosul, is supervised from Brazil. The high standard of quality achieved by the SCHULER plant in Brazil has opened up even the most demanding markets.

In the growing market of China, the SCHULER Group runs two joint venture corporations in cooperation with Chinese partners for the manufacture of mechanical presses and hydraulic presses.

Today, we stand on the threshold to a new millennium marked by increasing market globalisation and rapidly changing organizational and producing structures. Under these rapidly changing conditions, it is SCHULER's workforce which remains the single most important determining factor between success and failure. The technological orientation of the staff provides the innovative impetus which will secure the company's development as it moves into the 21st century.

This Metal Forming Handbook reflects the technical competence, the rich source of ideas and the creativity of the SCHULER Group's workforce. The book takes an in-depth look at the pioneering stage of development reached by today's presses and forming lines, and at related production processes, with particular emphasis on the development of control engineering and automation. Developments in the classical fields of design, mechanical engineering, dynamics and hydraulics are now being influenced to an ever greater degree by more recently developed technologies such as CAD, CAM, CIM, mechatronics, process simulation and computer-aided measurement and process control technology. In today's environment, the main objective of achieving enhanced product quality and productivity is coupled with lower investment and operating costs. In addition, questions of reliability, uptime, accident prevention, process accounting, economical use of resources and environmental conservation play also a central role.

In view of the fundamental importance of metal forming technology today, this Handbook offers the reader a reference work whose usefulness stretches to practically every branch of industry. The book provides an in-depth analysis of most of the important manufacturing technologies as a system comprising the three elements: process, production line and product.

2 Basic principles of metal forming

■ 2.1 Methods of forming and cutting technology

■ 2.1.1 Summary

As described in DIN 8580, manufacturing processes are classified into six main groups: primary shaping, material forming, dividing, joining, modifying material property and coating *(Fig. 2.1.1)*.
Primary shaping is the creation of an initial shape from the molten, gaseous or formless solid state. Dividing is the local separation of material. Joining is the assembly of individual workpieces to create subassemblies and also the filling and saturation of porous workpieces. Coating means the application of thin layers on components, for example by galvanization, painting and foil wrapping. The purpose of modifying material property is to alter material characteristics of a workpiece

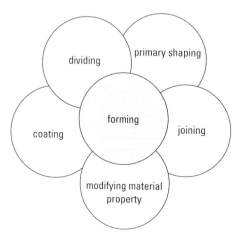

◀ **Fig. 2.1.1**
Overview of production processes

to achieve certain useful properties. Such processes include heat treatment processes such as hardening or recrystallization annealing.

Forming – as the technology forming the central subject matter of this book – is defined by DIN 8580 as manufacturing through the three-dimensional or plastic modification of a shape while retaining its mass and material cohesion. In contrast to deformation, forming is the modification of a shape with controlled geometry. Forming processes are categorized as chipless or non-material removal processes.

In practice, the field of "forming technology" includes not only the main category of forming but also subtopics, the most important of which are *dividing* and *joining through forming (Fig. 2.1.2)*. Combinations with other manufacturing processes such as laser machining or casting are also used.

■ 2.1.2 Forming

Forming techniques are classified in accordance with DIN 8582 depending on the main direction of applied stress *(Fig. 2.1.3)*:

– forming under compressive conditions,
– forming under combined tensile and compressive conditions,
– forming under tensile conditions,
– forming by bending,
– forming under shear conditions.

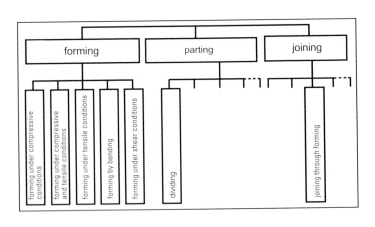

◀ **Fig. 2.1.2**
Production
processes
used in
the field
of forming
technology

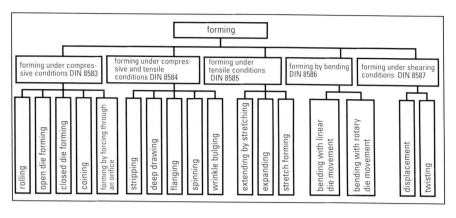

▲ **Fig. 2.1.3** Classification of production processes used in forming in accordance with DIN 8582

The DIN standard differentiates between 17 distinct forming processes according to the relative movement between die and workpiece, die geometry and workpiece geometry *(Fig. 2.1.3)*.

Forming under compressive conditions
Cast slabs, rods and billets are further processed to semi-finished products by *rolling*. In order to keep the required rolling forces to a minimum, forming is performed initially at hot forming temperature. At these temperatures, the material has a malleable, paste-like and easily formable consistency which permits a high degree of deformation without permanent work hardening of the material. Hot forming can be used to produce flat material of the type required for the production of sheet or plate, but also for the production of pipe, wire or profiles. If the thickness of rolled material is below a certain minimum value, and where particularly stringent demands are imposed on dimensional accuracy and surface quality, processing is performed at room temperature by cold rolling. In addition to rolling semi-finished products, such as sheet and plate, gears and threads on discrete parts are also rolled under compressive stress conditions.

Open die forming is the term used for compressive forming using tools which move towards each other and which conform either not at all or only partially to the shape of the workpiece. The shape of the workpiece is created by the execution of a free or defined relative movement

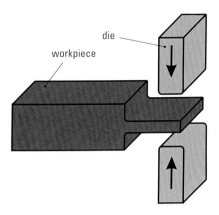

◀ **Fig. 2.1.4** Open die forming

between the workpiece and tool similar to that used in the hammer forging process *(Fig. 2.1.4)*.

Closed die forming is a compressive forming process, where shaped tools (dies) move towards each other, whereby the die contains the workpiece either completely or to a considerable extent to create the final shape *(Fig. 2.1.5)*.

Coining is compressive forming using a die which locally penetrates a workpiece. A major application where the coining process is used is in manufacturing of coins and medallions *(Fig. 2.1.6)*.

Forming by forcing through an orifice is a forming technique which involves the complete or partial pressing of a material through a forming die orifice to obtain a reduced cross-section or diameter. This technique includes the subcategories *free extrusion, extrusion of semi-finished products and extrusion of components* (cf. Sect. 6.1).

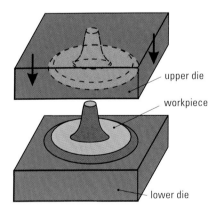

◀ **Fig. 2.1.5** Closed die forming

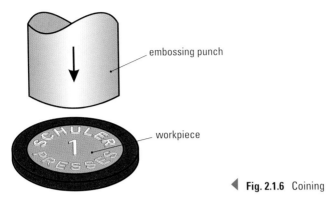

◀ **Fig. 2.1.6** Coining

During *free extrusion*, a billet is partially reduced without upsetting or bulging of the non-formed portion of the workpiece (*Fig. 2.1.7* and cf. Sect. 6.5.4). Free extrusion of hollow bodies or tapering by free extrusion involves partial reduction of the diameter of a hollow body, for example a cup, a can or pipe, whereby an extrusion container may be required depending on the wall thickness.

In extrusion of semi-finished products a heated billet is placed in a container and pushed through a die opening to produce solid or hollow extrusions of desired cross-section.

Cold extrusion of discrete parts involves forming a workpiece located between sections of a die, for example a billet or sheet blank (cf. Sects. 6.5.1 to 6.5.3 and 6.5.7). In contrast to free extrusion, larger deformations are possible using the extrusion method.

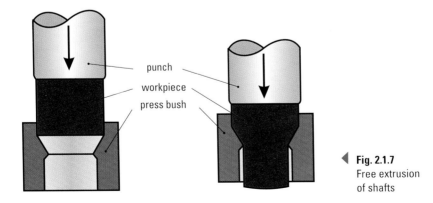

◀ **Fig. 2.1.7**
Free extrusion
of shafts

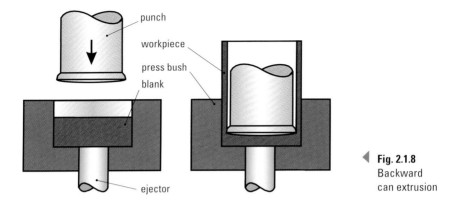

Fig. 2.1.8
Backward
can extrusion

Extrusion is used for the manufacture of semi-finished items such as long profiles with constant cross sections. Cold extrusion is used to produce individual components, e.g. gears or shafts. In both methods, forming takes place using either rigid dies or active media. In addition, a difference is drawn depending on the direction of material flow relative to the punch movement – i.e. forwards, backwards or lateral – and the manufacture of solid or hollow shapes *(cf. Fig. 6.1.1)*. Based on the combination of these differentiating features, in accordance with DIN 8583/6 a total of 17 processes exist for extrusion. An example of a manufacturing method for cans or cups made from a solid billet is backward cup extrusion *(Fig. 2.1.8)*.

Forming under combination of tensile and compressive conditions
Drawing is carried out under tensile and compressive conditions and involves drawing a long workpiece through a reduced die opening. The most significant subcategory of drawing is *strip drawing*. This involves drawing the workpiece through a closed drawing tool (drawing die, lower die) which is fixed in drawing direction. This allows the manufacture of both solid and hollow shapes. In addition to the manufacture of semi-finished products such as wires and pipes, this method also permits the production of discrete components. This process involves reducing the wall thickness of deep-drawn or extruded hollow cups by ironing, and has the effect of minimizing the material input, particularly for pressure containers, without altering the dimensions of the can bottom *(Fig. 2.1.9* and cf. Sect. 6.5.5).

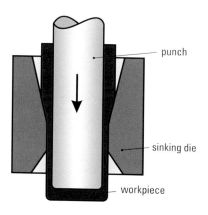

punch

sinking die

workpiece

◀ **Fig. 2.1.9** Can ironing

Deep drawing is a method of forming under compressive and tensile conditions whereby a sheet metal blank is transformed into a hollow cup, or a hollow cup is transformed into a similar part of smaller dimensions without any intention of altering the sheet thickness (cf. Sect. 4.2.1). Using the *single-draw deep drawing technique* it is possible to produce a drawn part from a blank with a single working stroke of the press *(Fig. 2.1.10)*.

In case of large deformations, the forming process is performed *by means of redrawing,* generally using a number of drawing operations. This can be performed in the same direction by means of a telescopic punch *(Fig. 2.1.11)* or *by means of reverse drawing,* which involves the second punch acting in opposite direction to the punch motion of the previous deep-drawing operation *(Fig. 2.1.12)*.

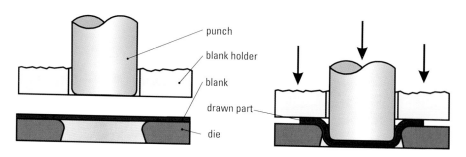

punch

blank holder

blank

drawn part

die

▲ **Fig. 2.1.10** Single-draw deep drawing with blank holder

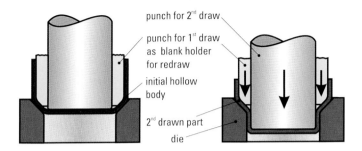

punch for 2nd draw

punch for 1st draw
as blank holder
for redraw

initial hollow
body

2nd drawn part

die

▲ **Fig. 2.1.11** Multiple-draw deep drawing with telescopic punch

The most significant variation of deep drawing is done with a rigid tool *(Fig. 2.1.10)*. This comprises a punch, a bottom die and a blank holder, which is intended to prevent the formation of wrinkles as the metal is drawn into the die. In special cases, the punch or die can also be from a soft material.

There are deep drawing methods which make use of active media and active energy. Deep drawing using active media is the drawing of a blank or hollow body into a rigid die through the action of a medium. Active media include formless solid substances such as sand or steel balls, fluids (oil, water) and gases, whereby the forming work is performed by a press using a method similar to that employed with the rigid tools. The greatest field of application of this technique is *hydromechanical drawing,* for example for the manufacture of components from stainless steel *(Fig. 2.1.13,* cf. Sects. 4.2.4 and 4.2.5).

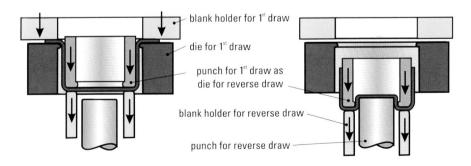

blank holder for 1st draw

die for 1st draw

punch for 1st draw as
die for reverse draw

blank holder for reverse draw

punch for reverse draw

▲ **Fig. 2.1.12** Reverse drawing

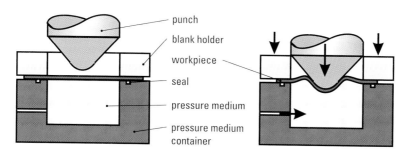

▲ **Fig. 2.1.13** Hydromechanical deep drawing

Flanging is a method of forming under combined compressive and tensile conditions using a punch and die to raise closed rims (flanges or collars) on pierced holes *(Fig. 2.1.14)*. The holes can be on flat or on curved surfaces. Flanges are often provided with female threads for the purpose of assembly.

Spinning is a combined compressive and tensile forming method used to transform a sheet metal blank into a hollow body or to change the periphery of a hollow body. One tool component (spinning mandrel, spinning bush) contains the shape of the workpiece and turns with the workpiece, while the mating tool (roll head) engages only locally *(Fig. 2.1.15)*. In contrast to shear forming, the intention of this process is not to alter the sheet metal thickness.

Wrinkle bulging or upset bulging is a method of combined tensile and compressive forming for the local expansion or reduction of a generally tubular shaped part. The pressure forces exerted in the longitudinal direction result in bulging of the workpiece towards outside, inside or in lateral direction *(Fig. 2.1.16)*.

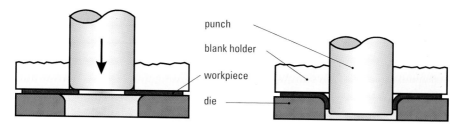

▲ **Fig. 2.1.14** Flanging with blank holder on a flat sheet

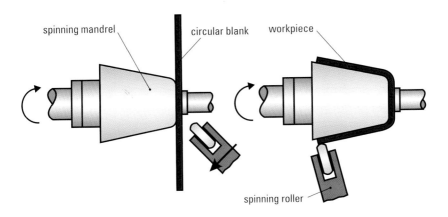

spinning mandrel   circular blank   workpiece

spinning roller

▲ **Fig. 2.1.15** Spinning a hollow body

Forming under tensile conditions
Extending by stretching is a method of tensile forming by means of a tensile force applied along the longitudinal axis of the workpiece. Stretch forming is used to increase the workpiece dimension in the direction of force application, for example to adjust to a prescribed length. Tensile test is also a pure stretching process. Straightening by stretching is the process of extending for straightening rods and pipes, as well as eliminating dents in sheet metal parts.

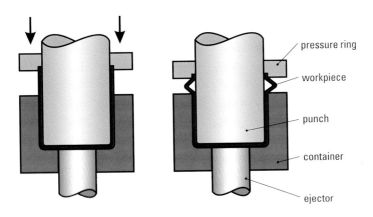

pressure ring

workpiece

punch

container

ejector

▲ **Fig. 2.1.16** Wrinkle bulging to the outside

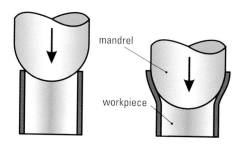

◀ **Fig. 2.1.17** Expanding by stretching

Expanding is tensile forming to enlarge the periphery of a hollow body. As in case of deep drawing, rigid *(Fig. 2.1.17)* as well as soft tools, active media and active energies are also used.

Stretch forming is a method of tensile forming used to impart impressions or cavities in a flat or convex sheet metal part, whereby surface enlargement – in contrast to deep drawing – is achieved by reducing the thickness of the metal.

The most important application for stretch forming makes use of a rigid die. This type of process includes also *stretch drawing* and *embossing*. Stretch drawing is the creation of an impression in a blank using a rigid punch while the workpiece is clamped firmly around the rim *(Fig. 2.1.18)*. Embossing is the process of creating an impression using a punch in a mating tool, whereby the impression or cavity is small in comparison to the overall dimension of the workpiece *(Fig. 2.1.19)*.

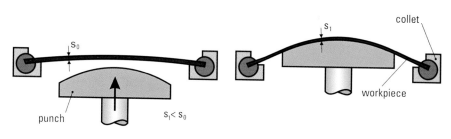

▲ **Fig. 2.1.18** Stretch forming

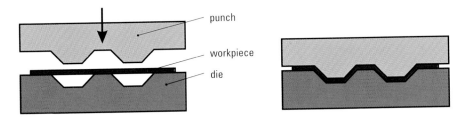

▲ **Fig. 2.1.19** Embossing

Forming by bending

In *bending with a linear die movement* the die components move in a straight line (cf. Sect. 4.8.1). The most important process in this sub-category is *die bending,* in which the shape of the part is impacted by the die geometry and the elastic recovery *(Fig. 2.1.20).* Die bending can be combined with die coining in a single stroke. Die coining is the restriking of bent workpieces to relieve stresses, for example in order to reduce the magnitude of springback.

Bending with rotary die movement includes roll bending, swivel bending and circular bending. During roll bending, the bending moment is applied by means of rolling. Using the roll bending process, it is possible to manufacture cylindrical or tapered workpieces *(Fig. 2.1.21).* The roll bending process also includes roll straightening to eliminate undesirable deformations in sheet metal, wire, rods or pipes *(Fig. 2.1.22* and cf. Sect. 4.8.3) as well as corrugating and roll forming *(Fig. 2.1.23* and cf. Sect. 4.8.2).

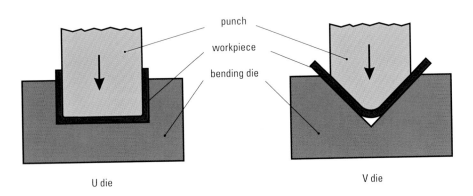

U die

V die

▲ **Fig. 2.1.20** Die bending

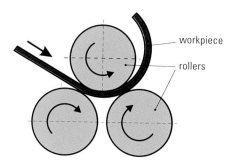

◀ **Fig. 2.1.21** Roll bending

Swivel bending is bending using a tool which forms the part around the bending edge *(Fig. 2.1.24)*. *Circular bending* is a continuous process of bending which progresses in the direction of the shank using strip, profile, rod, wire or tubes *(Fig. 2.1.25)*. Circular bending at an angle greater than 360°, for example is used in the production of springs and is called coiling.

Forming under shear conditions
Displacement is a method of forming whereby adjacent cross-sections of the workpiece are displaced parallel to each other in the forming zone by a linear die movement *(Fig. 2.1.26)*. Displacement along a closed die edge can be used for example for the manufacture of welding bosses and centering indentations in sheet metal components.

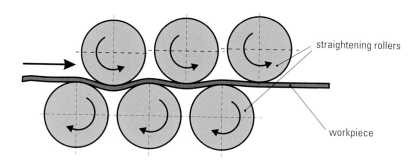

▲ **Fig. 2.1.22** Roll straightening

Twisting is a method of forming under shearing conditions in which adjacent cross-sectional surfaces of the workpieces are displaced relative to each other by a rotary movement *(Fig. 2.1.27).*

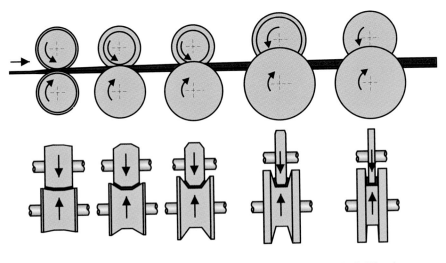

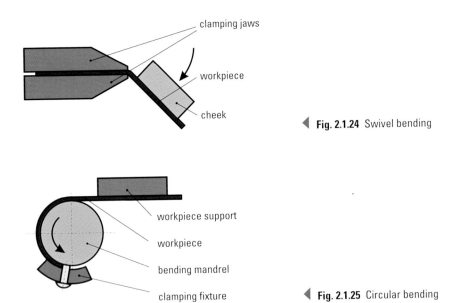

▲ **Fig. 2.1.23** Roll forming

◀ **Fig. 2.1.24** Swivel bending

◀ **Fig. 2.1.25** Circular bending

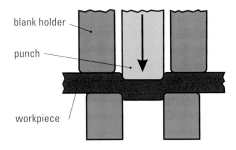

blank holder

punch

workpiece

◀ **Fig. 2.1.26** Displacement

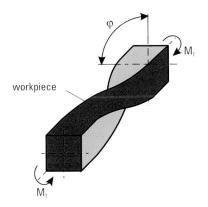

workpiece

◀ **Fig. 2.1.27** Twisting

■ 2.1.3 Dividing

Dividing is the first subgroup under the heading of parting, but is generally categorized as a "forming technique" since it is often used with
other complementary production processes *(cf. Fig. 2.1.2)*. According
to the definition of the term, dividing is taken to mean the mechanical separation of workpieces without the creation of chips (non-cut-
ting). According to DIN 8588, the dividing category includes the subcategories shear cutting, wedge-action cutting, tearing and breaking
(Fig. 2.1.28). Of these, the shear cutting is the most important in industrial application.

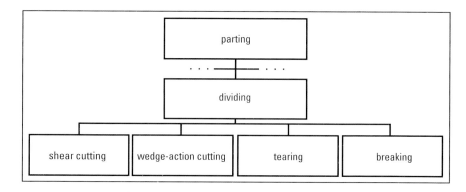

▲ **Fig. 2.1.28** Parting techniques classified under forming

Shear cutting – known in practice as shearing for short – is the separation of workpieces between two cutting edges moving past each other (*Fig. 2.1.29* and cf. Sect. 4.5).

During single-stroke shearing, the material separation is performed along the shearing line in a single stroke, in much the same way as using a compound cutting tool. Nibbling, in contrast, is a progressive, multiple-stroke cutting process using a cutting punch during which small waste pieces are separated from the workpiece along the cutting line.

Fine blanking is a single-stroke shearing method that uses an annular serrated blank holder and a counterpressure pad. Thus the generated blanked surface is free of any incipient burrs or flaws, which is frequently used as a functional surface (*Fig. 2.1.30* and cf. Sect. 4.7).

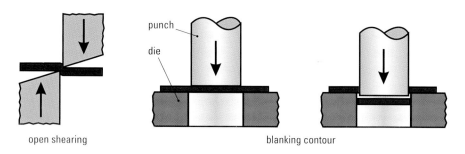

▲ **Fig. 2.1.29** Shearing

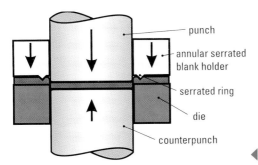

Fig. 2.1.30 Fine blanking

Wedge-action cutting of workpieces is generally performed using a wedge-shaped cutting edge. The workpiece is divided between the blade and a supporting surface. Bite cutting is a method used to divide a workpiece using two wedge-shaped blades moving towards each other. This cutting method is employed by cutting nippers or bolt cutters *(Fig. 2.1.31)*.

The processes *tearing* and *breaking* subject the workpiece either to tensile stress or bending or rotary stress beyond its ultimate breaking or tensile strength.

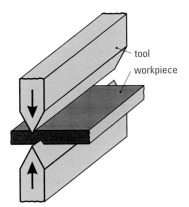

Fig. 2.1.31 Bite cutting

■ 2.1.4 Combinations of processes in manufacturing

Various combinations of different forming processes or combinations of forming, cutting and joining processes have been found to be successful over many years.

Stretch drawing and deep drawing, for example, assume an important role in the sheet metal processing industry (cf. Sect. 4.2.1). During stretch drawing, the blank is prevented from sliding into the die under the blank holder by means of a locking bead and beading rods or by applying a sufficiently high blank holder force *(Fig. 2.1.32).* As a result, the blank is subjected to tensile stress during penetration of the punch. So the sheet metal thickness is reduced.

Deep drawing, in contrast, is a process of forming under combined tensile and compression conditions in which the sheet is formed under tangential compressive stress and radial tensile stress without any intention to alter the thickness of the sheet metal *(cf. Fig. 4.2.1).*

For example when drawing complex body panels for a passenger car, stretch drawing and deep drawing may be conducted simultaneously. The tool comprises a punch, die and blank holder *(Fig. 2.1.32).* The blank holder is used during stretch drawing to act as a brake on the metal, and during deep drawing to prevent the formation of wrinkles.

Modern pressing techniques today permit the desired modification of the blank holder force during the drawing stroke. The blank holder forces can be changed independently at various locations of the blank holder during the drawing stroke. The blank is inserted in the die and clamped by the blank holder. The forming process begins with penetra-

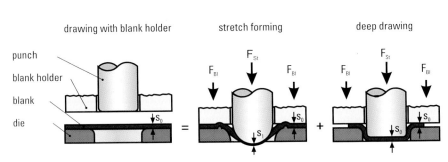

▲ **Fig. 2.1.32** Overlapping deep drawing and stretch forming in the sheet metal forming process

tion of the punch to perform a stretch drawing process in which the wall thickness of the stretched blank is reduced. The bottom of the drawn part is subsequently formed.

The deep drawing process begins once the required blank holding force has been reduced to the extent that the blank material is able to flow without generating wrinkles over the rounded sections of the die. At the end of the drawing process, the blank holder force is frequently increased again in order to obtain a reproducible final geometry by respecting the stretching portion of the drawing stroke.

In addition to deep drawing, body panels are additionally processed in the stamping plant by forming under bending, compressive and shearing conditions. A characteristic of the *bending* process is that a camber is forced on the workpiece involving angular changes and swivel motions but without any change in the sheet thickness. The springback of the material resulting from its elastic properties is compensated for by overbending (cf. Sect. 4.8.1). Another possibility for obtaining dimensionally precise workpieces is to combine compressive stresses with integrated restriking of the workpiece in the area of the bottom dead center of the slide movement.

Forming is almost always combined with *cutting*. The blank for a sheet metal part is cut out of coil stock prior to forming. The forming process is followed by trimming, piercing or cut-out of parts (cf. Sect. 4.1.1).

If neither the cutting nor the forming process dominates the processing of a sheet metal part, this combination of methods is known as *blanking*. Where greater piece numbers are produced, for most small and medium-sized punched parts a progressive tool is used, for example in the case of fine-edge blanking (cf. Sect. 4.7.3). However, solid forming processes often also combine a number of different techniques in a single set of dies (cf. Sect. 6.1).

The call for greater cost reductions during part manufacture has brought about the integration of additional production techniques in the forming process. Stacking and assembly of punched parts, for example, combines not only the classical blanking and forming processes but also joining for the manufacture of finished stator and rotor assemblies for the electric motor industry *(Fig. 2.1.33, cf. Fig. 4.6.22 and 4.6.23)*. Sheet metal parts can also be joined by means of forming, by the so-called hemming or flanging *(Fig. 2.1.34)*.

Dividing, coating and modifying material property technologies will substantially expand the field of application covered by forming technology in the future. This will allow finish processing in only a small number of stations, where possible in a single line, and will reduce costs for handling and logistics throughout the production sequence.

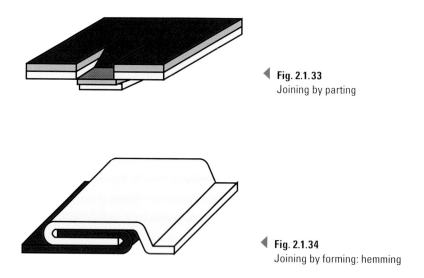

◀ **Fig. 2.1.33**
Joining by parting

◀ **Fig. 2.1.34**
Joining by forming: hemming

2.2 Basic terms

2.2.1 Flow condition and flow curve

Metallic materials may be shaped by applying external forces to them without reducing their structural cohesion. This property is known as the formability of metal. Deformation or flow occurs when the rows of atoms within the individual crystalline grains are able, when stressed beyond a certain limit, to slide against one another and cohesion between the rows of atoms takes place at the following atomic lattice. This sliding occurs along planes and directions determined by the crystalline structure and is only made possible by, for example, dislocations (faults in the arrangement of the atomic lattice). Other flow mechanisms such as twin crystal formation, in which a permanent deformation is caused by a rotation of the lattice from one position to another, play only a minor role in metal forming technology.

Flow commences at the moment when the principle stress difference $(\sigma_{max} - \sigma_{min})$ reaches the value of the flow stress k_f, or when the shear strain caused by a purely shearing stress is equal to half the flow stress, given by:

$$k_f = \sigma_{max} - \sigma_{min}$$

By neglecting the principle stress σ_2, this mathematical expression represents an approximate solution of the shearing stress hypothesis with the greatest principle stress σ_1 and the smallest principle stress σ_3:

$$k_f = \sigma_1 - \sigma_3$$

The value of the flow stress depends on the material, the temperature, the deformation or strain, φ, and the speed at which deformation or strain rate is carried out, φ̇. Below the recrystallisation temperature, the flow stress generally rises with increasing deformation, while the temperature and deformation rate exert only a minimal influence. Exceptions to this rule are forming techniques such as rolling and forging, in which extremely high deformation rates are used. Above the recrystallisation temperature, the flow stress is generally subject to the temperature and deformation rate, while a previous deformation history has only minimal influence. The flow stress generally drops with increasing temperature and decreasing deformation rate.

Accordingly, DIN 8582 differentiates between metal forming processes involving a lasting change in strength properties and those involving no appreciable change in strength properties, previously designated as cold and hot forming. In the temperature range between, deformation involves only a temporary change in the strength properties of the material. In this case, the deformation speed is higher than the recovery or recrystallisation rate. Recrystallisation starts only after completion of the forming process. The rules of metal forming with lasting change in the strength properties apply in this case.

The DIN 8582 standard also breaks down the process according to forming without heating (cold forming) and forming after the application of heat (hot forming). These terms simply specify whether heating devices are necessary. Unlike their former meaning, these terms are not physically related to the material concerned. The flow stress of the individual materials is determined by experiments in function of deformation (or strain) and deformation rate (or strain rate) at the various temperature ranges, and described in flow curves. One of the uses of flow curves is to aid the calculation of possible deformation, force, energy and performance.

■ 2.2.2 Deformation and material flow

Actual deformation φ, also called logarithmic or true strain, is given by:

$$\varphi_1 = \int_{h_0}^{h_1} \frac{dh}{h} = \ln \frac{h_1}{h_0}$$

in which φ_1 is deformation in one principle axis and φ_2 and φ_3 in the other two principle axes. This equation will give, for example, the amount of compression in a body with height h *(Fig. 2.2.1)*. φ is calculated from the compression relative to the starting measurement ε or from the relative deformation

$$\varepsilon_1 = \frac{h_1 - h_0}{h_0} = \frac{\Delta h}{h_0},$$

in which h_0 stands for the height of the body before compression and h_1 the final height of the body after compression:

$$\varphi_1 = \ln \frac{h_1}{h_0} = \ln(1 + \varepsilon_1)$$

In accordance with the law of volume constancy, according to which the volume is not altered by the deformation process *(Fig. 2.2.1)*, the sum of all deformation values is always equal to zero:

$$\varphi_1 + \varphi_2 + \varphi_3 = 0$$

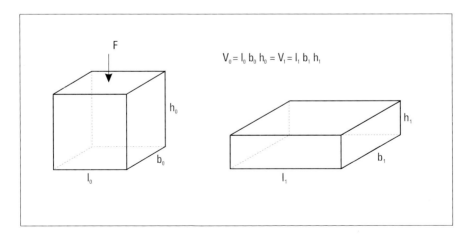

▲ **Fig. 2.2.1** Dimensional changes in frictionless upsetting of a cube

The greatest deformation, which is equal to the sum of the two other deformations, is designated the principle deformation φ_g:

$$\varphi_1 = -\left(\varphi_2 + \varphi_3\right) = \varphi_g$$

The principle deformation must be a known quantity, as it forms the basis for every calculation, for example of deformation force. It is easy to determine, as it carries a different sign to the other two. In the compression of a cubic body, for example, the increase of width ($b_1 > b_0$) and length ($l_1 > l_0$) results in a positive sign, while the decrease of height ($h_1 < h_0$) produces a negative sign *(Fig. 2.2.1)*. Accordingly, the absolute greatest deformation is along the vertical axis φ_1.

Similar to the sum of deformations, the sum of deformation rates $\dot{\varphi}$ must always be equal to zero:

$$\dot{\varphi}_1 + \dot{\varphi}_2 + \dot{\varphi}_3 = 0$$

The flow law applies approximately:

$$\varphi_1 : \varphi_2 : \varphi_3 = \left(\sigma_1 - \sigma_m\right) : \left(\sigma_2 - \sigma_m\right) : \left(\sigma_3 - \sigma_m\right),$$

with the mean stress σ_m given by

$$\sigma_m = \frac{\sigma_1 + \sigma_2 + \sigma_3}{3}$$

The material flow along the direction of the stress which lies between the largest stress σ_{max} and the smallest stress σ_{min} will therefore be small and will be zero in cases of plane strain material flow, where deformation is only in one plane.

■ 2.2.3 Force and work

In calculating the forces required for forming operations, a distinction must be made between operations in which forces are applied directly and those in which they are applied indirectly. Direct application of force means that the material is induced to flow under the direct application of an exterior force. This requires surfaces to move directly against one another under pressure, for example when upsetting and rolling. Indirect application of force, in contrast, involves the exertion of a force some distance from the actual forming zone, as for example when the material is drawn or forced through a nozzle or a clearance. Additional stresses are generated during this process which induce the material to flow. Examples of this method include wire drawing or deep drawing.

In the direct application of force, the force F is given by:

$$F = A \cdot k_w$$

where A is the area under compression and k_w is the deformation resistance. The deformation resistance is calculated from the flow stress k_f after taking into account the losses arising, usually through friction. The losses are combined in the forming efficiency factor η_F:

$$\eta_F = \frac{k_f}{k_w}$$

The force applied in indirect forming operations is given by:

$$F = A \cdot k_{wm} \cdot \varphi_g = A \cdot \frac{k_{fm}}{\eta_F} \cdot \varphi_g = A \cdot \frac{w_{id}}{\eta_F}$$

where A represents the transverse section area through which the force is transmitted to the forming zone, k_{wm} is the mean deformation resistance and k_{fm} the mean stability factor, both of which are given by the integral mean of the flow stress at the entry and exit of the deformation zone. The arithmetic mean can usually be used in place of the integral value. The referenced deformation work w_{id} is the work necessary to deform a volume element of 1 mm³ by a certain volume of displacement:

$$w_{id} = \int_0^{\varphi_g} k_f \cdot d\varphi \cong k_{fm} \cdot \varphi_g$$

The specific forming work can be obtained by graphic or numerical integration using available flow curves, and in exactly the same way as the flow stress, specified as a function of the deformation φ_g. *Figure 2.2.2* illustrates the flow curves and related work curves for different materials.

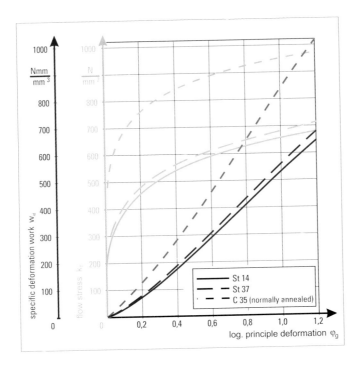

Fig. 2.2.2
Flow curves
and curves
showing the
specific defor-
mation work
for different
materials

If there is no flow curve available for a particular material, it can be determined by experimentation. A tensile, compressive or hydraulic indentation test would be a conceivable method for this. If the specific deformation work w_{id} and the entire volume V or the displaced volume V_d are known quantities, the total deformation work W is calculated on the following basis:

$$W = V \cdot \frac{w_{id}}{\eta_F} \cong V \cdot \varphi_g \cdot \frac{k_{fm}}{\eta_F} = V_d \cdot \frac{k_{fm}}{\eta_F}$$

■ 2.2.4 Formability

The identification of formability should only be based on those cases of material failure caused as a result of displacement or cleavage fractures where no further deformation is possible without failure. If, therefore, a material breaks before reaching maximum force as a result of a cleavage fracture, this characteristic may be taken as a point of reference in the determination of formability, for example during a tensile test. However, cases of failure in which the stability criterion between the outer and inner forces is indicative of the achievable deformation, cannot be used as a basis for determining formability. Such cases include for example the uniform strain of a material with marked necking. The formability of different materials differs even though other conditions are equal. Thus it is that some materials are described as malleable and others as brittle. These descriptions are usually based on the characteristics revealed in tensile testing for fractures due to shrinkage or elongation.

The formability of a material is not a fixed quantity, however, it depends on the mean hydrostatic pressure p_m exerted during the forming operation:

$$p_m = \frac{p_1 + p_2 + p_3}{3}$$

Thus, for example, a material can have a low formability for one type of forming operation where the mean hydrostatic pressure is low. However, if a different forming process is employed in which the mean hydrostatic pressure is higher, the same material can be formed without problems. Even marble can be plastically deformed if the mean hydrostatic pressure exerted is sufficiently great.

■ 2.2.5 Units of measurement

Since the implementation of the law governing standardised units of measurement, it is only permissible to use the variables prescribed by the statutory international unit system (SI units).

Table 2.2.1 provides a comparison of the units applied in the old technical system of measurement and the SI units. The given variables can be transferred to the international metric system using the factor 9.81. For approximate calculations, it is generally sufficient to use factor 10, for example:

$$1\ kp \approx 10\ N$$

Table 2.2.1: Conversion of technical to SI units of measurement

Unit system	Technical (m kp s)	SI (MKS)
Time: t	1 s	1 s
Length: l	1 m	1 m
Speed: v	1 m/s	1 m/s
Revolutions/no. of strokes:	rpm	rpm
Acceleration: a	1 m/s²	1 m/s²
Mass: G	9,81 kg = 1 kp s²/m	1 kg 1 t = 1000 kg
Density: ρ	9,81 kg/m³ = 1 kp s²/m⁴	1 kg/m³
Force: F	9,81 kg m/s² = 1 kp 1MP = 1000 kp	1 N (Newton) = 1 kg m/s² 1 kN = 1000 N
Pressure: p	9,81 N/m² = 1 kp/m² 0,0981 bar = 1 at 1,333 mbar = 1 Torr	1 N/m² = 1 Pa (Pascal) = 1 kg/m s² 10⁵ N/m² = 1 bar
Tension: σ	9,81 N/mm² = 1 kp/mm²	1 N/mm²
Torque: M	9,81 N m = 1 kp m	1 N m 1 kN m = 1000 N m
Mass moment of inertia: J	9,81 kg m² = 1 kp m s²	1 kgm²
Performance: P	9,81 W = 1 kp m/s 0,7355 kW = 1 PS	1 N m/s = 1 W (Watt) 1kW = 1000 W
Work: W	9,81 J = 1 kp m	1 J (Joule) = 1 N m = 1 W s 1kJ = 1000 J
Quantity of heat: Q	4,19 kJ = 1 kcal	1 kJ
Sound pressure level: L	1 dB(A)	1 dB(A)

Bibliography

DIN 8580 (Entwurf): Manufacturing Methods, Classification, Beuth Verlag, Berlin.
DIN 8582: Manufacturing methods, forming, classification, subdivision, Beuth Verlag, Berlin.
DIN 8583: Manufacturing methods, forming under compressive conditions, Part 1-6, Beuth Verlag, Berlin.
DIN 8584: Manufacturing methods, forming under combination of tensile and compressive conditions, part 1-6, Beuth Verlag, Berlin.
DIN 8585: Manufacturing methods, forming under tensile conditions, part 1-4, Beuth Verlag, Berlin.
DIN 8586: Manufacturing methods, forming by bending, Beuth Verlag, Berlin.
DIN 8587: Manufacturing methods, forming under shearing conditions, Beuth Verlag, Berlin.
DIN 8588: Manufacturing methods, terms, dividing, Beuth Verlag, Berlin.
Lange, K: Umformtechnik, Band 1: Grundlagen, Springer-Verlag, Heidelberg (1984).

3 Fundamentals of press design

3.1 Press types and press construction

The function of a press is to transfer one or more forces and movements to a tool or die with the purpose of forming or blanking a workpiece. Press design calls for special knowledge of the production process to be used. Depending on the intended application, the press is designed either to execute a specific process or for mainly universal use.

In a "specialized production line", in view of economic production, the output is the most important issue, while maintaining the required part quality. Material-related influences such as the maximum deep drawing speed or workpiece-related influences such as the suitability of parts for transportation, as well as aspects such as operating ergonomics and working safety all have to be taken into consideration.

The purpose of a *universal production line,* in contrast, is to offer flexibility and use a larger variety of dies covering as wide a part spectrum as possible.

Achievement of the maximum possible uptime is a determining factor in the design of any press, with the objective of reducing unproductive or downtime to a minimum – for example necessary for changing dies, maintenance or tryout. Furthermore, all presses are expected to ensure the longest possible service life of tools and dies. Therefore, for example, the precision guidance of the slide is required.

On the basis of the production techniques used, presses are divided into the following subcategories:

- sheet metal forming presses (cf. Sect. 4.4),
- blanking presses (cf. Sects. 4.6 and 4.7.4),

– presses for solid forming, such as forging, extrusion and coining
  presses (cf. Sect. 6.8),
– presses for internal high-pressure forming (cf. Sect. 5.6),
– pressure forming, stretching and stamping presses.

Shears are also commonly used, generally in separate lines, for the manufacture of sheet metal blanks (cf. Sects. 4.6.1 and 4.6.2).

3.1.1 Press frame

The function of the press frame is to absorb forces, to provide a precise
slide guidance and to support the drive system and other auxiliary
units. The structural design of the frame depends on

– the pressing force – this determines the required rigidity,
– the dimensions of dies influencing the size of the tool area,
– work area accessibility that determines on the shape of the press
  frame,
– the degree of guidance precision. This influences both the shape and
  the rigidity of the frame.

Presses with relatively low press forces, up to 2,500 kN, frequently make
use of the *open-front press* design *(cf. Fig. 4.6.14)*. This construction is
characterized particularly by the easy access to the tool area. However,
its drawback lies in the asymmetrical deflection of the frame, which
contributes to reductions in part accuracy and die life, particularly in
blanking applications. Inclined or horizontal designs permit faster part
ejection making use of gravity following the forming process, for example when forging or coining *(cf. Fig. 6.8.20)*. As a rule, open-front presses
are used in conjunction with single dies.
    Presses with a nominal pressing force over 4,000 kN are constructed
exclusively in a gantry-type design. These are known as *straight-side
presses (Fig. 3.1.1)*. In this press type, the press bed with the bed plate,
the two uprights and the crown form the frame. The application spectrum of straight-side presses ranges from small parts produced using
progression dies, compound progression dies or transfer dies through to
individual dies for varying part sizes (cf. Sect. 4.1.1). When using progression tools, the parts are conveyed by the sheet strip itself. Con-

versely, when using individual dies, an additional gripper rail is generally integrated in the press. A particularly economical production
process can be achieved by accommodating individual resp. transfer
dies in one and the same press. When automated, using a transfer system, this type of line is known as a *transfer press*.

Originally, press frames were produced of gray cast iron. Progress
made in the field of welding technology subsequently allowed thick

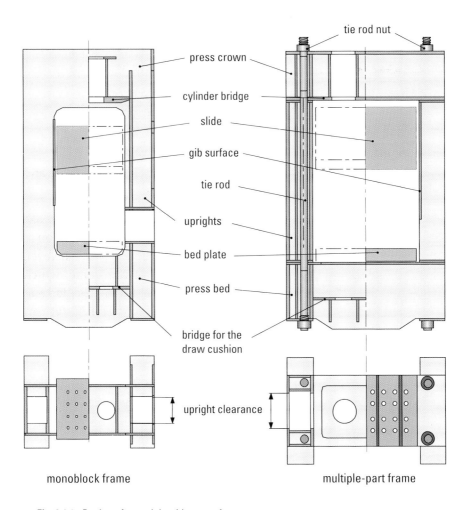

press crown
cylinder bridge
slide
gib surface
tie rod
uprights
bed plate
press bed
bridge for the
draw cushion
tie rod nut
upright clearance

monoblock frame   multiple-part frame

Fig. 3.1.1 Design of a straight-side press frame

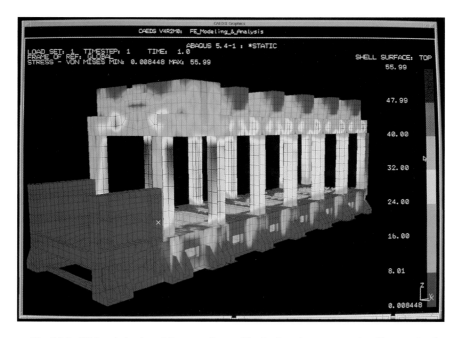

Fig. 3.1.2 FEM optimization of the press frame: Distribution of stresses under off-center load

plate metal to be reliably joined together, opening up a high degree of design flexibility. Thus, presses with welded frames can be constructed precisely for individual requirements.

Compound designs integrating welded and cast components are also possible. Small straight-side presses are built exclusively using the welded *monoblock frame* type, while larger designs use a *multiple-part frame* construction which effectively limit the size of components to facilitate both processing and transport. The individual components are assembled together using tie rods which generally reach along the entire height of the press, and pre-tensioned to a determined force *(Fig. 3.1.1)*. Careful *configuration of the frame* is essential, as it is subjected to stress both by the drive elements and auxiliary units, and is also required to ensure precise guidance of the press slide. Care must be taken not to exceed the load and elastic deformation limit of the frame. To satisfy these complex design requirements, press frames are optimized with the aid of the *Finite Element Method (FEM)*. This involves the computation and depiction of stresses also under eccentric loads *(Fig. 3.1.2)*. By ensuring the correct distribution of material and configuration of transitional radii between the bed and uprights, as well as between the uprights and press crown, it is possible to avoid exceeding the maximum permissible stress levels, also at rated force. Another benefit of FEM-calculated components is the ability to save on material not needed to ensure rigidity. On top of the *press bed* lies the bed plate (bolster plate) used to center and hold down the lower half of the die (*Fig. 3.1.1* and cf. Sect. 3.4). The bolster plates on the press bed and the slide are screwed into place to allow simple removal, for example to alter the T-notch configuration, to retrofit hydraulic clamping devices or to adapt the openings to new pressing requirements.

3.1.2 Slide drive

The press frame absorbs all the forces acting on the press, and guides the slide together with the upper die. The slide, which is moved in the vertical direction by a drive system acting in the *press crown*, transfers the forming or blanking force to the die. Alongside this type of *top driven press,* there are presses whose drive system is located in the press bed. This type of *bottom-driven press* is used primarily for solid forming, specifically as coining presses *(cf. Fig. 3.2.5).* A basic difference exists between *displacement-related, force-related and work-related presses.* The

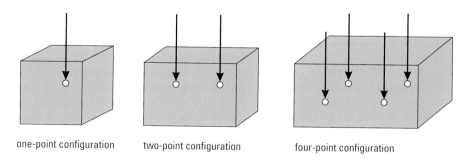

one-point configuration   two-point configuration   four-point configuration

Fig. 3.1.3 Connecting rod forces at the slide in one-, two- and four-point presses

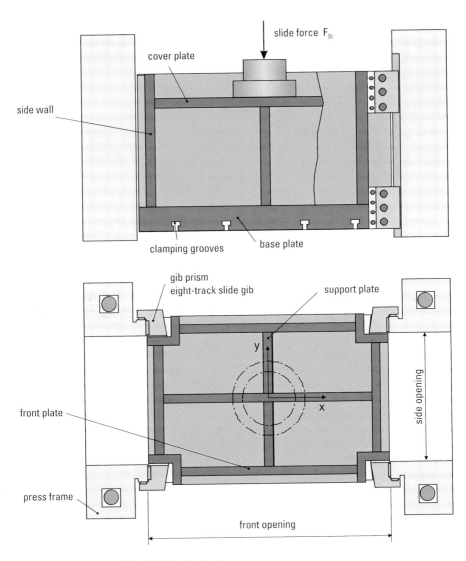

Fig. 3.1.4   Design of the slide in a hydraulic press

group of work-related presses includes screw presses and forging ham-
mers. Displacement and work-related presses are generally referred to as
mechanical/hydraulic presses.

In mechanical presses, force is transmitted to the slide by means of
connecting rods *(Fig. 3.1.9)*, while in hydraulic presses this transfer
takes place via the piston rods of the hydraulic cylinder *(Fig. 3.1.10)*.

Depending on the number of force transmitting elements, there are *one, two or four-point configurations in the slide (Fig. 3.1.3).* When an off-center load is exerted by the die on the slide, a multiple-point press is better able to counter the resulting tilting moment.

The slide, which generally has a welded box shape, is guided in the press frame *(Fig. 3.1.4).* In the case of off-center loads, the *slide gibs* also provide part of the supporting forces. *Figure 3.1.5* illustrates some common types of gibs. A multiple-point press, used in conjunction with a stable slide gib system, offers the best conditions for achieving good part quality and a long die life as a result of minimal tilting of the slide. For large-scale presses with two or four connecting rod drive systems, the gibs have long guidance lengths and relatively large vertical elongation of the frame. This calls for a readjustment feature. The gib clearance should not exceed 0.1 mm.

Standard deep drawing presses are equipped with easily adjustable and exchangeable 45° bronze gibs. In the case of blanking and transfer

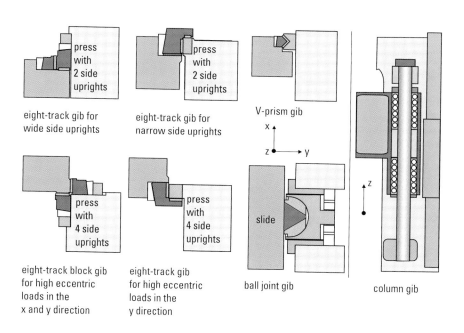

Fig. 3.1.5 Different gib configurations

presses, eight-track guides and exchangeable bronze gibs on the slide are the state of the art. A reliable system of central lubrication is essential to ensure optimum guidance. Instead of bronze slide gibs, plastic gibs are also used.

Nowadays, universal presses are frequently configured with eightfold roller gibs *(Fig. 3.1.6)*. This guidance system is pretensioned free of gib clearance by individual application of the rollers, which are each mounted in roller bearings. The rollers are lubricated and sealed for life and run on hardened guide rails on the press upright free of oil or grease.

Bottom-driven toggle presses feature a long press frame which calls for a special guidance system to compensate for necking, tilt and prevent high edge pressures *(cf. Fig. 3.2.5)*. Cardan universal ball joint bearings compensate for tilts and ensure that the supportive function is evenly spread over the gibs *(Fig. 3.1.5)*. These guidance components can be used in any position: inside, outside, sixfold, eightfold or in combination.

◀ **Fig. 3.1.6** Roller gib

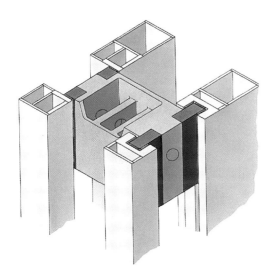

Fig. 3.1.7
Eight-track slide gib
in cross formation

The guidance ratio, i. e. the ratio of guide length to slide length, can be reduced to 1 : 3 in large-scale presses with a height restriction. Eccentric press forces can only be partially absorbed by the gibs in this case. For hot forming applications, the expansion of the slide due to temperature must be taken into consideration in the choice of slide gibs. In this case, guidance systems are used which permit expansion of the slide in one direction without generating additional stresses *(Fig. 3.1.7)*.

For ejection of parts that may stick to the upper dies, the slide can be fitted with pneumatic, hydraulic or mechanical *knock-out pins (cf. Fig. 3.2.12)*.

3.1.3 Drive systems for deep drawing presses

Single and double-action presses are used for sheet metal forming. The *double-action press* features, next to the drawing slide, also a separate blank holder slide *(Fig. 3.1.8, cf. Fig. 4.2.2)*. Both slides are actuated from above.

After the forming process, the part must, as a rule, be turned upside down for the next blanking and forming operation. This requires a

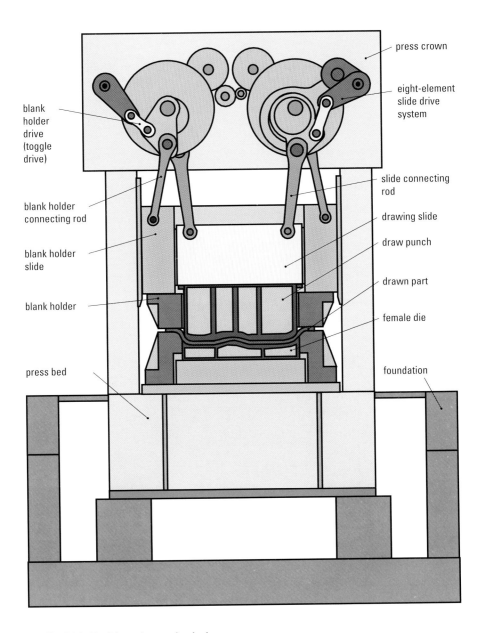

press crown

eight-element
slide drive
system

blank
holder
drive
(toggle
drive)

slide connecting
rod

blank holder
connecting rod

drawing slide

draw punch

blank holder
slide

blank holder

drawn part

female die

press bed

foundation

Fig. 3.1.8 Double-action mechanical press

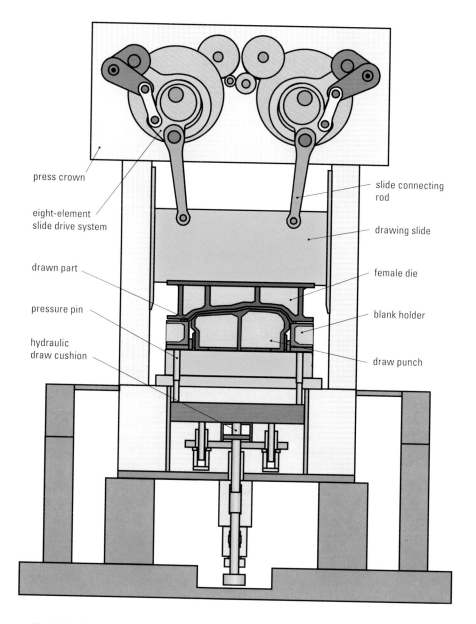

press crown

eight-element
slide drive system

drawn part

pressure pin

hydraulic
draw cushion

slide connecting
rod

drawing slide

female die

blank holder

draw punch

Fig. 3.1.9 Single-action mechanical press with draw cushion

bulky, cost-intensive turning device to be installed downstream from the double-action press. There are advantages in forming a part in the reverse position, i. e. the bottom of the part facing upwards. In this case part ejection is simpler through a slide cushion in subsequent forming stations. The alternative is to work with a draw cushion or nitrogen cylinders. Furthermore, the centering in subsequent stations is improved, since the parts can be positioned over their entire inner shape by the bottom die. In addition, the sensitive surfaces of the material – for example when working with automotive body panels – are less likely to sustain damage during transportation in the press and when centering in the die. It is also possible to install the blanking punches at a lower level for subsequent operations if the part has previously been turned over.

In the case of *single-acting deep drawing presses,* in contrast, the blank holder force is transmitted from underneath by a draw cushion (*Fig. 3.1.9* and *cf. Fig. 4.2.3*). Both the forming and the blank holder force are applied by the slide, as the slide forces the draw cushion downwards. After the drawing, without "turning over", the part is ideally positioned for the subsequent operations.

When used together with a hydraulically driven slide movement, the draw cushion can also be used for *active counter-drawing (Fig. 3.1.10* and *cf. Fig. 4.2.4).* From the point of view of energy savings, it is beneficial that the slide holds the blank while the draw cushion applies the forming force.

In order to achieve as great a degree of flexibility as possible, double-action presses can also be configured with a draw cushion.

3.1.4 Draw cushions

The function of the draw cushion is to hold the blank during deep drawing operations in mechanical and hydraulic presses (cf. Sect. 2.1.4). The cushion thus prevents the formation of wrinkles and, as a result of the ejection function, raises the parts to the transport level during the return motion of the slide. Modern draw cushions can be either hydraulically or pneumatically actuated. A major difference between the two is the larger force of the hydraulic draw cushion. The cushion force is initiated during downward travel of the slide by, for example, four hydraulic cylinders via the lifting bridges and com-

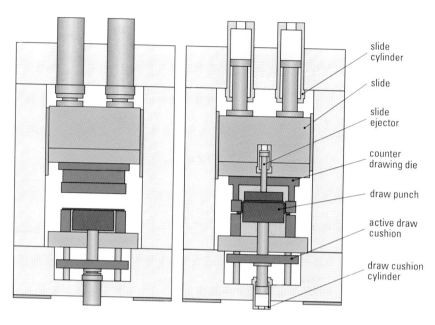

slide
cylinder

slide

slide
ejector

counter
drawing die

draw punch

active draw
cushion

draw cushion
cylinder

Fig. 3.1.10 Single-action hydraulic press with active draw cushion for counter drawing

pression columns into the corner areas of the blank holder on the die
(Fig. 3.1.11 and *cf. Fig. 4.4.39).* When the slide reaches the bottom dead
center, the draw cushion is moved back to the upper position either in an
uncontrolled, i.e. together with the return motion of the slide, or a controlled movement, i.e. after a prescribed standstill period, by the lift
cylinder acting in the center of the lifting bridge. Impact damping is
achieved by means of optimized proportional valve technology, ensuring
that the draw cushion reaches its upper end position without vibrations.

The benefit here is that the draw cushion force can be controlled on a
path-dependent and a time-dependent basis during the drawing process.
It is possible, for instance, not only to reduce and increase the overall
blank holder force, but the pressure levels of the four cylinders can also
be modified independently of each other due to the short pressure
response time within a range of 25 to 100% of the rated pressure. As a
result, the blank holder pressure can be individually controlled over the
entire deep drawing process at each of the four corner points of the die.
The precision control of the blank holder follows a prescribed force versus time profile *(Fig. 3.1.12).*

This offers the diemaker and the machine operator the capabilities for optimization of the drawing process and for the reduction of die start-up times. The stretch forming process, for example, can be initiated by briefly raising the blank holder force, or the material flow correspondingly facilitated by reducing the pressure in certain areas of the blank holder (cf. Sect. 2.1.4). To improve part quality, output and tool life as well as reducing noise, the draw cushion can be pre-accelerated. This means that the draw cushion is already set in motion before the slide impacts the blank. This reduces the relative speed between the slide and blank holder considerably as compared to an uncontrolled draw cushion. Due to the reduced impact of the blank holder, the num-

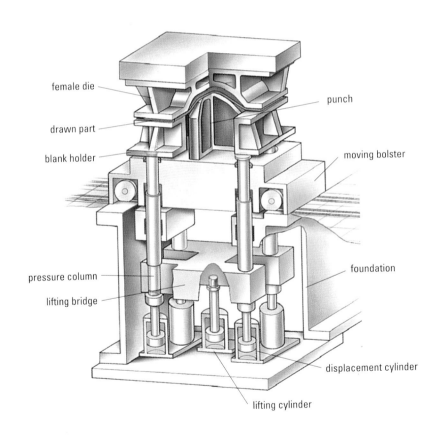

Fig. 3.1.11 Design of a hydraulic draw cushion

ber of press strokes can be increased without compromising part quality, while at the same time the noise level is reduced by as much as 8 dB(A).

An important characteristic of the hydraulic draw cushion with displacement cylinders is that the blankholding force, in contrast to the pneumatic draw cushion, is not fully applied instantanously on application of the slide, and is only built up completely with the start of the piston movement through displacement of the oil column in the cylinders. This allows the vibrations commonly encountered on impact of the slide, as indicated in *Fig. 3.1.13*, to be avoided. The negative influence of vibrations and noise, and also on the drawing process itself, are eliminated. This is of particular significance when producing large sheet metal components with high stroking rates. It is possible to largely avoid superimposed vibrations affecting the drawing process at slide impact speeds of, for example, 40% with a relatively large lead of the

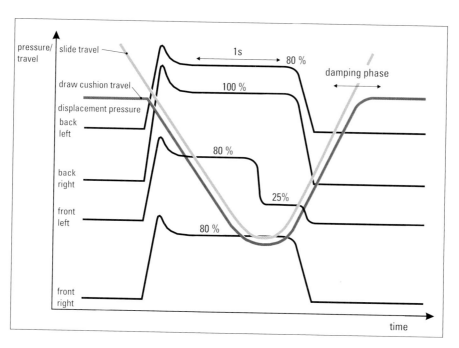

Fig. 3.1.12 Progression of blank holder force in a hydraulic draw cushion with a constant displacement pressure of 100 % back right, 80 % back left and front right and pressure changeover from 80 % to 25 % front left

pneumatic draw cushion *(Fig. 3.1.13, top)*. However, it is only marginally possible to achieve a constant blank holder force at high impact speeds *(Fig. 3.1.13, bottom)*. The conditions for the hydraulic draw cushion with displacement cylinders *(Fig. 3.1.11)* are significantly more favorable. Since the pressure build-up does not take place until the beginning of the piston movement, even at maximum impact speed the blank holder force peaks only minimally and is quickly dissipated *(Fig. 3.1.12)*. Accordingly, in general terms a lead of the blank holder in the die of some 30 mm is sufficient to ensure that the entire drawing process takes place at a constant blank holder force.

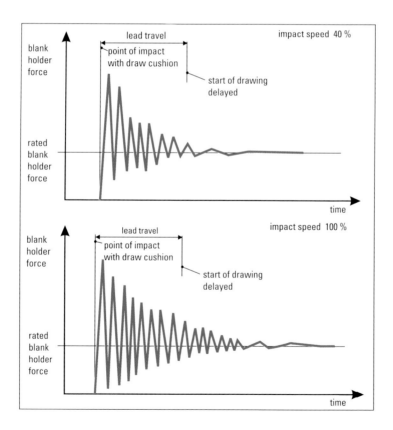

Fig. 3.1.13 Progression of blank holder force of a pneumatic draw cushion depending on impact speed

3.2 Mechanical presses

3.2.1 Determination of characteristic data

The following paragraphs will deal with basic principles concerning *force, work and power* of mechanical presses.

If a weight of 75 kg is suspended from a crane rope, a force of around 750 N (force F = Mass m · gravitational acceleration g) acts on the rope. As long as the weight is not raised, no work is performed, as the work W [Nm] is the product of force F [N] and distance h [m]:

$$W = F \cdot h$$

If this weight is now raised by 1 m, the work performed amounts to:

$$W = 750\,\text{N} \cdot 1\,\text{m} = 750\,\text{Nm}$$

This work can be obtained from the weight again when it drops by 1 m. However, the magnitude of the resulting force depends on the distance over which the work takes place:

$$F = W / h$$

If, therefore, a force of 750 N is uniformly exerted to raise the weight over a distance of 1 m, the same force will be expended if the entire work is performed when lowering the weight over a distance of 1 m. A different situation results if the weight initially drops freely for half the distance and then the entire work of 750 Nm is expended over the last 50 cm = 0.5 m. The force is then given by:

$$F = 750 \, \text{Nm} / 0.5 \, \text{m} = 1{,}500 \, \text{N}$$

Since the work has been performed over only half the distance, the force has doubled. If the entire work is performed over only one tenth of the distance (i. e. 0.1 m), the force rises tenfold:

$$F = 750 \, \text{Nm} / 0.1 \, \text{m} = 7{,}500 \, \text{N}$$

In Fig. 3.2.1, the force exerted over the relevant distance is indicated for all three cases. The results are three rectangles whose areas correspond to the *product of force F · distance h* and thus to the *work W*. If a curve is traced through the corners of the rectangles, provided the existing work of 750 Nm is performed evenly over a certain distance, it is possible to directly read the magnitude of the force applied. With a forming distance of 0.2 m, for instance, the force applied is 3,750 N *(Fig. 3.2.1)*.

In summary, it is generally valid that

– force and work are two terms which can be related to each other only by means of a third variable, i. e. distance.
– If a force is applied over a certain distance, work is performed. This corresponds in the force-distance graph to the area of the rectangle below the force curve.
– If a certain amount of work is performed, the distance over which it is performed determines the magnitude of the generated force.

The relations described here using the example of a weight, apply generally to the field of press construction. They are applied also in the *forging hammer*. The hammer has a certain weight and is raised over a certain distance. This gives it a defined quantity of potential energy. When it gives up the work stored in it, i. e. transfers its energy to the die, the magnitude of its force at the moment of impact depends on the displacement over which deformation takes place. This force can be very large if the displacement of the workpiece is very small, and can even result in damage to the tool and the press. On the other hand, in

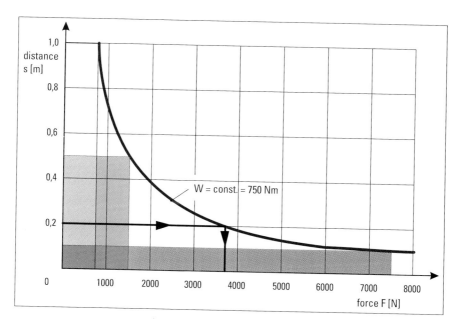

Fig. 3.2.1   Force-stroke diagram showing the relationship between force and stroke for constant work

case deformation takes place over large displacements, the potential energy may be insufficient, resulting in a need for repeated hammering or several blows.

These principles, illustrated here in the example of the forging hammer, also apply to mechanical presses. Here, instead of a raised weight, the energy is stored in the rotating mass of the flywheel (*Fig. 3.2.8*). However, while all the energy stored in the hammer must be expended, the energy in the rotating mass of the flywheel is only partially used during a press stroke. Thus, the electric motor driving the flywheel is not overloaded and does not need to have a large power capacity. In continuous operation, a flywheel slowdown between 15 and 20 % is estimated to be the greatest speed drop permissible. However, this gives no indication of the resulting forces and of the stress exerted on the components of the press.

This is illustrated more clearly by the example below (*Fig. 3.2.2*). The press is assumed to have the following parameters:

- rated press force: F_{N0} = 1,000 kN = 1,000,000 N at 30° before bottom dead center (BDC)
- usable energy per press stroke during continuous operation: W_N = 5,600 Nm
- continuous stroking rate: n = 55/min

Assuming a slowdown of 20 % during continuous stroking, the usable energy is 36 % of the total energy available in the flywheel. In the example given, therefore, the overall energy stored in the flywheel is:

$$W = W_N / 0.36 = 5,600\,\text{Nm} / 0.36 \approx 15,600\,\text{Nm}$$

The given *nominal load F_{N0}* in a mechanical press indicates that this value is based on the strength calculations of the frame and the moving elements located in the force flow such as crank shaft, connecting rod and slide. The nominal load is the greatest permissible force in operating the press. This limit can be defined on the basis of the permissible level of stress or by the "deflection" characteristics. In most cases, the

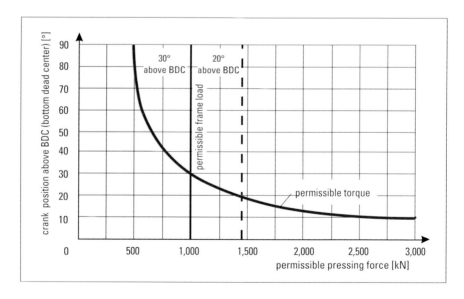

Fig. 3.2.2 Permissible press force of an eccentric press as a function of crank angle

stress on the frame is kept low to achieve the maximum possible rigidity. The nominal load is specified at 30° before bottom dead center to indicate that from here to the bottom dead center, the drive components that transmit the power, such as the driveshaft, the clutch, etc., have also been designed for the torque corresponding to the nominal press force. Therefore, between 90° and 30° before the bottom dead center, the movable parts must be subjected to smaller stresses to avoid overloading. The force versus crank angle curve, *Fig. 3.2.2*, indicates that the press in question may be subjected to a nominal load of 1,000 kN between 30° before bottom dead center and at bottom dead center, while at 90° before bottom dead center a slide load of only $F_{N0}/2 = 500$ kN is permissible.

If a press with the parameters specified here is used to perform an operation in which a constant force of $F = 1,000$ kN is applied over a distance of $h = 5.6$ mm, the energy used during forming is:

$$W = F \cdot h = 1,000,000 \, N \cdot 0.0056 \, m = 5,600 Nm$$

The press is then being used to the limits of its rated force and energy. If the same force were to act over a distance of only $h = 3$ mm $= 0.003$ m, the energy expended amounts to:

$$W = 1,000,000 \, N \cdot 0.003 \, m = 3,000 \, Nm$$

The force of the press is then fully used, while the available energy is not completely used.

The situation is much more unfavorable if the rated flywheel energy of 5,600 Nm is applied over a working distance of $h = 3$ mm. In this case, the effective force of the slide will be:

$$F = W_N / h = 5,600 \, Nm / 0.003 \, m \approx 1,867,000 \, N = 1,867 \, kN$$

As the maximum permissible press force is only 1,000 kN, the press is being severely overloaded. In spite of the fact that the flywheel slowdown is still within the normal limits and gives no indication of overloading, all elements that are subjected to the press force may be damaged, such as the press frame, slide, connection rods, etc. Serious overloading often occurs when press forming is conducted using high forces over small distances, for instance during blanking or coining. The great danger is that such overloading may go undetected. For this reason, *overloading safety devices* must be used to protect the press (cf. Sect. 3.2.8).

Another form of overloading results from taking too much energy from the flywheel. Such overloading can result in extremely high press forces if the displacement during deformation is too small. However, if the energy is applied over a large displacement, this type of overload is much less dangerous. For example, if the press described above is brought to a stop during a working distance of h = 100 mm, the entire flywheel energy of W = 15,600 Nm is utilized. Assuming that no peak loads have occurred, the mean press force exerted is only

$$F = W / h = 15,600 \, \text{Nm} / 0.1 \, \text{m} = 156,000 \, \text{N} = 156 \, \text{kN}$$

The press is therefore by no means overloaded although the flywheel has been brought to a standstill. In this case, only the drive motor suffers a very large slowdown and it is necessary to use a press of a larger flywheel energy capacity, although the permissible press force of 1,000 kN is more than adequate. Overloads of this type occur more frequently if deformation is done over a large distance, e. g. during deep drawing, open or closed die extrusion.

3.2.2 Types of drive system

Eccentric or crank drive
For a long time, eccentric or crank drive systems were the only type of drive mechanisms used in mechanical presses. The sinusoidal slide displacement of an eccentric press is seen in *Fig. 3.2.3*. The relatively high impact speed on die closure and the reduction of slide speed during the forming processes are drawbacks which often preclude the use of this type of press for deep drawing at high stroking rates.

However, in presses with capacities up to a nominal force of 5,000 kN, such as universal or blanking presses, eccentric or crank drive is still the most effective drive system. This is especially true when using automated systems where the eccentric drive offers a good compromise between time necessary for processing and that required for part transport. Even in the latest crossbar transfer presses, eccentric drive systems used in

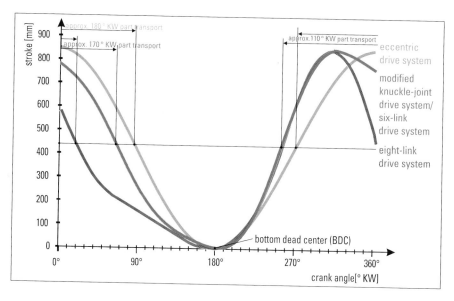

Fig. 3.2.3 Displacement-time diagram: comparison of the slide motion performed by an eccentric, a knuckle-joint and a link-driven press

subsequent processing stations, after the drawing station, satisfy the requirements for system simplification. Stroke lengths of up to 1,300 mm are achieved using eccentric gears with a diameter of 3,000 mm. The eccentric gears are manufactured from ductile iron in accordance with the highest quality standards (cf. Sect. 3.7).

Linkage drive

Requirements for improved economy lead directly to higher stroking rates. When using crank or eccentric drive systems, this involves increasing the slide speed. However, when performing deep-drawing work, for technical reasons the ram speed should not exceed 0.4 to 0.5 m/s during deformation. The linkage drive system can be designed such that in mechanical presses, the slide speed during the drawing process can be reduced, compared to eccentric drive by between a half and a third *(Fig. 3.2.3)*. The slide in a double-acting deep drawing press, for example, is actuated using a specially designed linkage drive *(Fig. 3.2.4 and cf. Fig. 3.1.8)*.

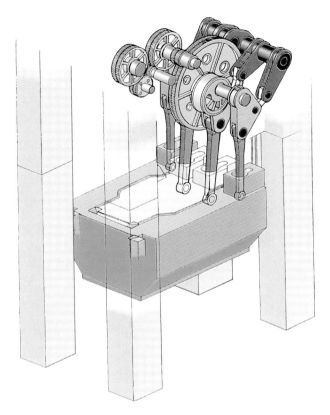

▲ **Fig. 3.2.4** Slide (eight-link drive system) and blank holder slide drive (toggle joint drive system) in a double-acting deep drawing press

This kinematic characteristic offers ideal conditions for the deep-drawing process. The slide hits the blank softly, allowing to build up high press forces right from the start of the drawing process, and forms the part at a low, almost constant speed. In addition, this system ensures smooth transitions between the various portions of the slide motion. During deformation, the drive links are stretched to an almost extended position. Thus, the clutch torque, the gear load and the decelerating and accelerating gear masses are between 20 and 30 % smaller than in a comparable eccentric press. In the case of single-acting machines, for instance, reducing the impact increases the service life of the dies, the draw cushion and the press itself.

This offers a number of important benefits in production: for the same nominal press force and the same nominal slide stroke, a link-driven press can be loaded substantially earlier in the stroke, because the press force displacement curve is steeper within the deformation range. Therefore, the linkage press has a more favorable force-displacement curve than on an eccentrically-driven press *(Fig. 3.2.2)*. Without increasing the impact speed, it is possible to achieve an appreciable increase in the stroking rate and output. Due to the improved drawing conditions, a higher degree of product quality is achieved; even sheet metal of an inferior quality can be used with satisfactory results. In addition, this system reduces stress both on the die and the draw cushion and also on the clutch and brake. Furthermore, noise emissions are reduced as a result of the lower impact speed of the slide and the quieter herringbone gears of the drive wheels.

In the case of large-panel transfer presses, the use of six or eight-element linkage drive systems is largely determined by deformation conditions, the part transport movement and the overall structure of the presses. As a result, in the case of transfer presses with tri-axis transfer system, a six-element linkage drive system is frequently used, since this represents the best possible compromise between optimum press geometry and manufacturing costs (cf. Sect. 4.4.7). In the case of crossbar transfer presses, either an eight-element linkage drive or a combination of linkage and eccentric drives are used in order to optimize the die-specific transfer movements (cf. Sect. 4.4.8). The linkage elements are made of ductile iron, the linkage pins are from steel with hardened contact surfaces and the bearing bushes are from highly stress-resistant non-ferrous alloys. A reliable system of lubrication and oil distribution guarantees safe operation of these highly stressed bearings (cf. Sect. 3.2.11).

Knuckle-joint drive system
This design principle is applied primarily for coining work. The knuckle-joint drive system consists of an eccentric or crank mechanism driving a knuckle-joint. *Figure 3.2.5* shows this concept used in a press with bottom drive. The fixed joint and bed plate form a compact unit. The lower joint moves the press frame. It acts as a slide and moves the attached top die up and down. Due to the optimum force flow and the favorable configuration possibilities offered by the force-transmitting

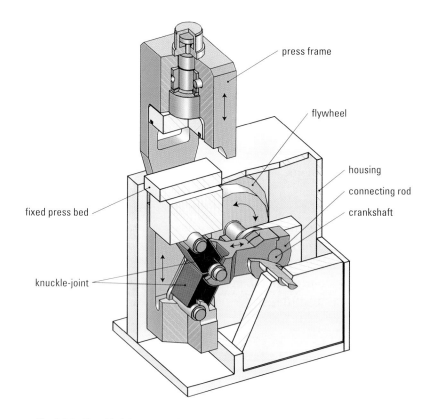

press frame

flywheel

housing

connecting rod

fixed press bed

crankshaft

knuckle-joint

Fig. 3.2.5 Knuckle-joint press with bottom drive

elements, a highly rigid design with very low deflection characteristics is achieved. The knuckle-joint, with a relatively small connecting rod force, generates a considerably larger pressing force. Thus, with the same drive moment, it is possible to reach around three to four times higher pressing forces as compared to eccentric presses. Furthermore, the slide speed in the region 30 to 40° above the bottom dead center, is appreciably lower *(cf. Fig. 6.8.3)*. Both design features represent a particular advantage for coining work *(cf. Fig. 6.8.21)* or in horizontal presses for solid forming *(cf. Fig. 6.8.7)*.

By inserting an additional joint, the kinematic characteristics and the speed versus stroke of the slide can be modified. Knuckle-joint and modified knuckle-joint drive systems can be either top *(cf. Fig. 4.4.10)* or bottom mounted. For solid forming, particularly, the modified top drive system is in popular use. *Figure 3.2.6* illustrates the principle of a press configured according to this specification. The fixed point of the modified knuckle-joint is mounted in the press crown. While the upper joint pivots around this fixed point, the lower joint describes a curve-shaped path. This results in a change of the stroke versus time characteristic of the slide, compared to the largely symmetrical stroke-time curve of the eccentric drive system *(Fig. 3.2.3)*. This curve can be altered by modifying the arrangement of the joints (or possibly by integrating an additional joint).

As a rule, it is desirable to reduce the slide velocity during deformation (e. g. reduction of the impact and pressing speed of the slide). Using this principle, the slide displacement available for deformation can be increased to be three or four times greater than when using eccentric presses with a comparable drive torque. This represents a further advantage when forming from solid (cf. *Linkage drive*).

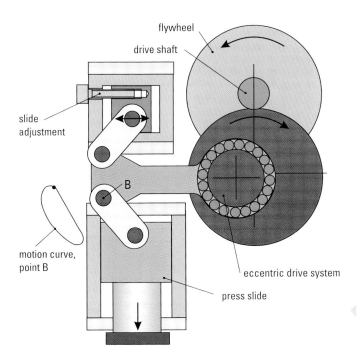

flywheel

drive shaft

slide adjustment

B

motion curve, point B

eccentric drive system

press slide

Fig. 3.2.6
Modified knuckle-joint drive system

The blank holder drive, sometimes used in double-acting deep draw presses, is a special case *(Fig. 3.2.4* and *cf. Fig. 3.1.8).* The required standstill of the blank holder during the deep-drawing phase is achieved in this type of machine by superimposing a double knuckle-joint system coupled with the eccentric or linkage drive of the slide. The standstill of the blank holder represents a crank angle of between 90 and 130° with a maximum residual movement of 0.5 mm. This residual movement does not, however, influence the deep-drawing process because the blank holder overload safeguard acts as a storage system and the elastic deflection of the entire press, which is a few millimeters, has an overriding effect. The slide drive, the blank holder drive and the wheel gear are integrated in the press crown.

3.2.3 Drive motor and flywheel

In larger-scale presses, the main drive system is usually powered by DC motors, primarily in order to provide a large stroking rate. Frequency-controlled threephase motors offer an alternative, particularly where a high protection class rating is called for. The output of a press, which is determined by force, slide displacement and speed, including lost energy, is provided by the main motor. Periodically occurring load peaks are compensated by the flywheel which stores energy. The highly elastic DC drive system is able to compensate for a drop of the flywheel speed of up to 20 % in every press stroke and to replace the consumed energy by acceleration of the flywheel prior to the subsequent stroke.

Useful energy should also be available during set-up operation with a reduced stroking rate, for example five strokes per minute. In this case, a speed drop of 50 % is admissible. An important criterion in configuring the energy balance is to ensure a short run-up time. Running up a large-panel press from its minimum to its maximum production stroking rate generally takes less than one minute under load.

Due to the required useful energy and the permissible flywheel slowdown of 20 %, the flywheel is generally designed to operate under the most energy-consuming condition, i. e. in the lower stroking range of the press. Although a high flywheel speed is beneficial here, this is limited by the admissible speed of the clutch, brake and flywheel itself.

In crossbar transfer presses, flywheels of up to 2,500 mm in diameter and a weight of 25 t are used. The mounting and lubrication of this type of flywheel in roller bearings are highly demanding, as a large quantity of lubricant and continuous temperature monitoring are required (cf. Sect. 3.2.11). The flywheel is driven by the main motor via a flat belt or high-performance V-belt. When the main drive system is switched off, the flywheel is brought to a standstill within a maximum of 30 s by reverse braking of the motor and by an additional pneumatic brake.

In isolated cases, normal car body presses and tri-axis transfer presses are equipped with mechanical creep speed drive systems of one stroke per minute. In this case, the drive is achieved by a threephase motor acting on a worm gear which actuates the drive shaft via the brake flange of the main brake. The transmitted torque generally falls short of the nominal drive torque. However, by means of a pneumatic or hydraulic servo system in the brake, the nominal force may be achieved for distances of up to 10 mm. This is generally sufficient for set-up operation.

3.2.4 Clutch and brake

One of the characteristics of mechanical presses is the *clutch* used to transmit the motor and flywheel torque to the gear shaft and, after clutch release, the *brake* which is used to decelerate the slide, the top die and the gear. Particularly when working in single-stroke mode, the masses in translational or rotational motion must be brought to a standstill after every stroke within an extremely short time: 200 to 300 ms for large-panel presses and 100 to 150 ms in universal presses. Conversely, after engaging the clutch, the same masses must be accelerated from zero to operating speed.

For safety reasons, braking is generated mechanically by spring power. The clutch torque is calculated from the nominal press force and the required working distance, generally 13 to 25 mm above bottom dead center.

Pneumatic single-disk clutch and brake combinations with minimum rotating masses have been in successful use for decades *(Fig. 3.2.7)*.

Pneumatic control systems with safety valves and damping devices are reasonable in cost and generally comply with requirements. One of the problems of pneumatic systems, however, is the limited switching frequency of single-stroke presses and the environmental damage caused by wear to the clutch and brake. In order to eliminate these drawbacks, hydraulic systems are being used increasingly in recent years. With the aid of cooling systems, these permit far higher switching loads per time unit and are practically immune to wear. In single-stroke presses, separate units which permit switching frequencies of up to 30/min have been used to permit more effective control and to avoid overshooting and undershooting by the clutch and brake. Large-panel transfer presses use clutch-brake combinations based on sintered disk sets with up to 20 friction surfaces and transmission torque levels of up to 500,000 Nm.

The compact drive system used in smaller mechanical presses, consists of a flywheel, clutch, brake and planetary gear in a complete unit. Such a system achieves a particularly high level of efficiency with an extremely low moment of inertia and small braking angle, coupled

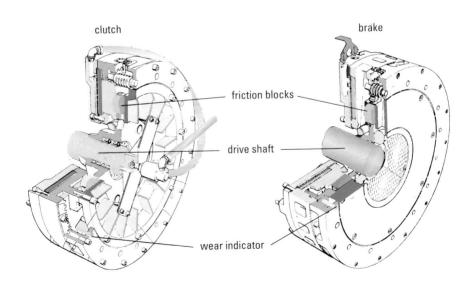

clutch

brake

friction blocks

drive shaft

wear indicator

Fig. 3.2.7 Single-disk clutch and brake in a large-panel press

with a long service life and low maintenance costs *(Fig. 3.2.8)*. The electric motor drives the eccentric shaft by means of a flywheel, a clutch-brake combination that can be either hydraulically or pneumatically actuated and a high-performance planetary gear *(cf. Fig. 4.4.4)*. This eliminates the need for the reducing gear generally required for universal presses. Compact drive systems using a hydraulic clutch-brake combination are particularly environmentally friendly, as they produce no abraded particles and run at an extremely low noise level.

3.2.5 Longitudinal and transverse shaft drive

The design of a mechanical press is determined mainly by the arrangement of the drive shafts and axles, and by the number of pressure points of the connecting rod on the slide *(cf. Fig. 3.1.3)*. The term longitudinal or transverse shaft drive is defined by the arrangement of the drive shafts relative to the front of the press, where the press operator stands.

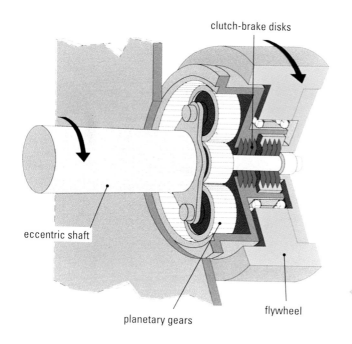

clutch-brake disks

eccentric shaft

planetary gears

flywheel

Fig. 3.2.8
Compact drive
system used in a
universal press

Longitudinal shaft drive is used mainly in single-acting one- and two-point presses with eccentric drive systems *(cf. Fig. 4.4.4 and 4.6.29).* These are generally blanking presses and universal presses in the lower force range of up to 8,000 kN. This classical drive arrangement is also used in long and narrow transfer presses where between two and six pressure points or connecting rods are located in the longitudinal direction. The connecting rods are driven by a continuous drive shaft. The transfer presses are designed with nominal pressing forces between 1,000 and 35,000 kN. They have up to four pairs of uprights and three slides (cf. Sect. 4.4.6).

The connecting rods are arranged in the area of the uprights and tie rods with the purpose of minimizing the bending stress on the press crown. This leaves the top of the work stations in the slide free for ejection functions. For simpler production and assembly, in larger-scale transfer presses, the longitudinal shaft drive is also replaced by a transverse shaft drive system. Knuckle-joint presses with single-point drive, covering a force range from 1,500 to 16,000 kN represent another special construction.

The *transverse shaft press* is designed to have one-, two- or four-point connections and pressing forces of between 1,600 and 20,000 kN. This design permits a greater bed depth, a longer stroke and also the use of a joint system. Double-acting deep drawing presses are only available in this configuration *(Fig. 3.2.4).*

Large-panel presses with tri-axis transfer in a range from 18,000 kN to 45,000 kN press force are also based on a transverse shaft drive system (cf. Sect. 4.4.7). To permit the distribution of pressing forces – generally over four connecting rods per slide – extensive gear units are required which are accommodated in multiple-walled press crowns. Thus, the gear drive systems are simplified and improved.

The latest development in the field of large-panel presses is the crossbar transfer press. Its functions include the production of passenger car side panels in dies measuring up to 5,000 × 2,600 mm (cf. Sect. 4.4.8). In a five-stage configuration, five transverse shaft drive units are joined to form a machine assembly similar to a press line, and all six drive shafts connected by clutches *(Fig. 3.2.9).* This transforms the complete press into a longitudinal shaft drive offering all the advantages of a transverse shaft drive.

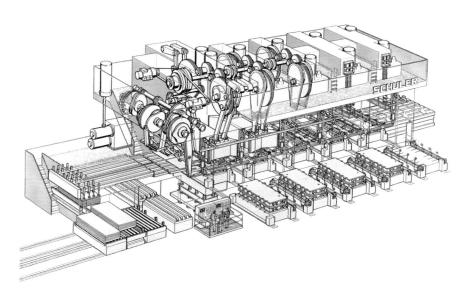

Fig. 3.2.9 Drive system of a transfer press with crossbar transfer

3.2.6 Gear drives

In contrast to blanking presses, whose continuous stroking rate corresponds to the speed of the flywheel, presses with one or two-stage reducing wheel gears are used for longer working strokes. In large-panel presses, the connecting rod and links are generally driven by a two-stage reduction gear. Depending on the necessary power distribution over two, four or, in the case of transfer presses, up to twelve connecting rods, the gear units with wheels and links form the nerve center of modern presses.

Eccentric wheels with a diameter of up to 3,000 mm and reducing gears of up to 1,600 mm in diameter are manufactured from high-strength ductile iron (cf. Sect. 3.7). The gear teeth on the shafts and pinions are hardened. Double helical gearing guarantees optimum power transmission and low running noise *(Fig. 3.2.4)*. The gears and quill pinion shafts are mounted on floating axles with friction bearing bushing for easy assembly. At the start of the power train is always the drive shaft with drive pinion, clutch and brake. Unlike other gear drives, roller bearings are used in the speed range from 300 to 500 rpm.

3.2.7 Press crown assembly

The connecting rod, links, gears and main drive system determine the capacity and the size of the press crown. The welded components weigh up to 150 t, as do the mounted gear drive elements contained in the crown of a large-panel press.

The dimensional limits of such large-scale parts are around 6,000 × 3,600 × 12,000 mm, whereby width, height and weight depend on the possibilities offered for road transport on low-bed trucks – which generally require special authorization. Waterways represent a less difficult method of transporting this type of components.

Stress-relief annealing of large welded assemblies and also machining on drilling centers is restricted to a maximum length of 12 m. The entire problem of transport logistics plays a major role in large-panel press design and must be taken into consideration right from the planning stage.

3.2.8 Slide and blank holder

The rotary motion generated by the press drive is transmitted to the linear motion of the slide and blank holder using a suitable speed and force progression. The most important functional units, particularly in large-panel presses, are the *pressure points* containing slide adjustment and an overload safety device.

Slide adjustment
In individual presses, pressure points with a slide adjustment of up to 600 mm are used in the main spindle in order to adjust the press shut height to various die heights *(Fig. 3.2.10)*. In universal and transfer presses, slide adjustment distances of 150 mm are sufficient. In this case, pressure points with a low overall height are useful. Using brake motors and worm gears, adjusting speeds of 60 mm/min are obtained. With multiple-stage presses used for solid forming, to achieve high stiffness, wedge adjustments are frequently used in the die set or at the slide clamping plate. In this case, a slide adjustment is no longer necessary.

Overload safety devices

The nominal force of the press is limited and safeguarded by the hydraulic overload safety device in the slide. If the nominal press force is exceeded, the pressure exerted on a hydraulic cushion integrated in the pressure point is quickly relieved, allowing an overload displacement.

For blank holders used in double-acting mechanical presses, the overload safety function is coupled to the adjustment function of the blank holder force. The overload safety function is activated only when the deflection of the four blank holder pressure points has exceeded a permissible magnitude of approx. 3 mm. The safety function is actuated by an air-hydraulic pressure scale. By changing the compressed air setting, the blank holder force can be adjusted to the requirements of the die at the four pressure points within a range of 40 to 100%.

Slide counterbalance

The slide weight is compensated by the slide counterbalance system. Thus, the weight of all moving parts such as joints, connecting rods, the slide and upper die, is balanced out by pneumatic cylinders *(Fig. 3.2.11)*. As a result, the drive system is largely free of gravity forces. There are no stresses acting on the slide adjustment and there is an addi-

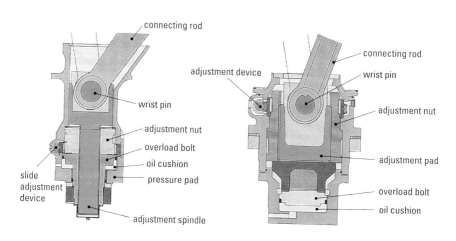

Fig. 3.2.10 Spindle (left) and pressure point (right)

tional safeguard against downward travel of the slide caused by gravity. Pneumatic weight compensation guarantees low-noise, vibration-free operation, smooth loading of the motor and short braking distances.

For counterbalancing weights up to 200 t, at 10 bar air pressure, cylinder sizes up to a 900 mm diameter are required. Long slide strokes of up to 1,300 mm and permissible pressure spikes of up to 25 % require substantial surge tanks. The arrangement of cylinders and tanks depends on the type of press and safety issues. In the case of counterbalance cylinders which are not externally enclosed, double piston rods are often required to safeguard against the risk of breakage *(Fig. 3.2.11)*. Conversely, if the system is integrated in the press crown, this additional safety feature is not necessary.

With every die change, the weight compensation system must be automatically adjusted to the new die weights for this purpose. Individual presses are fitted with special weighing and control functions. Program-controlled automatic systems used in transfer presses automatically set the program inherent pressure value when changing dies in order to reduce set-up times to a minimum.

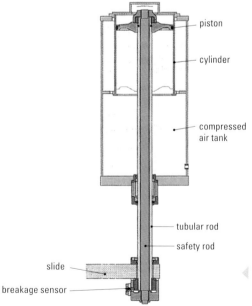

Fig. 3.2.11
Compensation of slide speed with double piston rod

Slide cushion

Some users of presses prefer to use slide cushions as an ejector for the top die. In these cases, pneumatic bellow cylinders are used to supply the required ejection forces *(Fig. 3.2.12)*. Here, ejection strokes can be anywhere between 100 and 200 mm, with pressure increases of up to 30%. The bellow cylinders act on an ejector plate which translate the force to the pressure pins of the cushion and of the die.

During the upstroke of the slide, the ejector plate and the pins are returned by force from the bellow cylinders to their lowest position. Damping elements are used to reduce the noise resulting from the hard impact.

Stroke adjustment

Today, the stroke of small mechanical presses can be adjusted by motor power *continuously* according to requirements of the formed parts with the aid of two interactive eccentrics *(Fig. 3.2.13)*. The stroke can, for example, be set in such a way that the required part height is reached during the deep drawing process and the part can be removed from the die without difficulties on completion of the deep-drawing operation.

The elastic clamp connection between the eccentric shaft and the eccentric bushing is hydraulically released. This involves expanding the

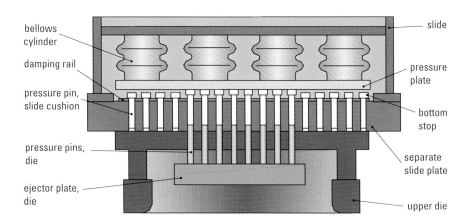

Fig. 3.2.12 Slide cushion with bellow cylinders

bushing against the clamp so that it can be rotated using a gliding motion in any direction on an oil cushion. Special seals are used to prevent oil leakage. The position of the eccentrics in relation to each other is modified from the control desk in accordance with the specific requirements of the die until the sum of the eccentricities reaches the required stroke. The positive connection between the eccentric shells is created as a result of pressure relief coupled with run-away of the oil through drainage grooves. For each die, it is possible to store the required stroke in the die data record. This information can be accessed again and used for automatic adjustment of the slide stroke.

3.2.9 Pneumatic system

The major pneumatic functions of mechanical presses are the clutch, brake, flywheel brake, slide counterbalance, slide cushion and bed cushion. In the case of transfer presses, pneumatic functions also include the application of air pressure to the transfer cam followers and transport by suction cups. Large volumes of air require compressed air tanks with a monitoring safety function and drainage system. The max-

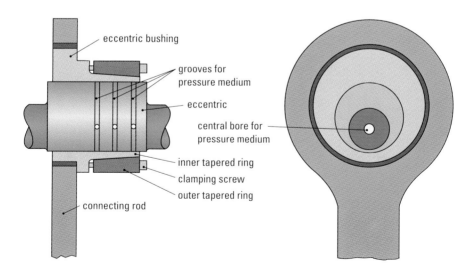

Fig. 3.2.13 Continuous stroke adjustment in an eccentric press

imum pneumatic pressure is generally 6 bar, although high-pressure networks or compressors producing a maximum of 16 bar are increasingly employed.

The engineering of these auxiliary units is determined by valid standards and legal requirements. In general, galvanized gas lines between 1" and 6" are used for installation purposes. Smaller lines are configured using steel piping and cutting ring or beaded unions. The installation required for a large-panel transfer press with a structural volume of around $40 \times 10 \times 15$ m can comprise up 40 t, including compressed air tank, pipes and appliances.

3.2.10 Hydraulic system

The use of hydraulics in mechanical press construction is relatively minimal. Standard use of hydraulic engineering in this type of press is restricted to the hydraulic overload safeguard, raising the moving bolster and hydraulic die clamping systems. All these functions are generally supplied from a central hydraulic unit. In the case of transfer presses, these basic functions are complemented by other functions such as engagement of the clutch between the press and transfer drive systems and the comprehensive system of scrap ejection flaps. In addition, various clamping and locking devices around the press are hydraulically driven.

The most important function in mechanical presses, engaging and braking the slide drive system, is being increasingly done by hydraulic means. The use of hydraulics permits higher switching frequencies in single presses and is also more environmentally friendly in terms of noise and air pollution.

The clutch and brake are supplied by a central hydraulic unit equipped with a cooling system. A dual-channel safety control with regulated damping facility and pressure filtration to 10 μm is required.

The increased use of deep drawing with single-acting presses places particularly stringent demands on the blankholding function and thus the draw cushion in the press bed (cf. Sect. 3.1.4), particularly in large-panel presses. Multiple-point bed cushions with pressure control capability are then only possible using a complex servo hydraulic system.

This requires separate supply units with oil tank, pump outputs of approx. 150 kW and oil flow rates of approx. 1,000 l/min for the draw cushion. Control is performed using proportional valves complying to DIN and ISO standards in the pressure range up to 300 bar.

3.2.11 Lubrication

The complex lubrication systems used to supply the lubrication points in mechanical presses are substantially simplified by using valve blocks *(cf. Fig. 3.3.4)*. The system most commonly used here consists of parallel-switching progressive distributors. Allocation of the required quantity of lubricant is aided by volume controllers, dividers and similar elements. The flow of oil is monitored at the distributor blocks. To ensure optimum operating conditions, a control system regulates the temperature of the lubricant within a relatively narrow bandwidth by using auxiliary heating and cooling functions.

To guarantee maximum operating safety, complex production installations for metal forming are always fitted with twin supply unit elements such as pumps and filters. If one pump fails, there is no need to interrupt production, as the second system cuts in automatically. Maintenance or exchange of the defective system does not result in production standstill, increasing uptime. The quality of the oil is safeguarded by supplementary filtration in the secondary circuit, increasing its service life.

In smaller presses, and in particular by knuckle-joint presses with bottom drive, central lubrication systems with multiple-circuit geared pumps are used. Up to 20 main lubrication points are continuously supplied directly by the gear stages with a constant supply of oil. During the short swivel movements executed by the thrust elements and joints, a constant flow of lubricant is available even under varying resistance levels in the individual user and supply lines.

3.3 Hydraulic presses

3.3.1 Drive system

Hydraulic presses operate on the physical principle that a *hydrostatic* pressure is distributed evenly through a system of pipes, and that a pressure p [N/m²], acting on a surface A [m²], produces a force F [N]:

$$F = p \cdot A$$

The force *acting on the slide* of the press depends on the forming or blanking process performed. For this reason, the pressure p acting on the piston surface is also a function of the deformation process conducted. Its measurement may be used directly to calculate the force acting on the slide. The maximum slide force can be selected by limiting the maximum hydraulic pressure through a relief valve at any position of the slide.

The *drive power* P [W] of a hydraulic press depends on the volumetric flow of hydraulic fluid $\dot{V}$ [m³/s] and therefore on the speed of the slide as well as on the hydraulic pressure p, the forces at the slide, and the mechanical and electrical losses η_{ges}:

$$P = \frac{\dot{V} \cdot p}{\eta_{ges}}$$

In contrast to mechanical presses, there is generally no reserve of energy available in hydraulic presses. Thus, the entire power must be applied when carrying out the forming operation. For this reason, hydraulic presses require a significantly higher drive power than in mechanical presses, with comparable force capacity (cf. Sect. 4.4.1).

Mechanical presses operate according to the throughfeed principle, and the output capacity is determined by the drive speed. In the case of hydraulic presses, in contrast, the flow direction of the hydraulic medium is reversed to close and open the press.

The *cycle time* is made up of a number of variable components, whereby there is an optimum press setting value for each die set. The total stroke, the drawing stroke, the forming force and blank holder force are optimized in accordance with the dies. Each of these setting values has an effect on the press cycle time, and thus on the press output. The drive power is calculated on the basis of the ram displacement versus time curve *(Fig. 3.3.1)* that is related to time constants of the press.

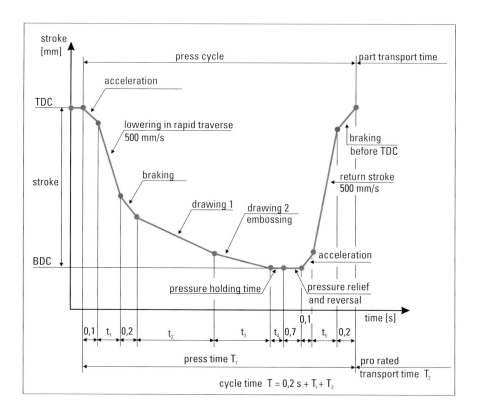

Fig. 3.3.1 Displacement-time diagram of a hydraulic press

Design of the drive system

The piston force of the *slide cylinder,* fastened to the press crown, acts directly on the slide *(Fig. 3.3.2* and *cf. Fig. 3.1.10).* The nominal press force is the sum of the active surfaces of all pistons multiplied by the hydraulic pressure. The movement sequence during the press cycle is executed by the slide cylinder with the slide control. The number of active hydraulic cylinders does not affect the press function. Whether the slide is driven at one, two or four points is determined on the basis of static calculations *(cf. Fig. 3.1.3).*

The function of the hydraulic system

Hydraulic systems convert electrical energy into hydraulic energy, which in turn is transformed into mechanical energy with the aid of the individual hydraulic cylinders inside the press. The hydraulic fluid in the cylinders actuates the press slide as follows:

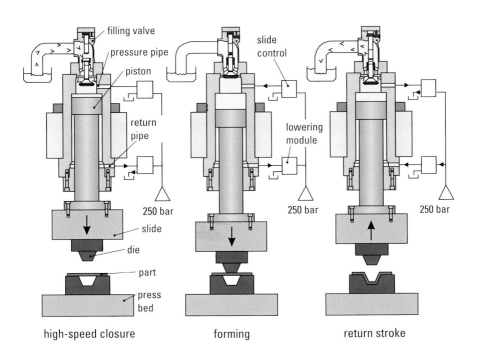

high-speed closure   forming   return stroke

Fig. 3.3.2 Functional principle of a hydraulic cylinder driving the slide of a hydraulic press

- Forces are generated through the action of the hydraulic fluid.
- Displacements are achieved by the supply of hydraulic fluid.
- Press (slide) speed is proportional to volumetric flow rates.
- Forward and reverse motions are controlled by the direction of fluid flow.

Design of hydraulic systems in presses

The hydraulic system is subdivided into several functional assemblies *(Fig. 3.3.3)*. The hydraulic fluid is stored in the *oil tank*. All suction pipes to the pumps or cylinders are supplied with fluid from the reservoir. The fluid tank is suspended at the press crown to ensure that the cylinders are filled when the press closes at high speed. The *hydraulic drive system*, comprised of pumps and motors, forms a structural unit together with the clutch and connecting flange. The pumps used are either immersion

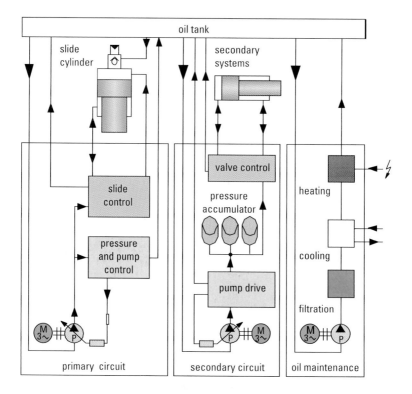

Fig. 3.3.3 Arrangement of a press hydraulic system

pumps suspended in the oil tank, or they are mounted on supports in the press foundation. The motors drive the pumps via an elastic clutch. The pumps are supplied by suction pipes, and the hydraulic fluid is conveyed to the valve control by means of high-pressure pipes. The *valve control system* guides the hydraulic fluid to the appropriate units. Depending on the function in question, the hydraulic cylinders are extended or retracted, moved actively or passively in the press. The return flow is guided back to the oil tank. The *consumers,* cylinders or hydraulic motors, move the slide or cushions, or adjust spindles. They generate forces, lift loads or clamp the dies. The service *unit* operates in a bypass circuit away from the main hydraulic flow. In this unit, the fluid is recirculated by means of filters and coolers in a separate circuit, purifying and cooling it to ensure that the press remains ready for production at all times.

Pressure accumulator drive system
If the pump conveys the fluid to a pressure accumulator, it is possible to smooth out the energy delivery from the main electrical net. Large quantities of hydraulic oil can be drawn from the pressure accumulator for a short time, which has to be replenished by the pump before the next press cycle. If the extraction of hydraulic oil is followed by a long idle period, the accumulator can be supplied by a small pump – for example when loading and unloading the die after retraction of the slide. However, for processes requiring large quantities of hydraulic fluid to be drawn in quick succession, a higher delivery power must be available – for example a drawing operation followed by the opening of the press. For the second example, a pressure accumulator-type drive system is not suitable, since not only a large accumulator system but also a large pump is required. Draw presses involve short cycle and transport times which often cannot be achieved using an accumulator drive system. Table 3.3.1 demonstrates additional differences between pump and accumulator drive systems.

3.3.2 Hydraulic oil

Hydraulic systems in presses are operated using standard hydraulic oil (HLP) according to DIN 51524 Part 2. Hydraulic oils are required to comply with strict standards:

Table 3.3.1: Comparison of hydraulic drive systems

Pump drive system	Pressure accumulator drive system
The installed pump capacity determines the achievable speeds with the motor output.	Speeds are determined by valve cross-sections and the pressure head.
Requirement-oriented hydraulic oil delivery results in peak loads on the electrical mains.	Continuous hydraulic oil delivery exerts an even degree of stress on the mains.
Work capacity is unlimited, as the pressing force is available over the entire stroke.	Work capacity is limited by the available useful volume of the accumulator.
The system pressure corresponds directly to the power requirement of the die.	Pump pressure corresponds to the accumulator boost pressure. In case of a lower energy requirement, excess energy is converted into heat.
The pressing force can be simply limited by means of pressure reducing valves.	It is not possible to regulate the pressure of an accumulator, which always operates at full capacity.

– lubrication and wear protection,
– constant viscosity level at temperatures between 20 and 60 °C,
– resistance to temperature,
– low compressibility,
– low foaming characteristics,
– low absorption of air,
– good rust protection,
– good filtration properties,
– low cost.

All hydraulic units are designed for use with hydraulic oils conforming to these standards. Characteristics, life expectancy and pressure ranges, as well as the choice of seals and gaskets correspond to the oil type used. Due to the *compressibility of hydraulic oil,* the volume of hydraulic fluid changes under pressure. As pressure is increased, the fluid column is compressed. This property affects the control and regulating functions in the press: the higher the pressure, the longer the response time. When calculating cycle times or determining the use of different drive systems, the compressibility factor must be taken into account, as this has a pronounced effect, particularly where short forming strokes are involved, for example when blanking or coining. In the case of mineral oils, a compressibility factor of 0.7 to 0.8 % per 100 bar oil pressure can be expected. This property of hydraulic fluids also affects the release of pressure in the press cylinder. If the area under pressure is opened too rapidly, this results in potentially damaging pressure relief spikes. Flame-retardant hydraulic fluids should only be used where this is absolutely mandatory and the appliance manufacturer has issued the relevant approval. These call for the use of special gaskets, in some cases lower operating pressure and also special measures during start up.

Between the pump drive system and the cylinders, the press functions are controlled by valves. Directional, pressure and non-return valves act in conjunction with proportional and servo valves. Depending on the system and position – in the primary or secondary circuit – valves of various dimensions are used. The execution of all the necessary press functions calls for a large number of hydraulic connections which are grouped together in a block configuration *(Fig. 3.3.4)*. *Block hydraulic systems* have gained in popularity, because:

– external connections only need to be directed from the pump and to the cylinder,
– the valves are surface-mounted and easy to exchange,

Fig. 3.3.4 Block hydraulic system

- their nature inevitably results in a standardization process: the valve blocks do not need to be redesigned for each different press,
- all the valves are brought together to create a functional assembly on a single block,
- integrated valves are sensibly combined with space-saving, low-cost auxiliary valves,
- troubleshooting is simplified by the presence of drilled-in functional and test connections.

3.3.3 Parallelism of the slide

The capacity of the press frame to absorb eccentric loads plays a major role. Eccentric forces occur during the forming process when the load of the resulting die force is not exerted centrally on the slide, causing it to tilt *(Fig. 3.3.5)*. The standard press is able to absorb a maximum eccentric moment of load of

$$M_x = 0.075\,m \cdot F_{N0}[kNm]$$

$$M_y = \frac{M_x}{2}[kNm]$$

via the slide gibs. This results in a maximum slide tilt of 0.8 mm/m. If a higher off center loading capability is desired, then the press design must be more rigid. In this case, the slide gibs will have greater stability, the press frame will be more rigid, and the slide will be higher.

If it is not economically possible to achieve the allowable amount of slide tilt within the required limits by increasing press rigidity, it is necessary to use hydraulic parallelism control systems, using electronic control technology *(Fig. 3.3.5)*, for example in the case of hydraulic transfer presses *(cf. Fig. 4.4.14)*. *The parallelism control systems* act in the die mounting area to counter slide tilt. Position measurement sensors monitor the position of the slide and activate the parallelism control system *(Fig. 3.3.5)*. The parallelism controlling cylinders act on the corners of the slide plate and they are pushed during the forming process against a centrally applied pressure. If the electronic parallelism monitor sensor detects a position error, the pressure on the leading side is increased by means of servo valves, and at the same time reduced on the opposite side to the same degree. The sum of exerted parallelism

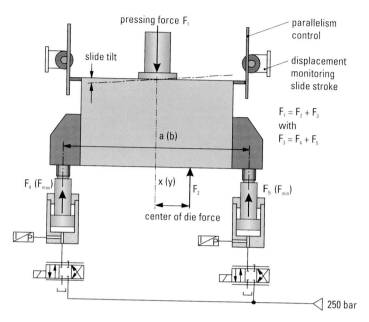

pressing force F_1

parallelism control

displacement monitoring slide stroke

slide tilt

$F_1 = F_2 + F_3$
with
$F_3 = F_4 + F_5$

a (b)

x (y)

F_2

center of die force

F_4 (F_{max})

F_5 (F_{min})

250 bar

Fig. 3.3.5 Control system for maintaining slide parallelism

control forces remains constant, and the slide tilt balance is restored. Depending on the deformation speed, a slide parallelism of 0.05 to 0.2 mm/m is achieved. A central device adjusts the system to different die heights by means of spindles at the slide.

Full-stroke parallelism control involves the use of parallel control cylinders, with their pistons permanently connected to the slide *(Fig. 3.3.6)*. These act over the entire stroke of the slide so that no setting spindles are required to adjust the working stroke. Two cylinders with the same surface area, arranged well outside the center of the press, are subjected to a mean pressure. The tensile and compressive forces are balanced out by means of diagonal pipe connections. The system is neutral in terms of force exerted on the slide. If an off-center load is exerted by the die on the slide, a tilt moment is generated. The slide position sensor detects a deviation from parallel and triggers the servo valve. The valve increases the pressure on the underside of the piston acting on the leading side of the slide, and thus also on the opposite upper side of the piston. At the same time, the pressure in the other connecting pipe is reduced. The

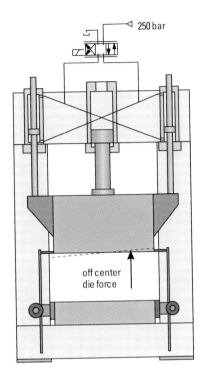

250 bar

off center
die force

Fig. 3.3.6
Full-stroke parallel control of the press slide

opposing supporting torques exerted on the two sides counteract the tilt moment. The maximum deviation measured at the parallel control cylinders is between 0.6 and 0.8 mm during drawing, bending and forming operations.

3.3.4 Stroke limitation and damping

The deformation depth, which depends on the slide position, is determined in hydraulic presses by the switchover point at the bottom dead center (BDC) *(Fig. 3.3.1)*. Depending on speed, the slide travels past this switchover point by a few millimeters. However, repeatability remains within close limits, meaning that the slide is reversed at the BDC with sufficient accuracy for most forming operations. However, if a tolerance of less than 1 mm is required, a mechanical stroke limitation is called for. In combination with a hydraulic damping system, this can achieve a marked reduction in the blanking shock involved in presswork *(Fig. 3.3.7)*. There

Fig. 3.3.7
Stroke limiting and damping device

are two variations in use. Standard damping is used for material thicknesses exceeding 3 mm and for light blanking which only utilizes some 30 % of the press capacity – for example trimming the edge of a workpiece in a transfer press. Where high-strength materials and sheet thicknesses below 3 mm are involved, a pre-tensioned damping system is required, since this is capable of absorbing the energy released instantaneously at the moment of material separation.

3.3.5 Slide locking

When switching off the press and when working in the die mounting area, the slide must be reliably safeguarded by a slide locking device. German accident prevention legislation (UVV 11.064 §12) specifies that hydraulic presses with a bed depth of more than 800 mm and a stroke length of more than 500 mm must be equipped with at least one fixture acting on the upper end position of the slide. This fixture must

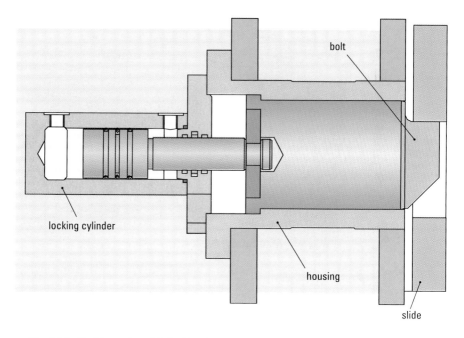

Fig. 3.3.8 Positive acting slide locking device

prevent the slide from dropping under its own weight when the control system is switched off. The *locking device at the top dead center* (TDC) is a positive-acting locking device at the upper point of the nominal stroke. A bolt guided in the press frame is pushed by a hydraulic piston into a recess in the slide *(Fig. 3.3.8)*. The bolt is configured to carry all moving weights, including that of the die superstructure.

The *infinitely variable locking device* is a non-positive acting slide safeguard in which a retaining rod at the slide is guided by the clamping head in the press crown *(Fig. 3.3.9)*. The pilot jaws hold the movable elements in any position – even in case of a power failure. This locking device, which is arranged in the slide area, is recommended particularly in presses featuring a blank holder and internal slide. The slide can be fixed in any position, eliminating the need for a separate safety device for the blank holder.

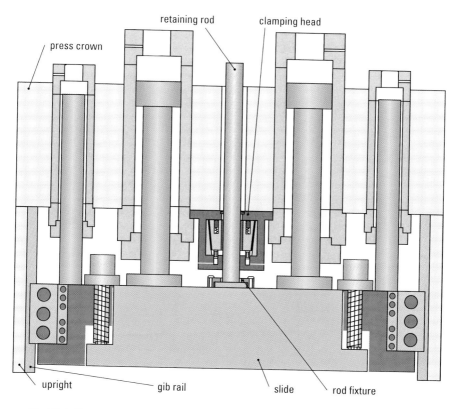

retaining rod   clamping head

press crown

upright   gib rail   slide   rod fixture

Fig. 3.3.9 Non-positive acting locking device for the blank holder and the drawing slide

3.4 Changing dies

3.4.1 Die handling

Devices used for the fast exchange of dies help to reduce set-up times, allowing the economical production of different components even where small batch sizes are required. Small and medium-sized forming and blanking dies are transported within the stamping plant using cranes, forklifts or other industrial trucks. The transportation of larger-scale dies into the press and their alignment call for special die changing fixtures, as the dies cannot be deposited directly on the press bed. Trouble-free die change is only possible if the upper and lower dies are perfectly centered. In addition, it must be possible to deposit the dies outside the press. The purpose of die changing fixtures is to simplify and accelerate transport into the die mounting area, to ensure simple handling of the pressure pins for the draw cushion, and to permit fast, reliable clamping of dies.

Swivelling brackets
A *swivelling bracket* is a low-cost die changing fixture for the transport of the die into the press. Such fixtures can be used up to a die weight of around 6 t. In the case of dies weighing up to around 2 t, insertable brackets suspended from retaining hooks located in the press bed are used. When higher die weights are involved, support feet, where possible fitted with rollers, are essential. Double-jointed swivelling brackets can be turned into a parked position after die change *(Fig. 3.4.1)*. It is possible for dies to be manually pushed into position by one person up

Fig. 3.4.1
Swivelling brackets

to a weight of 3 t provided the press bed and brackets are equipped with hard roller bars.

Motorized die change carts
Rail-bound carts for die changing are designed to facilitate the exchange of tools up to around 20 t in weight. The use of two adjacent tool tables represents a particularly economical solution, as the set of dies can be prepared for use when one table is located in the parked position while production is going on. If the press is fitted with protective grilles and automatic die clamps, automatic die change is possible. In the home position, the die is unclamped and the protective grilles open. The *push-pull drive system* extends, couples automatically to the die and pulls the die out of the press *(Fig. 3.4.2)*. Once the cart has been released, traversed and locked back into the push-in position, the new die can be pushed into the assembly area.

However, in the case of presses with forward-positioned uprights, this changeover technique becomes considerably more complex. In this case, a simple push-pull drive is not sufficient to bridge the distance between the exchange cart and the press bed. Using an additional mechanical *overrun drive* and brackets with roller bars at the press, the push-pull drive system transports the die all the way to the die stops. The automatic coupling between the drive and the die-set eliminates the need for intervention by operating staff.

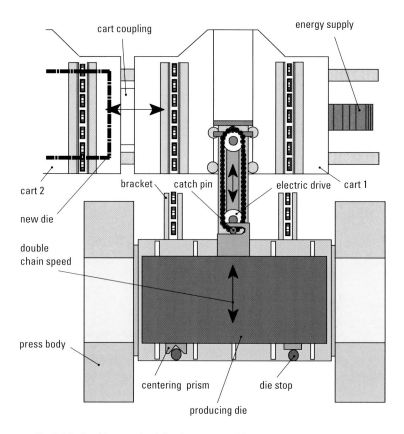

Fig. 3.4.2 Double motorized die charge cart with overrun drive

In cases where dies are positioned on roller conveyors, the force needed to displace the dies is minimized by using profile rails with roller elements in the T-slots of the clamping plate. Equipped with spring-mounted ball roller elements, these profile-rails offer a freedom of movement in all directions when handling lightweight dies. Heavier dies, in contrast, should be transported on hydraulically raisable rollers, as linear contact leads to a higher surface pressure. The surface of the die body coming into contact with the rollers must be hardened.

Die centering on the press bed plays a determining roll when changing dies. Manually displaceable dies can be aligned with sufficient accuracy using a central groove in the clamping plate or by means of centering aids in the T-slots or clamping plates. Dies or presses with automatic transfer

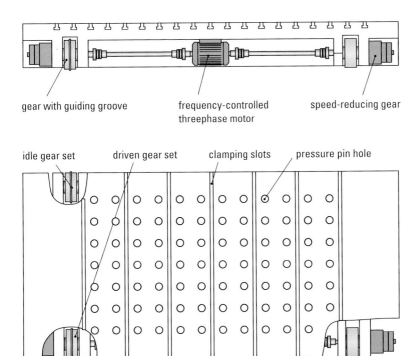

Fig. 3.4.3 Moving bolster

devices, in contrast, must be more precisely positioned in order to ensure exact positioning of the parts in each die station *(cf. Fig. 4.4.34)*.

Moving bolsters

Quick die changing systems for large presses comprise traversable moving bolsters, automated die clamping devices, gripper rail couplings and devices for automatic setting of path, pressure and time axes from the data memory. They can be ideally adjusted to the individual production conditions of the part being formed and can be automatically accessed from the data memory and edited when changing dies.

The upper and lower die of the set are moved out of the press on moving bolsters, and the bolsters bearing the new die sets are moved in *(Fig. 3.4.3* and *cf. Fig. 4.4.38)*. Preparation of the die-set for the next pressed part is performed outside the press while production continuous. By automating the die changing process, the dies can be changed, depending on the die transport system used, within a period of 10 to

30 min – in modern presses in even less than 10 min. The die changing sequence for large-panel transfer presses is described in Sect. 4.4.7. Among moving bolster type tool changing systems, those with left/right-hand extending rails and T-tracks have proven to be most popular *(Fig. 3.4.4)*. The former require more space, but achieve the shortest possible changeover times, as both sliding bolsters can be extended and retracted simultaneously.

Changing individual dies

If the service life of the die used in any particular press station is shorter than that required for the entire production run, the die must be changed during production. Most individual dies of a progressive tool

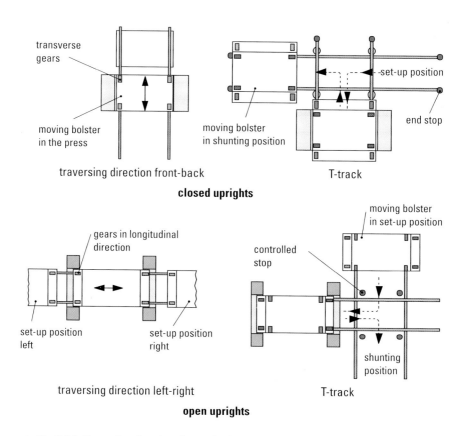

Fig. 3.4.4 Traversing directions for moving bolsters in presses with closed and open uprights

are changed in solid forming *(cf. Fig. 6.7.1)*. For dies weighing under 10 to 15 kg, this is generally performed manually. For heavier dies, when changing individual die blocks or for hot forming, die changing arms with a hoisting power of 50 to 800 kg are used *(Fig. 3.4.5)*. The up and down movement is powered by electric motor in mechanical presses, while hydraulic presses are fitted with hydraulic lifting cylinders. The swivel or traversing movement into the die space can be executed manually, or controlled by pushbuttons to the preselected die position. Alternatively, fully automatic systems are also available.

3.4.2 Die clamping devices

The clamping of dies onto the bed and slide clamping plates must be performed as quickly and reliably as possible. Clamping devices are either electrically or hydraulically powered, in some cases also in combination with manual actuation by means of spring power, threaded spindles, toggle lever, wedge action or self-locking mechanisms. Whichever means is used, safety is always a prior concern *(Fig. 3.4.6)*. Different clamping systems are often used for the upper and lower die, since the lower dies

Fig. 3.4.5 Die changing arm for solid forming

are pushed over the press bed, while the slide is lowered vertically on the upper dies. The task of classifying the different clamping systems is excessively complex, as the selection depends heavily on the required clamping force, the type of die and press involved and the degree of press automation.

Small dies are generally held by push-in clamps or clamping rails. The rails are fixed on the clamping plate, while push-in clamps are inserted into the T-slots on the clamping plate. The advantage of working with this type of small, lightweight fixture is that it restricts the usable table surface only minimally and permits easy handling. If the clamping force of the push-in clamps is not sufficient when connected to the hydraulic system of the press at 250 bar, it can be increased to 50 kN by raising the line pressure to 400 bar.

T-slot clamps are frequently used to secure medium-sized dies *(Fig. 3.4.7)*. A pressure of 250 bar is applied to the clamping cylinders by the press hydraulic system. As a result, a clamping force of up to 200 kN can be generated. The clamps can be released by means of spring power after a drop in the oil pressure. The safety of the fixture is guaranteed by positive locking action in the T-slot and monitored by pressure sensors which bring the press to an immediate standstill in case of a drop in pressure. For large dies with differing widths, it is necessary to use

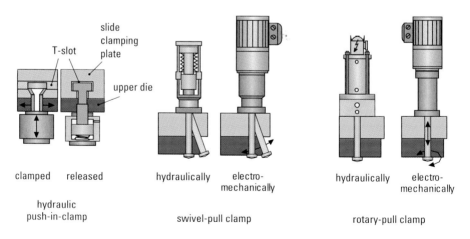

Fig. 3.4.6 Different die clamp designs

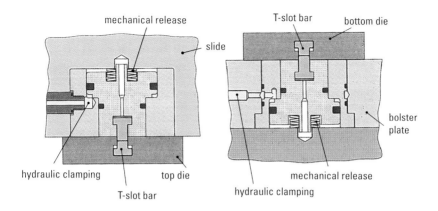

Fig. 3.4.7 T-slot clamp for upper and lower tool

clamps that can be adjusted automatically to the dimensions of the die used *(Fig. 3.4.8)*. These adjustable fixtures can be combined with all types of clamps – hydraulic, electric and hydromechanical. If there are no adjusting devices available, the clamps are mounted outside the clamping plates, and the dies must be adjusted to the clamp spacing by using adapter plates.

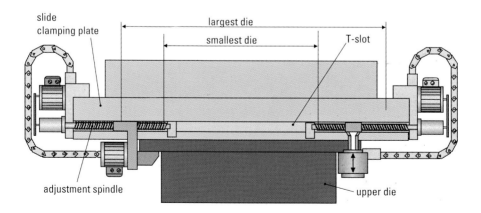

Fig. 3.4.8 Automatic die clamping device with adjusting facility: electromechanical clamp (left), hydraulic clamp (right)

3.5 Press control systems

3.5.1 Functions of the control system

The functions performed by press control systems include the reading of signals and data transmitted from the control elements and sensors located on the machine and their conversion into control commands to the machine's actuators. The generation of control commands depends on the selected operating mode such as "set-up", "automatic die change" and "automatic continuous operation".

Functions to be monitored or initiated include, for example, safety devices, clutch/brake, main drive system, lubrication, draw cushions, die clamps, moving bolsters, part transport and automatic die change. In addition, the control system executes machine functions and fault diagnostics.

3.5.2 Electrical components of presses

The electrical components of presses and their peripherals are broken down into three main areas:

- operating and visualization system,
- electrical control (located centrally in the switch cabinet or decentrally distributed in the unit),
- sensors and actuators in the press.

3.5.3 Operating and visualization system

Process visualization is an instrument which makes the production sequences of machines and plants ergonomically accessible. It supports the operation, observation and monitoring of technical plant and equipment. The control units are positioned in various locations around the machine and are basically broken down according to three categories:

- free-standing operator consoles,
- operator stations permanently integrated in the machine *(Fig. 3.5.1)* and
- movable pendant operator stations.

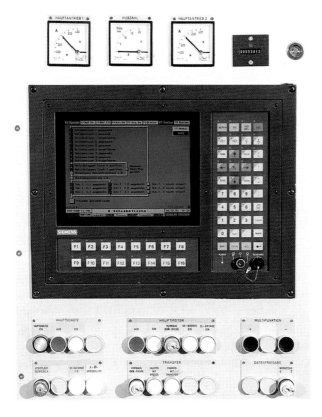

Fig. 3.5.1
Operator station

Control functions can be subdivided into central and local systems: central control functions refer to higher-level press functions, while local control functions are used for manual operation of individual functions directly at the press.

Control units are comprised of input and display units for operation and visualization and can range from simple keys with pilot lamps through to industrial PCs with high-resolution color monitors. The main function of the visual display system is to support the press operator on start-up and repeat start-up of the press, for example following a die change or machine fault, and to permit the input of various operating parameters *(Fig. 3.5.2)*. In addition, the visual display system can be used to store die and machine data and for logging functions. Information system functions which are integrated into the visual display system offer the user support both for maintenance work and troubleshooting. There is an increasing tendency for this type of system to make use of multi-media technology (cf. Sect. 4.6.7).

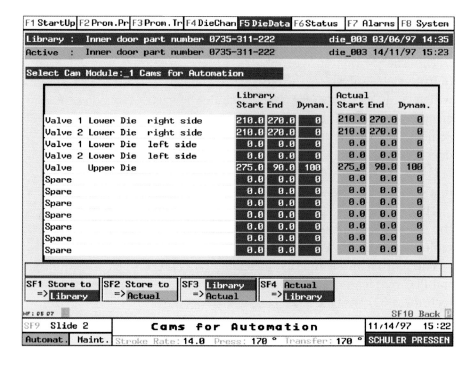

▲ **Fig. 3.5.2** Tooling data input mask for cam values

3.5.4 Structure of electrical control systems

Electrical control systems are broken down into four main areas:

– the conventional part,
– the programmable logic controller (PLC),
– input and output systems and
– the communication system.

The conventional part of the electrical system consists of the power supply, drive controllers, the individual electrical and electronic control devices and the low-voltage switchgear for power control monitoring and attaching the components in the control cabinets and the machine. The safety control represents a particularly important component of the conventional part for all functions relevant to safety such as the emergency stop, monitoring of safety gates or safety doors and the engagement of the clutch/brake in mechanical presses.

In addition, the electrical control system must reliably prevent the press from starting up or running through independently, following the occurrence of any individual fault. In case of failure of any component, it must ensure that no further press closing movement can be initiated.

The safety function of the conventional part of a hydraulic press monitors the valves responsible for reliable standstill of a movement which could lead to injury. These include the valves located in the lower cylinder area which control die closure. Hydraulic redundancy is achieved by means of two monitored valves. Failure of one of these does not lead to the execution of a hazardous movement. If one valve fails, the press's valve monitoring circuit automatically brings the press to a standstill.

The PLC controls and monitors all machine functions not covered by the conventional control circuits and automatic die change, performs fault diagnostics and provides the interface to other machine sections.

The sensors and actuators of the machine are connected to the central PLC by means of the input/output system. There are two different technologies being used: first the central input/output modules which are integrated in the PLC racks or in additional units of the control cabinet; second, on an increasing basis, local input/output modules within the press connected via fast communication systems to the central PLC unit *(Fig. 3.5.3)*.

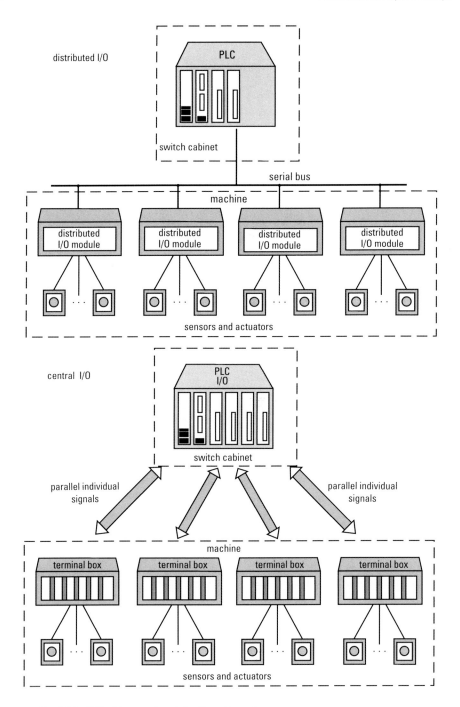

Fig. 3.5.3 PLC with central and distributed input/output systems

The communication system connects the PLC, and provides data exchange with the visual displays within the machine and also communication with external systems.

3.5.5 Functional structure of the control system

Depending on the type of machine or press, the control system is broken down according to the following functional groupings:

- feed,
- main drive system,
- safety control,
- slide functions (interlocking, overload safeguard, weight compensation, die clamping system, adjustment),
- draw cushion or blank holder functions (pressure control, adjustment),
- moving bolster functions (traversing, lifting/lowering, clamping),
- functions of the workpiece transport devices and dies,
- hydraulic, pneumatic, lubrication functions,
- special die changing functions.

Cam operated limit switches play a special role in press control systems. These press-specific functional units generate switching signals for certain functions depending on the position of the crank angle. A difference must be drawn between electromechanical rotating cam limit switches *(Fig. 3.5.4)*, which are used for safety-relevant functions, and electronic cam switches that are programmable. The latter comprise a measuring system for the crank angle of the press and an electronic evaluating module.

In contrast to mechanical presses, valves for the downstroke and upstroke of the slide are separate in a hydraulic press, so eliminating the need for a cam operated safety device.

3.5.6 Major electronic control components

Safety control systems used in Europe are generally still the dual-channel, self-monitoring conventional type using contactors. In contrast, since the mid-eighties, the safety control systems used in large-panel

⬥ **Fig. 3.5.4** Electromechanical rotary cam switch

presses for example in the US market have been frequently designed to use two PLC systems, each monitoring the other. American PLC manufacturers have offered suitable solutions for this type of application for some time. In Germany too, press safety control solutions based on redundant PLC systems are becoming increasingly popular.

The systems used are standard commercially available PLCs *(Figs. 3.5.5* and *3.5.6).* They are adapted to the functional requirements of the press by means of intelligent modules such as rotating cam limit switches, fast positioning modules, fast and conventional communication modules and, where required, by independently operating task-specific control systems.

Operations involving critical real-time functions and complex data processing tasks which are not suitable for the PLC are processed by independent subsystems. These include the control of the draw cushions and multiple-axis part transport devices or fast feed systems, acquisition, processing and display of sensor outputs associated with the drawing process as well as acquisition and evaluation of machine and production data. Hydraulic presses, for example, are equipped with quick-acting pressure systems to control the parallelism of the slide.

Digital rectifiers and DC or three-phase asynchronous motors are used for main, transfer and ancillary drive systems *(Fig. 3.5.7).* Workpiece handling systems are being increasingly equipped with dynamic digital servo drive systems, in which control intelligence is often integrated into the drive system electronics *(Fig. 3.5.8).*

Fig. 3.5.5 PLC European manufacturer

3.5.7 Architecture and hardware configuration

The control system is divided into modules according to functional groupings. This applies to the circuit diagram, the arrangement of the switch cabinet and structure of the PLC software. The functional modules do not necessarily correspond to the mechanical structural units – in some cases similar functions such as all emergency stop circuits, all protective grilles and the lubrication are grouped together.

Criteria used to determine the control structure include the logical and functional coherence between different system elements and the minimization of interfaces between the individual groups of functions. A well designed architecture of the control system enhances clarity and transparency of the electrical documentation and thus reduces the time spent in design and start-up as well as in troubleshoooting during production.

3.5.8 Architecture of the PLC software

The PLC software is designed to correspond to the functional groupings of the circuit diagram. Functional sequences, in particular those used

Fig. 3.5.6
PLC American
manufacturer

for workpiece handling and die changing, are programmed using
sequential programming methods. Where fast response times are called
for, the relevant PLC input signals are defined as interrupt inputs
which, as the name suggests, interrupt cyclically processed control software functions and trigger a response program which responds to external events in an extremely short time.

3.5.9 Future outlook

Programmable logic controllers are ideally suited to their respective
application as single-purpose systems. They are particularly well suited
for the processing of logic operations. The trend is towards equipping
systems with ever more extensive, cost-intensive processing capability.
Development of programming languages, on the other hand, has
slowed down almost to a halt. Programs are still being written in the
form of ladder diagrams and instruction lists, which are comparable to

Fig. 3.5.7 Switch cabinet with components for main press drive

an assembler language. These originally application-oriented programming languages are now being used for achieving visual display, data
processing, data management and communication functions.

The consequences of this trend are high software development and
maintenance costs, non-transparent programs – which not even program design standardization can make any easier to understand – and
long lead times to put the press into operation. To permit the integration of this type function, the PLC has to be equipped with additional
expensive hardware modules.

In comparison, industrial computers, for example PCs, offer the benefits of graphic user interfaces, data processing capability and facility
for object-oriented programming through the use of powerful programming tools. The basic requirements imposed on a future-oriented
press control system, i. e. the ability to access as many functions and as
much information as possible directly at the machine, are only achievable in the future through the use of greater intelligence and more computing capacity. In addition, modern automation engineering calls for

Fig. 3.5.8 Components of a servo drive system

control systems with greater networking capability through standard networks for the exchange of control, production and process data as well as the capability for diagnosis and intervention in the control system through public telecommunication networks. The PLC alone is not able to cope with these demands, although its continued existence as a subordinate, fast machine control concept is still justified.

Practically all the components required for realization of the above functions are available in the industrial computer. Alongside its computing performance, the industrial computer also profits from the open-ended software world. Furthermore, despite the steady rise in performance limits, the costs of the technology are falling in many cases. In view of this trend, control structures featuring distributed intelligence and higher-level industrial computers interlinked with fast communication systems are certain to gain increasingly in popularity. Added to this is the increased use of distributed input/output modules and sensor/actuator communication systems with a view to reducing the necessary installation and wiring input. Methods of programming the control software will become more standardized and simplified in the future, for example by the introduction of the international standard IEC 1131.

Other approaches exist in the form of programming through the use of status graphs, application ("object") oriented programming and user language. In this case, reusable modules with any degree of complexity can be generated and stored in a user-specific macro library. The prepared objects in the library are in parametric form and can be grouped with a minimum of complication to cover a large number of the press or machine functions. All that is necessary to then join the interfaces for data exchange.

Information technology is being integrated to an ever greater degree into the plants or machines themselves.

As internationally stable "information highways" become available, new forms of fault diagnostics and maintenance back-up will naturally develop, allowing possible faults and operating errors to be ascertained and mechanical faults to be traced using video technology or microphones, so realizing a system of temporary or continuous on-line monitoring.

3.6 Press safety and certification

3.6.1 Accident prevention

Due to their mode of operation and the energy input levels involved, presses represent – compared to many other machines – a higher potential hazard to the operator. However, as a result of considerable effort, notably the revision of the German accident prevention regulations for presses VBG7n5.1 dated April 1, 1987, the number of serious accidents occurring on or at presses has been substantially reduced. In addition to complying with legislation on safety requirements based on the "state of the art", manufacturers must also have the capability to recognize and implement sensible safety measures. The manufacturer's own sphere of responsibility in this field is a major portion of European safety legislation. As part of the service package offered by manufacturers, the operating instruction manual contains special safety provisions and instructions.

A major goal is to make the press operator recognize the necessity for safety measures. The unpredictable nature of human reaction in surprise or emergency situations must be taken into account here. Highlighting the accident risk is the first and most important step when training machine personnel on safety matters. As long as the operator or maintenance personnel are not convinced of the necessity for safety measures, they will be inclined to find ways of bypassing them. Excessive or superfluous safety measures by the manufacturer are also certain to create greater willingness on the part of operating staff to attempt to override safety functions. Reasons that justify negligence with regard to safety measures become quickly insignificant if an accident does actually occur.

3.6.2 Legislation

Alongside legislation and regulations defining requirements imposed on technical products – such as the German Pressure Vessel and Appliance Safety Acts – legislation has also been adopted which regulates cases of human injury caused by technical equipment. If the machine manufacturer adheres to the technological standards defined by the so-called characteristic requirements, it may be assumed that he will be exempt from liability. An accident taking place despite these precautions is defined as a "statutory accident", in which case the companies providing industrial accident insurance are responsible for settling any financial claims arising as a result of the damage.

In Europe, these characteristic requirements are defined by the EC Machine Directive, the EMC (electromagnetic compatibility) Directive and the Directive Governing Simple Pressure Vessels. Product liability is also treated on a standardized basis in Europe through the EC Directive for Harmonization of Product Liability Legislation.

In the case of particularly dangerous machines, state agencies or institutions acting on behalf of state agencies are commissioned to check and certify equipment provided the technical requirements of the equipment have been correctly implemented. Foreign trade in technical products is obstructed to a considerable degree by differences of opinion on the best ways to achieve safety and on the targeted standards of safety in the individual countries. Machine manufacturers operating on a global basis need to study the numerous different safety requirements and standards in force in each different export market. Thus, they must become familiar with the different safety philosophies, and must use locally approved components.

3.6.3 European safety requirements

Since January 1, 1995, new safety regulations have been in effect in the EEC (European Economic Community) which are briefly outlined below *(Fig. 3.6.1)*. EC legislation is based on a system of decrees and directives. EC decrees apply directly in the respective member state, while directives are taken up by the national governments, which have to convert them within a reasonable period into national law with or

EC law governing technology:

Objectives:
– Harmonization of safety standards in Europe
– No compromises on safety in a member state
– Free trade in machinery

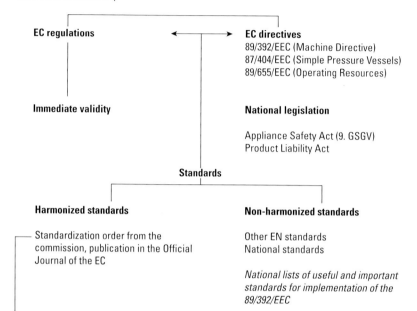

EC regulations   ⟵   ⟶   **EC directives**
                                                                                     89/392/EEC (Machine Directive)
                                                                                     87/404/EEC (Simple Pressure Vessels)
                                                                                     89/655/EEC (Operating Resources)

Immediate validity   **National legislation**

                                                                                     Appliance Safety Act (9. GSGV)
                                                                                     Product Liability Act

                                               Standards

Harmonized standards   **Non-harmonized standards**

Standardization order from the   Other EN standards
commission, publication in the Official   National standards
Journal of the EC
                                                                       National lists of useful and important
                                                                       standards for implementation of the
                                                                       89/392/EEC

→ **Type A standard**

General principles

EN292 (General principles for design)

→ **Type B standard (horizontal standard)**

Definition of safety devices, stipulations of general relevance
with different machines
EN294 (Safety distances to prevent)
EN574 (Two-hand operated devices)
EN953 (Insulating safety devices)

↳ **Type C standard (vertical standard)**

Stipulations relating to a particular machine category
EN692 (Mechanical presses)
EN693 (Hydraulic presses)

▲ **Fig. 3.6.1** Chart showing European safety legislation:
                  EC Machine Directives 89/392/EEC (ECMD) CE Mark

without their own additions. A series of directives was issued relating to the characteristic requirements of technical products. For machine tool manufacture, the directives listed in Table 3.6.1 together with the end of the transitional period are of particular importance.

The EC Machine Directive (ECMD) 89/392/EEC establishes the major health and safety requirements imposed on machines. Thus it is the most important directive for the machine manufacturer and was converted into national German law by the 9th decree relating to the Appliance Safety Act. The product liability law, which has been in effect since November 15, 1989, was derived from the EC Directive on Harmonization of Product Liability Legislation. Both the EC directives and EC decrees are based on harmonized EN standards formulated on behalf of the Commission and published in the EC Official Bulletin after harmonization. These standards allow the machine manufacturers to comply with the relevant directives. The manufacturers can, but are not obliged to, comply with the standards. The manufacturer can have recourse to the presumption that the requirements of the directive are fulfilled by application of the harmonized standards.

In the standards, concrete solutions are put forward for achievement of the prescribed safety standard: safety grilles on presses, for example, are required to have a closing force of no more than 150 N. Otherwise, a danger of crushing exists which can, however, also be avoided through the use of a contact strip. In case of failure to comply with a harmonized

Table 3.6.1: European standards

Directive	Content	Valid without restriction from
89/392	ECMD	1/1/95
93/68	Amendment to ECMD (Identification)	1/1/95
93/44	Amendment to ECMD (Safety components)	1/1/97
91/368	Amendment to ECMD	1/1/95
87/404	Simple pressure vessels to 30 bar	1/1/93
89/336	Electromagnetic compatibility	1/1/96
73/23	Low voltage directive	1/1/97
86/188	Machine noise	1/1/90
89/655	Operating resources directive	1/1/97

standard, the manufacturer can fulfill the health and safety standards using alternative methods. In contrast, however, failure to comply with the ECMD will result in prosecution of the manufacturer, withdrawal of the product from the European market and decommissioning of the hazardous press.

EN standards are classified under three categories, each with its own function and purpose. General guidelines and safety targets are set down in so-called A standards. B standards impose requirements on certain facilities, for example safety equipment used for different types of machines. C standards define machine categories and the requirements imposed on these. For a given machine category, the C standards lay down the way in which risks are likely to occur and how they should be evaluated and safeguarded against. Due to the generality with which machines are grouped to create machine categories – the mechanical press category, for example, ranges from the notching machine to the large-panel transfer press – it can, in certain cases, be expedient to deviate from the standard while still complying with the requirements of the ECMD.

As documentary evidence of compliance with health and safety requirements, the machine manufacturer issues the owner with a Certificate of Conformity *(Fig. 3.6.2)* according to annex II A, and attaches the CE mark to the machine. No machine has been commissioned in the EEC without a Declaration of Conformity or CE mark since January 1, 1995.

The ECMD differentiates between independently working machines and machines intended for installation as part of an overall production line. A production line comprised of several components constitutes a machine which has only one CE mark and only one declaration of conformity. For all production line components which are incapable of operating independently, the subcontractor issues a manufacturer's declaration in compliance with annex II B. Under certain circumstances, component machines of this type do not permit safeguarding unless integrated in the overall line. An individual destacker without the press, for instance, permits free access to hazardous movements and is thus incapable of complying with the ECMD. The party responsible for the overall line, whose function it is to issue the declaration of conformity, must therefore ensure that the manufacturer of the components

- stipulates in the manufacturer's declaration that the component complies as far as possible with the ECMD,
- has compiled and archived the required documentation as laid down by annex V, point 3,
- stipulates the applied safety standards and
- has made reference in the operating instructions to the safety regulations for the manufacturer of the line into which his components are to be integrated.

All those involved assume that the prescribed minimum statement, as indicated in annex II B, is insufficient and the machine cannot be commissioned until the overall line has been proven to comply with the ECMD. Frequently, and in particular in the case of large-scale presses, part of the equipment originates from the user and part of it from the purchaser of the press. As the ECMD requires static stability and safe access to all points of possible intervention, with the declaration of conformity the production line manufacturer assumes responsibility also for the structural work carried out for the press, for example the construction of the necessary foundations. The identity of the manufacturer or construction foreman as defined by ECMD must be contractually laid down in the case of this type of production line. The responsible party then has the other contractual partners confirm compliance with health and safety standards to the extent of their respective performance.

3.6.4 CE marking

The flow chart in *Fig. 3.6.3* indicates what is required of the manufacturer wishing to apply the CE mark to his machine. An underlying difference is drawn between two cases. According to annex IV of the ECMD, presses intended for manual feed operation have been classified as hazardous machines since January 1, 1995. In this case, a C standard – the EN692 for mechanical and EN693 for hydraulic presses – must exist, and the press must be constructed without restrictions in accordance with this standard. If this is not the case, or if the relevant C standard does not yet exist, the manufacturer is required to commission a type test of the machine by a notified body (i. e. an appropriate testing

---

EG-Declaration of Conformity
per specifications of the machine directive 89/392/EEC, annex II A

Applied EC-directives :

[X]   machine directive 89/392/EEC and 91/368/EEC
[X]   electromagnetic compatibility directive 89/336/EEC and 93/31/EEC
[X]   low-voltage directive 73/23/EEC
[ ]   ....

We herewith declare that the following identified machine / line

....

(make, type designation, year of manufacturing, order number)

as marketed by us meets the essential health and safety requirements of the EC machinery
directive 89/392/EEC relating to the design and construction. This declaration will become
invalid in case of modifications of the machine / line, not accepted by us.

Applied harmonized standards:

[X]   DIN EN 292
[X]   DIN EN 294
[X]   DIN EN 60204 T1
[X]   DIN EN 692
[X]   DIN EN 418
[X]   E DIN EN 953
[X]   DIN EN 954
[X]   DIN EN 982
[X]   DIN EN 983
[X]   DIN EN 1037

Applied national German standards and technical specifications :

[ ]   ....

A technical construction file has been compiled and will remain available.

An instruction manual for the machinery has been composed and delivered in :

[X]   community language of the manufacturers country (original version)
[X]   community language of the user country
[ ]   ....

Place, date   ....

Manager design services   ....
(title and designation of the undersigned)   (Legally binding signature
                                       of the manufacturer)

▲ **Fig. 3.6.2** Sample Declaration of Conformity

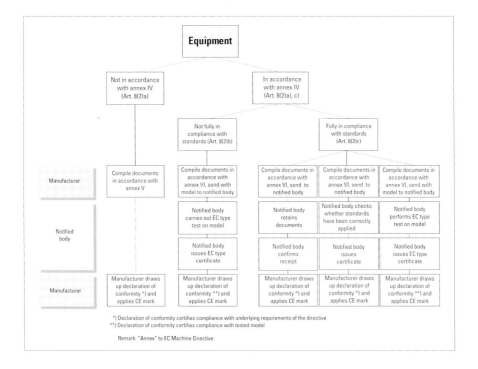

Fig. 3.6.3 Sequence for the conformity declaration and the CE mark

authority) in accordance with annex VI. The manufacturer is only enti-
tled to issue the declaration of conformity and apply the CE mark on
the basis of an EC type test certificate. If the press has been constructed
on the basis of the type C standard, the manufacturer may choose any
one of three further procedures. Either he sends the technical documentation to the testing authority, which only confirms receipt, or he
arranges for a review of the technical documentation, or he commis-
sions an EC type test of the machine.

In certain machines it is necessary to reach directly with the hands
between the two halves of the die after each stroke. These machines are
categorized as machines with manual feed or removal.

Even machines in which parts are inserted manually during set-up
are not categorized as annex IV machines. In the case of machines not
designed to permit manual feed or removal, the manufacturer issues
the declaration of conformity and applies the CE mark at his own risk.
Consequently, a press intended exclusively for automatic operation

cannot be used for manual feed operation. This possibility is excluded in the operating instructions – as far as possible, it should be possible for the control system to prevent manual production.

Certificates issued by the manufacturer himself, however, must satisfy certain conditions. In addition to adhering to the major health and safety requirements of the ECMD, the manufacturer is required to draw up a document prior to issuing the declaration of conformity in which he provides written proof of compliance with these safety requirements. The statutory retention period for this document is 10 years. The documentation, intended for the customer, comprises the operating instruction manual, which contains the necessary information for the machine user. The focal point of the documentation is the hazard analysis and risk assessment (Table 3.6.2). This involves a systematic evaluation of the machine by the manufacturer concerning potential hazards, an evaluation of the risk involved, and the appropriate measures for their minimization. These must be described in the documentation. The outcome takes on the form of a list of solutions undertaken to comply with the points derived from annex I of the ECMD. In addition, the documentation includes plans and diagrams – circuit diagram, hydraulic and pneumatic diagrams and functional characteristics –, calculations, test results, certificates and manufacturer's declarations from suppliers in as far as these are necessary for verification of the required safety standard. This documentation does not need to be physically

Table 3.6.2: Extract from a hazard analysis with risk assessment

1 Hazardous situation, danger area	Type of danger	Number of hazard in accordance with EN414	Part event	Continuous operation	Set-up	Maintenance	Repairs	Degree of severity	Overall frequency category	Bypassable	Protection category	Safety gear / remark	Doc. No.	89/392, annex I	Applied standards EN692	Others
2 Active area, die mounting area	Crushing, shearing	1.1, 1.2	Reaching in from front and other hazards in active area during operation	x				3	3	0	4	Protective shield 1st safeguarding to category 4, n.o. and n.c. contacts, early opening tS, safety distance 320 mm to match stopping time 0.2 s to EN999. Together with swivelling gripper box prevention of access to danger zones (EN294) (see drawing)	1, 2	1.1.2a,f, 1.3.7, 1.3.8, 1.4.2.2B, 1.4.1, 1.6.1	5.3.3, 5.3.5., 5.3.6, 5.3.10, 5.3.12, 5.3.15, 5.5.6, 5.5.7, 5.5.8	EN294 EN999 EN954 EN953 EN1088

present, but comprises an index which makes reference to the various sections by means of an identification number.

In the case of a hazard analysis, safety is monitored over each phase of the machine's life, from assembly through to commissioning, utilization, repair and final scrapping. Protective measures must apply equally to machine operators, set-up and maintenance staff. Depending on the risk, safety must also be guaranteed in case of machine malfunctions. As, for example, the failure of a brake valve and thus the unchecked movement of the slide could lead to serious injury, dynamically monitored press safety valves must be used unless it is possible to provide an equivalent degree of safety by keeping safety devices closed. The monitored valves are redundant (double valves) and, depending on the requirements of their respective functions, they are self-monitoring. This means that despite a brake valve failure, the "braking" function is still performed, accompanied by a message indicating whether the two valves have operated correctly or not. A positive "function OK" signal is used for self-monitoring of the signalling module (cf. Sect. 3.5).

In analysing potential hazards, the manufacturer must also consider "foreseeable unusual situations", e.g. surprise situations of the type previously mentioned. The operating instructions must accordingly include a statement on machine utilization in accordance with the intended purpose of the machine, and also prohibit any foreseeable cases of violation, for example operation of the press while observers are located behind the closed protective gear in order to observe the machine operation. Indications of possible violation of operating regulations are often only possible after observation of operating, set-up and maintenance staff. The results of the hazard analysis are reflected in the design of protective gear, remarks relating to residual risk at the machine and in the operating instruction manual. Thus, they are made known to the machine user.

3.6.5 Measures to be undertaken by the user

When purchasing a machine, the user must establish by means of a contractual agreement the identity of the manufacturer as defined by the ECMD and the position with regard to items of equipment provided by the user. When purchasing machines, whether new or used, from foreign countries outside of Europe, the user should contractually stipulate compliance with the current revision of 89/EEC/392. European manufacturers are required in any case to comply with this legal requirement, making a contractual provision to this effect unnecessary.

When the user himself performs no conversion work on the press, he is only required to be acquainted with the provided operating instructions and the accident prevention regulations applicable for press operation. In Germany, the following accident prevention regulations are among the most important in this context:

- VBG 7n5.1 for eccentric and related presses,
- VBG 7n5.2 for hydraulic presses,
- VBG 7n5.3 for spindle presses,
- VBG 5 for power-driven operating equipment.

Directive 89/655/EEC on the provision of equipment obliges the user to retrofit all machines to a defined minimum standard by January 1, 1997. It is safe to assume that users who converted their machines in line with accident prevention regulations valid up until December 31, 1993, will be unaffected by this ruling.

If the user intends to perform his own conversion work on the machine, he assumes full responsibility for his actions. He must accordingly be acquainted with all the valid directives, regulations and standards. Conversion of an annex IV machine, i. e. a press intended for manual feeding and unloading, requires notification of the testing body which performed the EC type test. The conversion of an automatic machine for manual feeding or unloading calls for EC type testing unless the complete construction has been performed in compliance with the C standard. Although the ECMD does not apply to used machinery, i.e. machines coming into circulation in the EEC prior to December 31, 1994. These machines do not bear the CE mark and do contain certain provisions which are of importance in this context. Used machines from foreign countries outside Europe must always bear the CE mark and be accompanied by a conformity declaration. The conversion of used machines must be performed with a view to the health and safety requirements of annex I of the ECMD. However, in case of major modifications, functional changes or increased performance, the entire machine or producton line must be subjected to a conformance check by the construction foreman, after which it will receive the CE mark.

3.6.6 Safety requirements in the USA

The following brief analysis of safety requirements in the USA provides a representative comparison with safety regulations applicable outside Europe. Alongside the ANSI (American National Standards Institute) Standard B11.1-1971, the regulations of the OSHA (Occupational Safety and Health Administration) apply.

ANSI national standards cover a number of areas including safety equipment, although application of these standards is voluntary. In contrast, compliance with the OSHA regulations is a statutory requirement.

This stipulates that a Lock-Out Procedure to §1910.147 be used on all machines. This requirement specifies a description of how each type of energy and each energy source can be rendered harmless. This must be performed in four stages: shut-down of the energy source, dissolution of residual energy, a check of energy-free status and safeguarding the energy-free status against unintentional restart. This step-by-step approach for the isolation of energy sources has been adopted by the EN1037 to apply to European standards. As a minimum requirement, main switches must be capable of locking in the energy-free status. The OSHA represents the philosophy that every staff member should carry padlocks usable only by himself or herself which safely eliminate any possibility of restart by third parties. Only he or she can cancel the disabled status.

Like the ANSI standard, the OSHA requires from the manufacturer the compilation of instructions ensuring safe machine operation and maintenance. According to OSHA, equipment must be delivered with a lock-out plate in accordance with *Fig. 3.6.4* indicating the devices required to render the equipment harmless using the four steps listed above for each of the energy sources used. The warning pictograms are also attached at the relevant points of the machine. As mentioned earlier, the OSHA places particular emphasis on the training of operators, set-up personnel and maintenance staff in particular. The objective of the training is to indicate possible sources of accidents, promote safety-conscious work and point out the purpose and function of the safety gear. Employers should provide training on initial transfer and commissioning of the equipment and thereafter once a year, and have this confirmed by means of a signature. Application of the OSHA regulations varies from region to region, as the OSHA orients its interpreta-

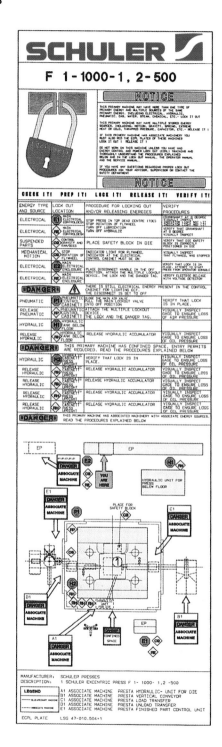

Fig. 3.6.4
Lock-out plate in accordance with
US regulations

tion of the safety measures to the regulations already existing in the respective companies. In the USA, the first burden of proof lies with the owner of the equipment who accordingly has an interest in equipment safety. For this reason, it makes sense for details of safety engineering to be coordinated with the safety officer of the company running the machine.

In view of the almost unpredictable amounts reached by damage claims in the USA, qualified safety engineering and good documentation are gaining in significance.

Characteristic requirements imposed on mechanical presses are defined in §1910.217, and those imposed on forging machines in §1910.218. Compliance Safety and Health Officers from the Department of Labor should be given access to monitor compliance with safety requirements. Much of this responsibility lies with the equipment user.

3.7 Casting components for presses

The proportion of cast components in presses in terms of their weight differs considerably. In individual presses such as minting and coining presses (cf. Sect. 6.8.7) or automatic notching machines (cf. Sect. 4.6.4), some 80 % of the total press weight is made up of gray cast iron, in general in nodular cast iron. This proportion is made up predominantly of large components such as uprights, press frames and press bodies (cf. Fig. 3.1.1). In the case of large presses with individual part weights exceeding 20 t, however, welded constructions predominate. Freedom from scale, good machining properties, non-leakage of oil and adherence to mechanical values laid down in the drawing specifications are criteria which govern the quality of castings. They determine the compliance of castings with general mechanical engineering requirements. Due to the size of the castings and their low production volume, these parts are typically made using the manual molding technique, i.e. without the use of mechanical aids.

The casting industry is challenged to a far greater degree by the production of components to drive mechanical presses, in particular transfer presses. Eccentric wheels, for example, cam wheels, reducing gears, connecting rods and clutch components represent a particular challenge to the castings producer due to the special quality requirements involved (cf. Sect. 3.2.6). The focal point of this type of drive system is the eccentric wheel, which can have a part weight ranging anywhere from 5 to 25 t (Fig. 3.7.1). For this type of component, ductile castings in GGG-70 quality are increasingly used, while in the past it was customary to use steel castings. This change necessitated extensive experimentation and continuous further development from the point of view

▲ **Fig. 3.7.1** Eccentric wheel of a large-panel transfer press

both of metallurgical and casting technology. Maximum care in the production process is an underlying requirement. According to the current state of the art in the field of foundry practice, the following stringent requirement criteria can be covered with GGG-70:

– high resistance to wear, i. e. the microstructure must be purely perlitic even with wall thicknesses of up to 200 mm,
– guaranteed hardness of 240 to 300 HB in all wall thicknesses – also at the tooth root,
– elongation over 2 % in order to avert fractures and to increase resistance to overloading.

In its raw cast state, i.e. without transformation annealing, these mechanical requirements are achieved using a selective alloying technique. The application of cooling elements made of iron, steel and silicon carbide, either cooled or uncooled, also serves to enhance hardness.

However, the casting technician can only alloy and cool within restricted limits, as elongation drops with increasing hardness. Due to this type of metallurgical-technological operation, and because of their geometrical shape, castings such as eccentric wheels are inevitably exposed to high degrees of residual stress. As a result, stress relief annealing at 560 to 580 °C is an absolute necessity.

In the foundry, quality standards must be assured in order to avert the risk of cracks and fractures in the castings. The marked tendency of the material to form scale substantially impedes the production of components made of GGG-70. The porosity risk therefore is addressed by feeding and selective solidification. All components, naturally including also large parts for press drive systems, are subject to highly stringent quality controls, including ultrasound, X-ray, crack detection and hardness testing, as well as testing of mechanical values using test-bar specimen, cast together with produced component.

Bibliography

Anselmann, N. (editor): Europäisches Recht der Technik, Beuth Verlag, Berlin, Vienna, Zurich (1993).

ANSI B11.1-1971: Safety requirements for the construction, care, and use of mechanical power presses, American National Standards Institute Inc., New York (1971).

Becker, U., Ostermann, H.-J.: Wegweiser Maschinenrichtlinie, Bundesanzeiger-Verlag, Cologne (1995).

Bodenschatz, W., Fichna, G., Voth, D.: Produkthaftung, Maschinenbau-Verlag, Frankfurt/Main, 4th edition. (1990).

DIN EN 692 (Draft): Mechanische Pressen – Sicherheit, Beuth Verlag, Berlin (1992).

DIN EN 693 (Draft): Hydraulische Pressen – Sicherheit, Beuth Verlag, Berlin (1992).

DIN EN 1037 (Draft): Sicherheit von Maschinen, Trennung von der Energiezufuhr und Energieabbau, Vermeidung von unerwartetem Anlauf, Beuth-Verlag, Berlin (1993).

Hoffmann, H., Schneider, F.: Schwingen verhindert-Hydraulische Zieheinrichtung für die Fertigung mit hoher Qualität in einfachwirkenden Pressen, Maschinenmarkt 95 (1989) 51/52.

Hoffmann, H.: Vergleichende Untersuchung von Zieheinrichtungen, Bänder Bleche Rohre (1991) 12 and (1992) 1.

Kürzinger, F.-X.: Kraftneutrale Parallelregelung über gesamten Hub, Bänder Bleche Rohre (1995) 9.

Occupational Safety and Health Administration: OSHA Regulations 29 CFR Ch XVII, Part 1910, Office of the Federal Register, Washington D.C., 7-1-92 Edition (1992).

Schmidt, K.: Die neue Druckbehälterverordnung, WEKA Fachverlag, Augsburg (1995).

Schwebel, R.: Geringe Parallelitätsabweichungen - Anwendungsspektrum hydraulischer Pressen, Industriebedarf (1994) 9.

Süddeutsche Metall-Berufsgenossenschaft: Unfallverhütungsvorschriften, Carl Heymanns Verlag KG, Luxemburg (1994).

TÜV Hannover-Sachsen-Anhalt: Kurzinformation EG-Maschinen-Richtlinie, Eigenverlag (1995).

4 Sheet metal forming and blanking

■ 4.1 Principles of die manufacture

■ 4.1.1 Classification of dies

In metalforming, the geometry of the workpiece is established entirely or partially by the geometry of the die. In contrast to machining processes, significantly greater forces are necessary in forming. Due to the complexity of the parts, forming is often not carried out in a single operation. Depending on the geometry of the part, production is carried out in several operational steps via one or several production processes such as forming or blanking. One operation can also include several processes simultaneously (cf. Sect. 2.1.4).

During the design phase, the necessary manufacturing methods as well as the sequence and number of production steps are established in a processing plan *(Fig. 4.1.1)*. In this plan, the availability of machines, the planned production volumes of the part and other boundary conditions are taken into account.

The aim is to minimize *the number of dies to be used* while keeping up a high level of operational reliability. The parts are greatly simplified right from their design stage by close collaboration between the Part Design and Production Departments in order to enable several forming and related blanking processes to be carried out in one forming station.

Obviously, the more operations which are integrated into a single die, the more complex the structure of the die becomes. The consequences are higher costs, a decrease in output and a lower reliability.

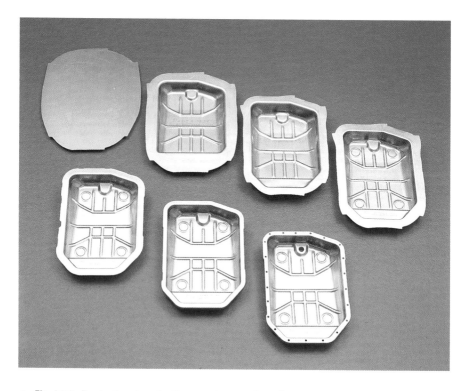

▲ **Fig. 4.1.1** Production steps for the manufacture of an oil sump

Types of dies

The type of die and the closely *related transportation of the part between dies* is determined in accordance with the forming procedure, the size of the part in question and the production volume of parts to be produced.

The production of large sheet metal parts is carried out almost exclusively using *single sets of dies*. Typical parts can be found in automotive manufacture, the domestic appliance industry and radiator production. Suitable transfer systems, for example vacuum suction systems, allow the installation of double-action dies in a sufficiently large mounting area. In this way, for example, the right and left doors of a car can be formed jointly in one working stroke *(cf. Fig. 4.4.34).*

Large size single dies are installed in large presses. The transportation of the parts from one forming station to another is carried out mechanically. In a press line with single presses installed one behind the other, feeders or robots can be used *(cf. Fig. 4.4.20 to 4.4.22)*, whilst in large-panel transfer presses, systems equipped with gripper rails *(cf. Fig. 4.4.29)* or crossbar suction systems *(cf. Fig. 4.4.34)* are used to transfer the parts.

Transfer dies are used for the production of high volumes of smaller and medium size parts *(Fig. 4.1.2)*. They consist of several single dies, which are mounted on a common base plate. The sheet metal is fed through mostly in blank form and also transported individually from die to die. If this part transportation is automated, the press is called a transfer press. The largest transfer dies are used together with single dies in large-panel transfer presses *(cf. Fig. 4.4.32)*.

In *progressive dies*, also known as *progressive blanking dies,* sheet metal parts are blanked in several stages; generally speaking no actual forming operation takes place. The sheet metal is fed from a coil or in the form of metal strips. Using an appropriate arrangement of the blanks within the available width of the sheet metal, an optimal material usage is

▲ **Fig. 4.1.2** Transfer die set for the production of an automatic transmission for an automotive application

ensured *(cf. Fig. 4.5.2 to 4.5.5)*. The workpiece remains fixed to the strip skeleton up until the last operation. The parts are transferred when the entire strip is shifted further in the work flow direction after the blanking operation. The length of the shift is equal to the center line spacing of the dies and it is also called the step width. Side shears, very precise feeding devices or pilot pins ensure feed-related part accuracy. In the final production operation, the finished part, i. e. the last part in the sequence, is disconnected from the skeleton. A field of application for progressive blanking tools is, for example, in the production of metal rotors or stator blanks for electric motors *(cf. Fig. 4.6.11 and 4.6.20)*.

In *progressive compound dies* smaller formed parts are produced in several sequential operations. In contrast to progressive dies, not only blanking but also forming operations are performed. However, the workpiece also remains in the skeleton up to the last operation *(Fig. 4.1.3 and cf. Fig. 4.7.2)*. Due to the height of the parts, the metal strip must be raised up, generally using lifting edges or similar lifting devices in order to allow the strip metal to be transported mechanically. Pressed metal parts which cannot be produced within a metal strip

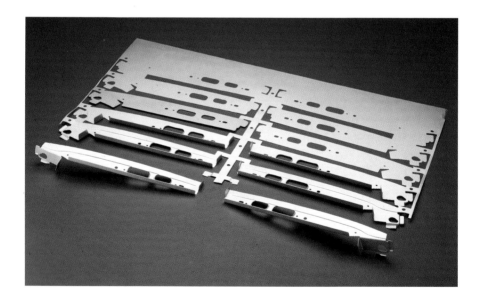

▲ **Fig. 4.1.3** Reinforcing part of a car produced in a strip by a compound die set

because of their geometrical dimensions are alternatively produced on transfer sets.

Next to the dies already mentioned, a series of *special dies* are available for special individual applications. These dies are, as a rule, used separately. Special operations make it possible, however, for special dies to be integrated into an operational sequence. Thus, for example, in *flanging dies* several metal parts can be joined together positively through the bending of certain metal sections *(Fig. 4.1.4* and *cf. Fig. 2.1.34)*. During this operation reinforcing parts, glue or other components can be introduced.

Other special dies locate special connecting elements directly into the press. *Sorting and positioning elements,* for example, bring stamping nuts synchronised with the press cycles into the correct position so that the *punch heads* can join them with the sheet metal part *(Fig. 4.1.5)*. If there is sufficient space available, forming and blanking operations can be carried out on the same die.

Further examples include bending, collar-forming, stamping, fine blanking, wobble blanking and welding operations *(cf. Fig. 4.7.14* and *4.7.15)*.

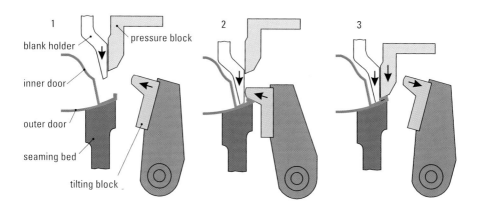

▲ **Fig. 4.1.4** A hemming die

◀ Fig. 4.1.5 A pressed part
with an integrated
punched nut

■ 4.1.2 Die development

Traditionally the business of die engineering has been influenced by
the automotive industry. The following observations about the die
development are mostly related to body panel die construction. Essential statements are, however, made in a fundamental context, so
that they are applicable to all areas involved with the production of
sheet-metal forming and blanking dies.

Timing cycle for a mass produced car body panel
Until the end of the 1980s some car models were still being produced for
six to eight years more or less unchanged or in slightly modified form.
Today, however, production time cycles are set for only five years or less
(Fig. 4.1.6). Following the new different model policy, the demands on
die makers have also changed fundamentally. Comprehensive contracts
of much greater scope such as Simultaneous Engineering (SE) contracts
are becoming increasingly common. As a result, the die maker is often
involved at the initial development phase of the metal part as well as
in the planning phase for the production process. Therefore, a much
broader involvement is established well before the actual die development is initiated.

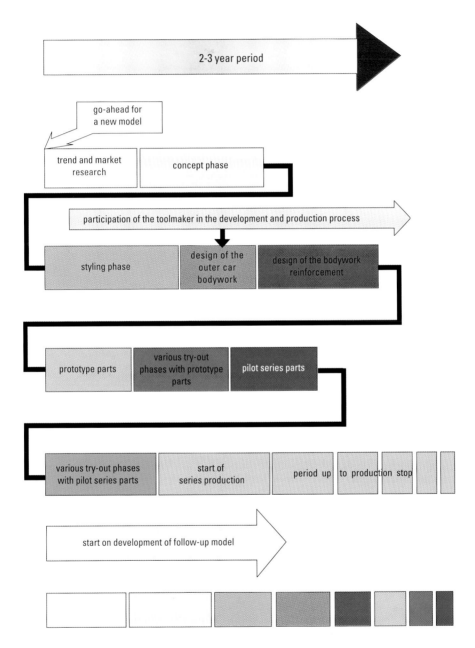

2-3 year period

go-ahead for
a new model

trend and market
research

concept phase

participation of the toolmaker in the development and production process

styling phase

design of the
outer car
bodywork

design of the bodywork
reinforcement

prototype parts

various try-out
phases with prototype
parts

pilot series parts

various try-out phases
with pilot series parts

start of
series production

period up to production stop

start on development of follow-up model

▲ **Fig. 4.1.6** Time schedule for a mass produced car body panel

The timetable of an SE project

Within the context of the production process for car body panels, only a minimal amount of time is allocated to allow for the manufacture of the dies. With large scale dies there is a run-up period of about 10 months in which design and die try-out are included. In complex SE projects, which have to be completed in 1.5 to 2 years, parallel tasks must be carried out. Furthermore, additional resources must be provided before and after delivery of the dies. These short periods call for precise planning, specific know-how, available capacity and the use of the latest technological and communications systems. The timetable shows the individual activities during the manufacturing of the dies for the production of the sheet metal parts *(Fig. 4.1.7)*. The time phases for large scale dies are more or less similar so that this timetable can be considered to be valid in general.

Data record and part drawing

The data record and the part drawing serve as the basis for all subsequent processing steps. They describe all the details of the parts to be

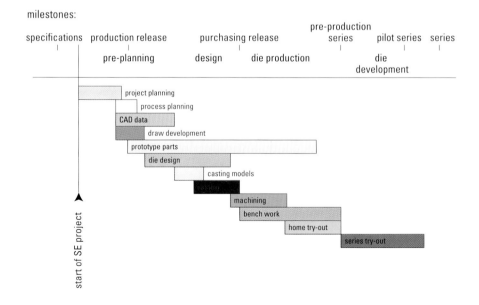

▲ **Fig. 4.1.7** Timetable for an SE project

produced. The information given in the part drawing includes: part identification, part numbering, sheet metal thickness, sheet metal quality, tolerances of the finished part etc. *(cf. Fig. 4.7.17)*.

To avoid the production of physical models (master patterns), the CAD data should describe the geometry of the part completely by means of line, surface or volume models. As a general rule, high quality surface data with a completely filleted and closed surface geometry must be made available to all the participants in a project as early as possible.

Process plan and draw development
The process plan, which means the operational sequence to be followed in the production of the sheet metal component, is developed from the data record of the finished part *(cf. Fig. 4.1.1)*. Already at this point in time, various boundary conditions must be taken into account: the sheet metal material, the press to be used, transfer of the parts into the press, the transportation of scrap materials, the undercuts as well as the sliding pin installations and their adjustment.

The draw development, i.e. the computer aided design and layout of the blank holder area of the part in the first forming stage – if need be also the second stage –, requires a process planner with considerable experience *(Fig. 4.1.8)*. In order to recognize and avoid problems in areas which are difficult to draw, it is necessary to manufacture a physical analysis model of the draw development. With this model, the forming conditions of the drawn part can be reviewed and final modifications introduced, which are eventually incorporated into the data record *(Fig. 4.1.9)*.

This process is being replaced to some extent by intelligent simulation methods, through which the potential defects of the formed component can be predicted and analysed interactively on the computer display.

Die design
After release of the process plan and draw development and the press, the design of the die can be started. As a rule, at this stage, the standards and manufacturing specifications required by the client must be considered. Thus, it is possible to obtain a unified die design and to consider the particular requests of the customer related to warehousing of standard, replacement and wear parts. Many dies need to be designed so that they can be installed in different types of presses. Dies are frequently

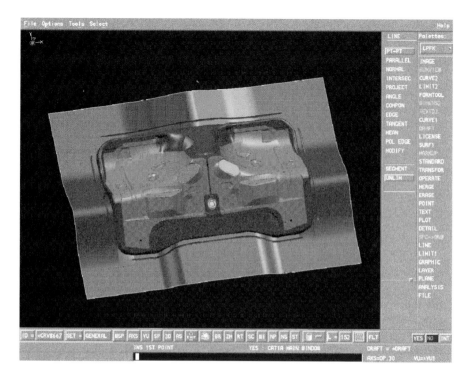

▲ **Fig. 4.1.8** CAD data record for a draw development

installed both in a production press as well as in two different separate back-up presses. In this context, the layout of the die clamping elements, pressure pins and scrap disposal channels on different presses must be taken into account. Furthermore, it must be noted that drawing dies working in a single-action press may be installed in a double-action press (cf. Sect. 3.1.3 and *Fig. 4.1.16*).

In the design and sizing of the die, it is particularly important to consider the freedom of movement of the gripper rail and the crossbar transfer elements (cf. Sect. 4.1.6). These describe the relative movements between the components of the press transfer system and the die components during a complete press working stroke. The lifting movement of the press slide, the opening and closing movements of the gripper rails and the lengthwise movement of the whole transfer are all superimposed. The dies are designed so that collisions are avoided and a minimum clearance of about 20 mm is set between all the moving parts.

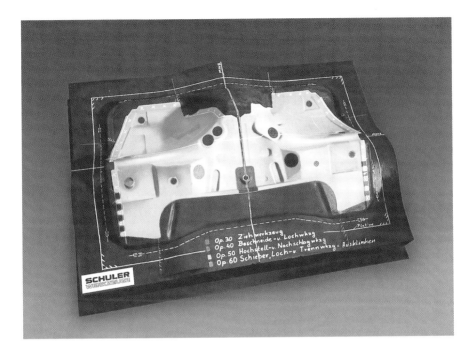

▲ **Fig. 4.1.9** Physical analysis model of a draw development

Guides of the upper and lower dies must carry the shearing forces and be sized according to the calculated and anticipated forces and functions. Large scale gibs show nearly no lateral displacement and a consistent, unchanging pressure pattern of the operational die (cf. Sect. 4.1.5). Furthermore, they reduce the tilting of the die and the press slides *(Fig. 4.1.10)*. Thus, a consistent overall high quality of pressed parts is achieved.

In order to avoid errors and potential problem areas later in production, the so-called FMEA (Failure Mode and Effects Analysis) is used with proven success during the design stage with the close collaboration of all those involved with the project. The design of the pressed parts and their arrangement and positioning in the dies have a great influence on production safety. In sheet metal forming well-designed press components will lead to simple, production-friendly and low cost tools. Difficult to form press parts, on the other hand, involve expensive, complicated and high cost dies, which are less easily integrated

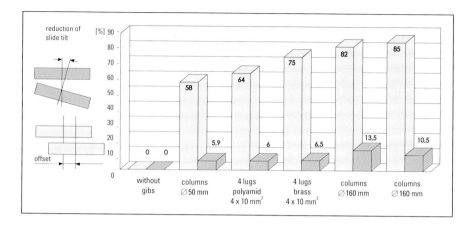

▲ **Fig. 4.1.10** Influence of the guidance system on the reduction of slide tilting and offset

into the production process. Tilting of the parts in the dies can compensate the negative aspects of an unfavorable part geometry. For example, for flanging and blanking a sheet metal part can be so rotated that the work direction corresponds to the direction of the flanging operation e.g. the movement of the blanking punch *(Fig. 4.1.11)*. However, it is often then necessary to increase the number of dies, which results in production cost increases.

To meet deadlines, the forming dies must be designed first so that the casting patterns and the finished castings are available for use as soon as possible. After machining and try-out, dies can produce stampings that are laser trimmed and used in prototype pre-production assemblies *(Fig. 4.1.7)*.

The use of CAD systems for die design has become increasingly common *(Fig. 4.1.12)* even though conventional drawing board design may often be more cost effective. The reasons for this are that the drawing system is not yet perfected, data banks are incomplete and workstation screens only allow a perspective view in the ratio of 1 : 5 up to 1 : 10. However, considering the total production costs, high quality CAD produced die designs and die casting pattern data are provided as a by-product. Shrinkage and expansion allowances can be determined by computer and used for NC machining of casting models, an economic alternative to NC machining of castings. A further advantage of CAD design, in so far as it is based on a volume model, is the possibility

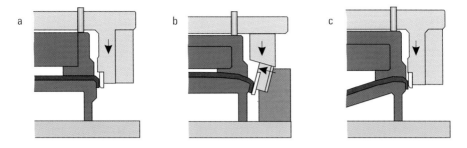

▲ **Fig. 4.1.11** Improvement to the positioning of the part by means of swivel movement in the folding operation:
    a   simple die construction for press parts with simple geometry;
    b   complex high cost die construction with a sliding element for difficult press part geometry;
    c   simple die construction via appropriate swivelling of the pressed part in an improved working position

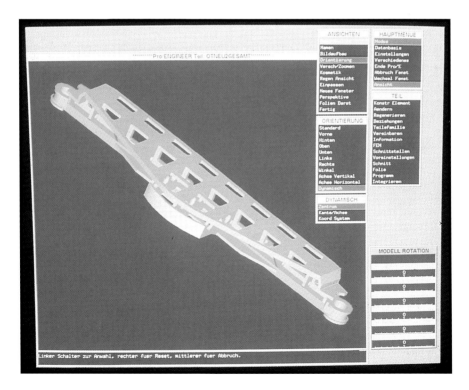

▲ **Fig. 4.1.12** CAD design of pivoting blanking tool

of ongoing collision testing, which clearly contributes to the reduction of errors. Jaw layout plans, die construction plans and single part drawings can be made available by this system. The general use of CAD/CAM systems in tool making have many inherent advantages that are linked to the process. Therefore, in the medium term, CAD/CAM techniques will be increasingly used to increase the total cost effectiveness of part production.

Mechanical processing
Today's technical possibilities in NC processing require ever more easy to manufacture designs from the die designer. In particular the use of five-axis milling machines allows to machine completely complex castings with a single fixturing operation. In the past, because of restricted conditions, extended operating and assembly activities were carried out with fixtures and additional clearances. Today, for example, three-dimensionally located sliding and guiding components can be cast *(Fig. 4.1.17)*.

High speed milling (HSM) is used increasingly for machining of sculptured surfaces in which economical cutting speeds of 2,000 m/min are attainable. Even strongly patterned surfaces for inner parts can still be machined with medium feed rates of around 5 m/min. Cermets and special PVD-coated carbides are capable of attaining surface qualities of $R_t = 10$ mm, which reduces the need for secondary manual polishing to a minimum.

Assembly and try-out
More accurate machining and finishing operations together with the above-mentioned advantages of a CAD supported die design, simplify the assembly task considerably. The general processing of data gives access to further theoretical and error-free blanking and contour information without having to resort to the use of costly aids. The expert knowledge of the qualified diemaker is, however, still necessary. In particular, in the pre-run or try-out phases when the die developer tries to produce a good part after the installation of the dies in the press, it becomes evident whether or not the theoretical design effort has been correct (cf. Sect. 4.1.5). Prototype tools or simulation methods provide, of course, considerable help. There are, however, additional factors which contribute towards the quality of the stampings obtained with

the production dies. The machining of the die surfaces and the gibs, the milling of the free form surface using HSM operations and the try-out on a precisely guided try-out press with adjustable hydraulic draw cushions should all be carried out efficiently *(Fig. 4.1.19* and *4.1.20)*.

Process control
In order to assure that the part dependent process plan and the draw development are correct, prototype dies or soft tools are produced *(Fig. 4.1.13)*. In so doing, the forming behavior of the sheet metal material, especially when the part geometry is complex and more prone to failure, must be investigated at the preliminary stages of the die design. However, the practise of using soft tools dedicated exclusively for this purpose is becoming less significant as computer-aided simulation processes are being increasingly used for die design and development.

Prototype dies
The preparation of prototype or soft dies is, however, worthwhile if further objectives need to be attained. At a very early point in the production process, the sheet metal parts are made available, which can be used, for instance, to construct vehicles for assembly and crash test pur-

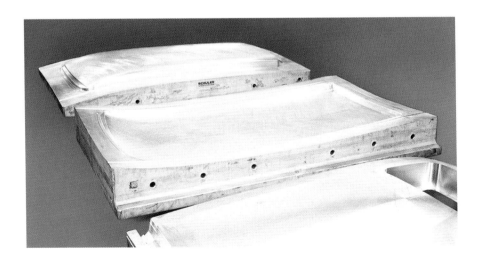

▲ **Fig. 4.1.13** Prototype drawing die in aluminium for the production of the roof of a station wagon

poses. In certain cases, these parts may be used as pre-production series *(Fig. 4.1.7)*. A further consideration which justifies the production of prototype dies is when the timing of their manufacture is close to the schedule for the beginning of series production. Thus, information on the formability and tolerances of the parts can be obtained, which contributes to goal oriented and cost-effective die manufacture.

The following are some of the points which must be taken into account here:

- the die material
- the complexity of the sheet metal part
- the blank material
- the design of the draw development
- the geometrical parameters such as degree of shrinkage, cambering, prestressing
- the process-related variable such as lubrication, blank holder forces, work-flow direction, draw beads
- the size and geometry of the blanks
- special factors and parameters which influence the results of the stamping operation

The type and design of the prototype dies are influenced by different factors. The choice of prototype material will be in line with the sheet metal specification, thickness and the number of parts to be produced. Today, plastics, aluminium alloys *(Fig. 4.1.13)*, different qualities of cast metals as well as rolled steel can be used as prototype or soft die material. A summary of the choice and application criteria is given in Table 4.1.1.

Methods of process simulation
Computer-aided simulation techniques, e.g. the analysis of the forming process without any physical dies, will increasingly replace the use of prototype and soft dies. This development is becoming more practical through rapidly growing hardware developments which allow the processing of ever increasing amounts of computerised data in ever shorter time spans. In addition, the computer software used for process simulation has become more reliable and more user friendly.

Table 4.1.1: Allocation of prototype dies

Die type	Plastic		Aluminium alloys		Iron alloys	
	without additives	*alloyed or fiber-reinforced*	*Zamak*	*Cerrotru*	*GS/GG cast iron*	*Plate steel*
Manufacturing process	complete structure in plastic, cast in layers or with back as the front casting	cast steel frame as basic structure for plastic panels (punch), casting of female die shape	model production for punch and female die (possibly blank holder), casting of models, reworking of functional surfaces	precise model manufacture, casting the punch, insertion of sheet metal thickness, casting of female die (no reworking)	model manufacture for punch and female die (possibly blank holder), casting of models, reworking of functional surfaces, integration of gib elements	construction of die from plates and jaws, screwed and pinned, reworking of functional surfaces, integration of gib elements
Repair and possibility for modification	machinable shape or new front casting	punch shape machinable, inserts are seamlessly glued in, casting of female die	weldable, inserts machinable	melt down existing die and recast	welding on, screw-mounted inserts, reworking by means of milling	welding on, new jaws or inserts
Range of parts	outer bodywork parts in steel and aluminium	outer bodywork parts and reinforcements in steel and aluminium	outer bodywork parts and reinforcements	outer bodywork parts and easily formable parts	reinforcement and chassis parts with complex shapes	reinforcement and chassis parts with flat shapes
Sheet metal thickness	sheet steel up to 2 mm, aluminium up to 4 mm	up to 2.5 mm	up to 1.5 mm	up to 1.2 mm	up to 3 mm	up to 3 mm
Piece numbers	with max. sheet metal thickness, extremely low piece numbers, otherwise suitable for small or pre-series	for outer bodywork parts, up to 1,000 pieces	easily deformable parts up to 100 pieces, complex parts up to 50 pieces	easily deformable parts up to 50 pieces, complex parts up to 20 pieces	depending on stress level and sheet metal material used, possibly with series capability	
Product. period	2 to 3 weeks	4 to 6 weeks	3 to 4 weeks	2 to 3 weeks	7 to 10 weeks	5 to 7 weeks

For the evaluation of the geometry data, which has been established by the processing plan and the draw development as well as the production parameters, CAE solutions are available which are based upon mathematical and plasto-mechanical equations or on experimentally proven data. Analytical and numerical simulation belong to these solutions. The numerical methods, and here in particular the Finite Element Method (FEM), deliver today the most accurate results. Stampings can be almost completely interactively analysed on a workstation in relation to the most important failure criteria. The results are the distribution of stress (tears) *(Fig. 4.1.14)*, the tendency towards wrinkling *(Fig. 4.1.15)* and the variation of the sheet metal thickness. The necessary steps and the prerequisites for a simulation run are as follows:

– preparation of CAD data for the punch, female die and blank holder as well as the total geometry of the draw development in the working direction *(Fig. 4.1.8)*,
– determination and input of the blank size (cf. Sect. 4.2.1 and 4.5),
– input of the process parameters such as sheet metal material, rolling direction, blank holder force, frictional conditions,
– definition i.e. establishment of the FE mesh (partly automatic) and
– initiation of the calculation process.

By varying the process and die parameters, the simulation can be repeated as often as required until a failure-free drawing part is produced. The advantage of this process lies in the fact that any objective and necessary component changes can be introduced at any time prior to the actual manufacture of the dies. Specific measures for the controlling metal flow, for example through the shape of the blank or through the use of draw beads, can be simulated. Similarly, calculations can be made to predict and compensate for spring back (cf. Sect. 4.8.1).

Through process simulation, process planners and die designers receive reliable information at an early stage in the die development, which really contributes to the production of fully operational and reliable dies.

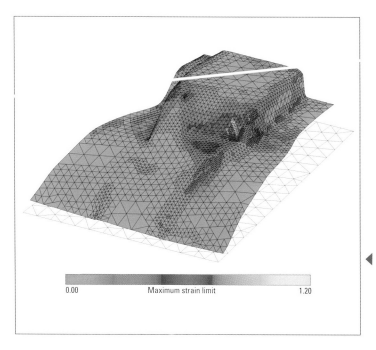

0.00   Maximum strain limit   1.20

◀ **Fig. 4.1.14**
Representation
of stress
distribution by
means of FEM
simulation

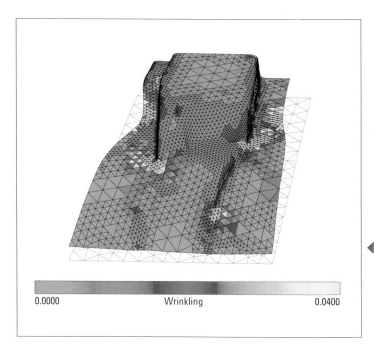

0.0000   Wrinkling   0.0400

◀ **Fig. 4.1.15**
Representation
of possible
wrinkle
formation by
means of FEM
simulation

■ 4.1.3 Die materials

The processing of high strength sheet metal, which is becoming more widely used in stamping technology, calls for higher pressing forces. This means higher loads on the dies and greater overall wear on the tools. In order to meet the requirements for long die life, while maintaining acceptable component quality, special materials – most of which have been hardened or coated – must be used in the construction of the dies. The basic structure of different types of dies and their most important components are shown in *Figs. 4.1.16* and *4.1.17.*

Table 4.1.2 lists the various components of dies used for drawing, blanking, folding and flanging as well as the materials used to manufacture them. Table 4.1.3 lists various surface hardening processes and coating methods for dies. Coatings, for example through hard chromium plating, can improve the sliding characteristics in deep drawing, so that the need for lubrication is greatly reduced.

■ 4.1.4 Casting of dies

The casting process is one of the steps in the overall process of manufacturing production dies for sheet metal forming where the highest standards are required by die makers. In contrast to castings used in general machinery manufacture, in die manufacture, multiple usage of a model is not possible, as no two tools are exactly the same. A disposable pattern made from polystyrene (polyester, hard foam) is therefore used for the die casting process *(Fig. 4.1.18)*.

The casting is carried out as a full form casting process: the polystyrene pattern stays in its resin-hardened quartz sand form. The liquid metal, which is poured in, evaporates or rather melts the pattern. With the full form casting process, the foundry industry has developed a production method which offers a large measure of design-related, technical, economic and logistical improvements in die making. For the designer, there is the possibility of precise measurement and functional control, as the full form pattern is identical in shape to the piece being cast. Product development is made easier, as the pattern can be altered without any problems by means of machining or gluing processes. Furthermore, the costs of core box, core production, core transport and core installation are reduced.

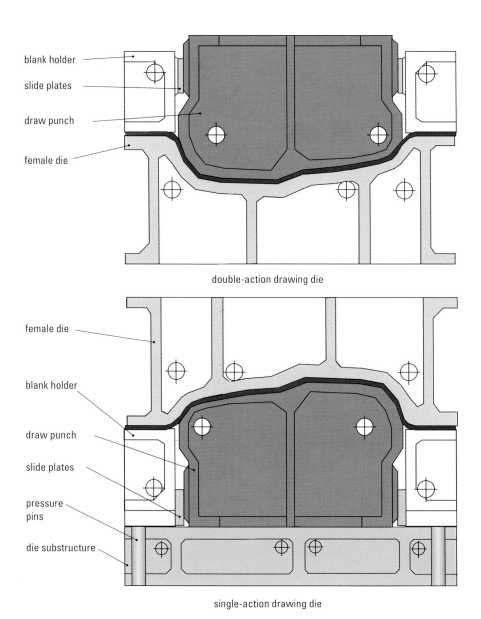

blank holder

slide plates

draw punch

female die

double-action drawing die

female die

blank holder

draw punch

slide plates

pressure pins

die substructure

single-action drawing die

▲ **Fig. 4.1.16** Basic design of different draw dies

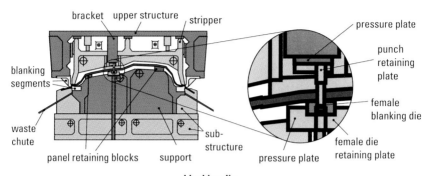

bracket upper structure stripper

pressure plate

punch retaining plate

blanking segments

female blanking die

waste chute

sub-structure

female die retaining plate

panel retaining blocks   support

pressure plate

blanking die

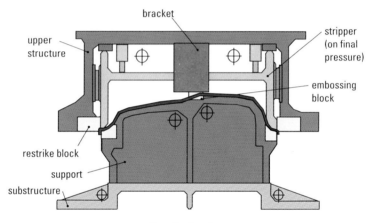

bracket

upper structure

stripper (on final pressure)

embossing block

restrike block

support

substructure

restrike edge bending die

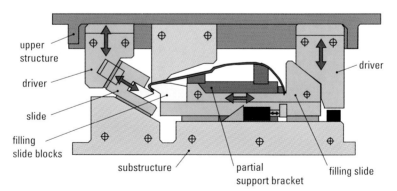

upper structure

driver

driver

slide

filling slide blocks

substructure

partial support bracket

filling slide

flanging die

▲ **Fig. 4.1.17** Basic design of different blanking, edge bending and flanging dies

Table 4.1.2: Die materials and applications

Die material No.	1.0046	1.0555	1.7140	1.2067	1.2769	1.2601	1.2363
Designation (DIN 16006)	GS-45	GS-62	GS-47 Cr Mn 6	G-100 Cr 6	G-45 Cr Ni Mo 42	G-X 165 Cr Mo V 12	G-X 100 Cr Mo V 51
R_m [N/mm²]	450	620	tempered, stress relieved 820-930	850	tempered to 1000-1030	800-900	800-900
Heat treatment	weldable, not hardenable	hardening conditionally possible	weldable surface hardenable	hardenable	surface hardenable, glow nitrided	hardenable	hardenable
Hardness after heat treatment	–	max. HRC 56 + 2 3 mm deep	HRC 56 + 2	HCR 60 + 2	working hardness HRC 50+/–2	HRC 60+/–2	HRC 60+/–2
Application	thin-walled blank holders, upper structures and substructures for small dies	blank holders, upper structures subject to high loads	forming punch, blanking segments, female blanking dies	forming punch, female die inserts, forming blocks, counterpressure pads	forming punches, female die inserts, forming blocks, blank holders and retainers with end pressure	blanking strips, blanking punches, coining punches, draw punches, female drawing dies, forming punches	blanking strips, blanking punches, coining punches, draw punches, female drawing dies, forming punches

Die material No.	1.2379	0.6025	0.6025 Cr Mo	0.7050	0.7060	0.7070	0.7070 L
Designation (DIN 16006)	G-X 155 Cr V Mo 12 1	GG-25	GG-25 Cr Mo	GGG-50	GGG-60	GGG-70	GGG-70 L
Basic hardness HB	max. 250	160-230	200-230	170-220	190-240	220-270	220-270
R_m [N/mm²]	860	250-350	250	500	600	700	700
Heat treatment	hardenable, TIC-TIN coating	conditionally weldable, glow nitriding	surface hardening, glow nitriding	surface hardening, temperable	conditionally weldable, surface hardening, glow nitriding	conditionally weldable, surface hardening, glow nitriding	surface hardening, glow nitriding
Hardness after heat treatment	HRC 60 + 2	approx. 1200-1400 HV2	approx. 53 HRC	approx. 54 HRC	approx. 56 HRC	approx. 56 HRC	approx. 56 HRC
Application	hemming blocks, edge bending blocks, coining punches for sheet metals up to 2,9mm, blanking cutters, bottom blanking dies, blanking punches	standard material for cast die components, blank holders adapters, upper structures and substructures, mounted parts, drives, slides	blank holders, draw punches, female drawing dies	instead of GG 25 and GS 45 with thin-walled materials in particular in case of compressive stress, e.g. counterpressure pads and strippers	blank holders, draw punches, female drawing dies, brackets	blank holders, female drawing dies	blank holders for parts with drawing depths >120 mm of outer bodywork parts on transfer presses

Table 4.1.3: Specifications and applications for surface hardening and coating of dies

| | SURFACE HARDENING PROCESSES | | | COATING PROCESSES | | | | | |
| | electrical | physical | | CVD process | | hard chromium plating | PVD process | | |
	induction hardening	flame hardening	glow nitriding	Ti C	Ti C - Ti N	hard chromium plating	Ti N	Cr N	Ti C N
Method	hardening temperature through high-frequency induced eddy currents	hardening temperature through acetylene oxygen flame	inclusion of nitrogen through electrical discharge in a vacuum	combination of hardening and coating process: hardening of the base material to approx. 60-62 HRC, coating > 1 000 °C, austenitizing, tempering		application of chrome coat in vacuum, diffusion or electrical precipitation at around 100 °C	heating to coating temperature below 500 °C, vacuum coating, vacuum cooling, distortion-free process		
Usable workpiece material	unalloyed and alloyed steels 0.2 - 0.6 % C e.g. Ck 45, 50 Cr Mo 4	unalloyed and alloyed steels 0.2 - 0.6 % C e.g. Ck 45, 50 Cr Mo 4	all ferrous materials e.g. 1. 2379, 1.2842, 1.2601, 1.0062, 0.6025, 1.7131	ledeburitic chromium steels, carbides preferably 1.2379		cast iron, steel, non-ferrous metals	cold and hot work steels, HSS, carbides, rust-proof steels e.g. 1. 2379, 1. 2601, 1. 2369, 1. 2378		
Workpiece requirement	metalically bright surface	metalically bright surface	finish machined, metalically bright surface	metalically bright and polished surface, roughness $R_z < 3$ μm, non-functional surfaces (fitting surfaces) with allowance, specification of tolerance field in the drawing		metalically bright surface, no material, processing or surface faults, as a rule prior to surface hardening	electrically conductive, non-magnetic condition, no shrinkage in case of soldered parts, metalically bright surface, roughness $R_z = 2 - 4$ μm, no burr, no assembled parts		
Layer characteristics	53 - 59 HRC, hardening depth 1,5 - 3 mm, wear-resistant surface, workpiece core with high toughness, following finish	wear-resistant surface, workpiece core with high toughness, danger of scaling, following finish	approx. 1200-1400 HV2, improvement of sliding properties, increase of strength, increase of surface roughness $\Delta R_a = 0.1-0.3$ μm	approx. 4 000 HV, max. charging temperature 300 °C, brittle, layer thickness up to approx. 9 μm, grey metal colored	approx. 3 000 HV, max. charging temperature approx. 450 °C, layer thickness up to approx. 9 μm, grey metal colored	600 - 1 100 HV0.1, rustproof surface, increased wear resistance, good coefficients of friction, max. layer thickness 100 μm	approx. 2 400 HV, max. charging temperature approx. 500 °C, chemical resistance, layer thickness up to 3 μm, gold colored	approx. 2 000 HV, max. charging temperature approx 600 °C, good chemical resistance, layer thickness up to 50-70 μm, silver colored	approx. 3 000 HV, max. charging temperature approx. 400 °C, layer thickness up to 3 μm, brown-violet
Range of parts	cast blanking cutters, gib rails, partial draw punches and female dies	lifting bolts, gib elements, cast blanking cutters, partial draw punches and female dies	all wear active die parts and gib elements	*active parts of forming and drawing dies* primarily for thicker uncoated sheet steels	primarily for aluminium plated and galvanized sheet metals	draw punches, female dies, blank holders	*active parts of blanking and separating dies and machining tools* primarily unalloyed deep-drawn sheets	non-ferrous metals and Cr Ni steels	primarily for low and high alloyed steels, e.g. VA sheet metals

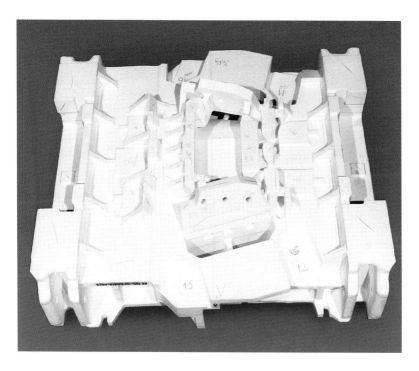

▲ **Fig. 4.1.18** Polystyrene pattern of a die

Compared to the use of other pattern materials such as wood, metal, or plastic, the designer is not restricted by the limitations of the conventional casting technology. As the pattern remains inside the form, there is no need to consider the parting line, undercuts, conicity and the tapers for pattern removal.

All ferrous materials are suitable as casting material for production dies. The residual matter left over from the vaporisation of the pattern impairs the finish of the surface quality of the die only to a minimal degree. However, in order to ensure that the operating surface of the die can have the best surface quality, this surface is placed face down in the casting box. The maintenance of the pattern during fabrication presents no problems – with a specific weight of 20 kg/m^3, even large polystyrene patterns can be transported without having to use large cranes. Compared to conventional preparation methods such as welding, casting techniques have the following advantages:

- ability to work with complicated components with small manufacturing tolerances,
- better distribution of forces in the cast piece via curvatures and smooth blending of surfaces,
- better damping characteristics of cast iron compared to steel plate construction; as a result, quiet and low vibration proven components that can be highly stressed and provide low noise level in production,
- less tendency to distortion with cast iron, which ensures that a high level of precision is delivered from the production system,
- good wear and sliding characteristics,
- die materials that can be hardened; thus, the cast component can fullfill various functional requirements without any additional attachments,
- oil-tight containers, etc.,
- no corners and cracks in the seams, which can lead to absorption of moisture and corrosion,
- good machinability, through which savings in the order of 20 to 50% on machining costs can be achieved,
- as a rule, no stress-relieving process is necessary.

■ 4.1.5 Try-out equipment

Try-out presses
Equipment for die try-out plays an important role in modern presswork operations. Dies are tested on a try-out press before starting production. In this way, the die set-up time on a production line press is reduced to a minimum. Press running times for all parts are consequently limited to set-up and production time only and this increases the profitability of the production line.

In making dies for sheet metal forming, the sculptured surface of the die cavity will be machined to form the male and female dies using the CAD-generated data for the finished part. As a general practice, both parts of the die are normally produced from cast patterns (cf. Sect. 4.1.4). During die development on a try-out press, it is possible to ascertain for the first time, where any reworking needs to be carried out on the geometry. The operational characteristics of the dies and the contact pattern are checked out. The flow of the material is evaluated to see how it compares with the theoretical calculations. The formed parts are inspected for presence of any wrinkling, tears or bulges. In particular, high material thicknesses do occur in the corner areas of the blank holder surface because of tangential compressive stresses, which lead to undesirable higher surface pressures and can adversely affect results of the deep drawing process. To rectify this, it is necessary to remove material and rework specific areas of the blank holder.

In a try-out press, the die geometry is modified by first forming a part and then grinding the punch, the die and the blank holder at the most critical points in order to obtain the required part quality. This procedure, e. g. forming and regrinding, is repeated until good quality parts are formed. The aim is to perfect the die on the try-out press, so that costly modifications for a series run on a production press are minimized. Try-out operations can only be considered complete when the required part quality is achieved, i. e. when the first good part is produced, and production on the production presses can then start.

If the draw cushion of a single-action press or for that matter, the blank holder of a double-action press, is equipped with a multi point controller, the try-out costs can be greatly reduced (cf. Sect. 3.1.4 and 3.2.8) – the production press onto which the die will later be installed and the try-out press should certainly be equipped, as far as possible, with a similar type of control system. Instead of working on the specific die area, or using draw beads *(cf. Fig. 4.2.6 and 4.2.7)*, the required deep drawing result can be achieved simply by making adjustments of individual control cylinders. An additional device, which controls the pressure during the draw stroke *(cf. Fig. 3.1.12)* contributes significantly to reducing try-out costs and achieving the highest part quality on the production press.

During the try-out process, the parameters of the forming operation, especially those which can be adjusted on the actual press being used, play a decisive role. Hydraulic and mechanical try-out presses test the dies under conditions which are as close as possible to those found in normal production runs. In order to achieve a faster and more reliable production start-up, a high degree of compatibility of the main press characteristics – such as the rigidity and transverse bending of the bed and slide, etc. – between the try-out press and the stations of the production press is an advantage. The closer the characteristics of the try-out press are to those of the production line press, the shorter will be the time required to be spent during production start-up and the sooner the production of good parts can be achieved. For example, the most successful way of developing dies for mechanical large-panel transfer presses is to use mechanical try-out presses, which have nearly identical characteristics. In particular, this means that for the first drawing stage, the try-out press, apart from having the same construction, also has a similar draw cushion to that of the single-action press *(Fig. 4.1.19, left)*. One or more single-action mechanical presses without a draw cushion, however, are used to develop the dies for the subsequent stations of a transfer press *(Fig. 4.1.19, right)*.

A much less expensive solution is provided by hydraulic presses *(Fig. 4.1.20)*. Admittedly, their kinematic characteristics and, in particular, the forming speed can only match those of a mechanical press to a limited degree *(cf. Fig. 3.2.3 to Fig. 3.3.1)*; but their energy capacity does not depend upon the stroking rate. Their flexibility makes any desired slide movement possible irrespective of whether the dowl edges should be carefully tested or any pre-production series of stampings are pro-

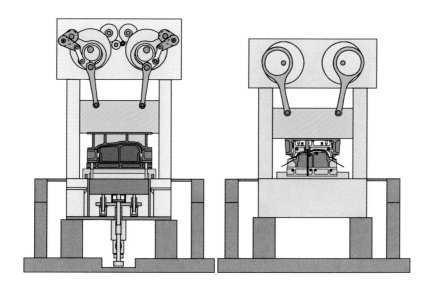

▲ **Fig. 4.1.19** Mechanical try-out presses for die try-out:
         with draw cushion for the first forming process (left) and
         without draw cushion for the subsequent processes (right)

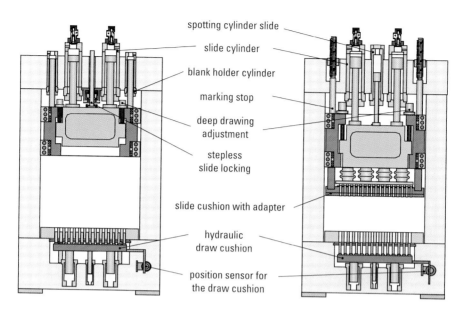

spotting cylinder slide
slide cylinder
blank holder cylinder
marking stop
deep drawing adjustment
stepless slide locking
slide cushion with adapter
hydraulic draw cushion
position sensor for the draw cushion

▲ **Fig. 4.1.20** Design of a hydraulic try-out press

duced. In addition, a spotting fixture for the slide and blank holder can be applied to the contours of the die with a precisely controlled speed. The die mounting area, the stroke and the force on all try-out presses are so designed that the press can accomodate the total range of large size tools. Force and speed can be selected in the ratio of at least 1 : 10, so that the dies can be tried out by starting with a very low ram speed. As a rule, both the punch and the die must be reworked at the same time during die try-out.

In order to facilitate the rework and save time, hydraulic try-out presses can be equipped with a pivoting slide plate *(Fig. 4.1.21)*. At first, the moving bolster carrying the lower die moves out of the press. When pivoted, the top die is rotated through 180° and lowered and centered onto a movable pallet truck. After this, the slide, the clamping plate and top die are lowered to a pre-defined height and locked into the roll-over device. The quick-action clamps release the slide plate from the slide, and the slide moves to the top dead center position. After the slide plate and top die have been turned over, the slide places the slide plate with the top die in the centering of the pallet truck and moves again to the

◄ Fig. 4.1.21
Rotating the slide of a hydraulic
try-out press

upper stop position. After being taken out of the back of the press on
the pallet truck, the top die is accessible from all sides. This allows convenient and time-saving reworking to be carried out. After reverse rotation, the die does not need to be repositioned as it is already positioned
by the clamping plate on the pallet truck.

The electrical system of a try-out press includes a sensor system for
measuring the setting parameters for each die. Sensors for measuring
various process parameters and a data acquisition system allow to perform statistical evaluation. The optimal setting parameters are saved
and made available as input into the control system of the production
press. In order to achieve production-like conditions in hydraulic try-
out presses, any elastic deformations or off-center movements of the
slide, which are detected at the die and the production line must be
considered by the control system.

During run-in, the dies need to be frequently moved in and out of
the press. A moving bolster allows this work to be carried out faster and
more safely *(cf. Fig. 3.4.3)*. In a try-out center for example, several try-

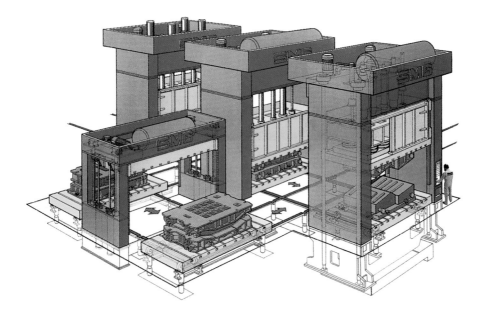

▲ **Fig. 4.1.22** Try-out center: try-out presses linked by T-tracks

out presses may be linked together with T-tracks *(Fig. 4.1.22* and *cf. Fig. 3.4.4)*. This arrangement allows comprehensive testing of the progressive forming operations so that die sets are more quickly and efficiently run-in.

Try-out presses and above all try-out centers are also suitable for producing short run parts or carrying on production on a smaller scale in the event of the breakdown of a large production press. Depending on the task they have to perform, such universal presses may be used, for example, for production as well as for try-out purposes.

Parallel lift mechanisms
For any repairs and reworking operations, the top die has to be lifted off the bottom die. For ergonomic (overhead working) and scheduling reasons (working at the same time on the top and bottom dies) both dies are laid down separately. Dies which cannot be handled manually can be moved with the help of a hydraulic parallel lift mechanism. This apparatus ensures quick clamping for repairs and maintenance of dies.

The mechanism, seen in *Fig. 4.1.23*, lifts the top die parallel to the lower die, and after a 180° turn places it next to the lower die. The mechanism ensures that both parts of the die are lifted safely and with consistent accuracy without any tilting.

■ 4.1.6 Transfer simulators

Transfer simulators are used for the adjustment and pre-setting of part-dependent tooling of the gripper rails and the vacuum suction crossbars of large-panel transfer presses *(cf. Fig. 4.4.29* and *4.4.34)*. The lower die and the part-dependent tooling are installed in the simulator; prior to production these will be coupled with the permanently installed transfer components of the press *(Fig. 4.1.24)*.

The movements of the transfer, part and die must be synchronized in relation to one another so that there can be no overlapping at critical points and consequently no collision between the transfer, the part itself and the die. Thus, data on gripper and suction cup positions, transport paths, transport height, part positioning and flow directions is stored in a programme. This programme is transferred onto the production line. The possibility of collisions of part and die, which was already verified as part of the CAD design process, is checked out in the simulator. The primary task for the simulator is, however, the matching of the part-dependent fixturing to the sheet metal part. A test is per-

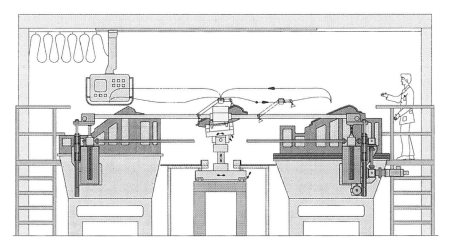

▲ **Fig. 4.1.24** Transfer simulator for a large-panel transfer press with crossbar transfer system

formed to ascertain whether the part is positioned precisely in its universal station and in the die. In addition, the energy supply and control system of the transfer rails are checked for correct functioning. Modern systems can also simulate the changing of part-specific templates used in the universal station.

■ 4.2 Deep drawing and stretch drawing

■ 4.2.1 Forming process

In deep drawing operations forming is carried out with tools (rigid and flexible), active media (e. g. sand, steel pellets, liquid) or active energy (i.e. a magnetic field). The following section covers possible applications for stretch and deep drawing processes with rigid tools as a first and subsequent operation *(cf. Fig. 2.1.10 to 2.1.12)*. Sections 4.2.4 and 4.2.5 will cover deep drawing with active media *(cf. Fig. 2.1.13)*.

Unlike in deep drawing, in *stretch forming* the sheet metal cannot flow because it is gripped by clamps or held in a blank holder. The forming process takes place with a reduction in the thickness of the sheet metal. Pure stretch drawing is a forming process conducted under tensile stresses *(cf. Fig. 2.1.18)*. Mostly large-scale work pieces such as, for example, body panel components used in aircraft or automotive manufacturing, are produced on special production lines. This is certainly true in producing low volume or large size parts, as, for example, with the wing elements of an aircraft. The advantage of this process is that the tooling and machining costs are relatively modest in proportion to the size of the part being produced.

Deep drawing is a compression-tension forming process according to DIN 8582. The procedure with the greatest range of applications is *deep drawing with rigid tooling,* involving draw punches, a blank holder and a female die *(cf. Fig. 2.1.10)*. The blank is generally pulled over the draw punch into the die, the blank holder prevents any wrinkling taking place in the flange. In pure deep drawing, the thickness of the sheet metal remains unchanged since an increase in the surface area does not occur, as it occurs in a stretch drawing process.

The material flow in the drawing process can be controlled, as required, by adjusting the pressure of the blank holder in between pure stretch drawing, in which no sheet metal can flow out of the flange area into the draw die, and deep drawing, in which there is a wrinkle-free and unrestricted material flow *(cf. Fig. 2.1.32)*.

In deep drawing, as a first step, a cup is produced from the flat blank. This cup can then be further processed, for example by an additional drawing, reverse drawing, ironing or extrusion process. In all deep drawing processes, the pressing force is applied over the draw punch onto the bottom surface of the drawn part. It is further transferred from there to the perimeter in the deformation zone, between the die and the blank holder.

The workpiece is subjected to radial tension forces F_R and tangential compression forces F_T *(Fig. 4.2.1)*. The material is compressed in the tangential direction and stretched in the radial direction. As the draw

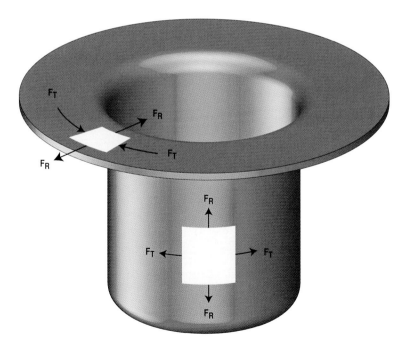

▲ **Fig. 4.2.1** Pressing forces in deep drawing of a round cup with a blank holder

depth is increased, the amount of deformation and the deformation resistance are also increased. The sheet metal is most severely stretched in the corner of the draw punch. Failure normally occurs at this point (base fracturing).

When drawing with a blank holder, the press must also transfer the blankholding force onto the die. This is achieved either from above via a blank holder slide or from underneath via a draw cushion. There are basically three possible ways of applying blank holder force.

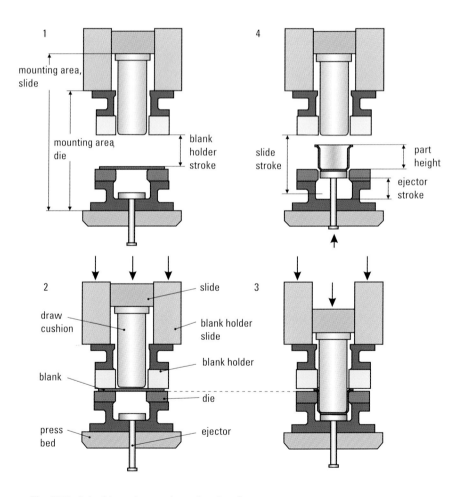

▲ **Fig. 4.2.2** A double-action top down drawing die

In *double-action drawing operations* the press has two slides acting from above: the drawing slide with the draw punch and the blank holder slide with the blank holder *(Fig. 4.2.2 and cf. Fig. 3.1.8)*. The blankholding slide transfers the blankholding force via the blank holder onto the blank and the draw die. The die and the ejector are located in the lower die on the press bed. During forming, the blank holder brings the sheet metal into contact against the die, the punch descends from above into the die and shapes the part while the sheet metal can flow without any wrinkling out of the blankholding area. In this case, the drawing process is carried out with a fixed blank holder and moving punch. In double-action drawing operations, the drawing slide can only apply a pressing force.

A disadvantage in double-action drawing is the fact that the parts for further processing in successive forming and blanking dies need to be rotated by 180° in an expensive and bulky rotating mechanism (cf. Sect. 3.1.3). With outer body parts the danger exists, furthermore, that the surface of the part may be damaged in the rotating operation.

Single-action drawing with a draw cushion works the other way round: the forming force is exerted by the slide above through the die and the blank holder onto the draw cushion in the press bed *(Fig. 4.2.3 and cf. Fig 3.1.9)*. The draw punch and the blank holder of the drawing tool are both located in a base plate on the press bed. Pressure pins, which come up through the press bed and the base plate transfer the blank holder force from the draw cushion onto the blank holder. The female die and the ejector are mounted on the press slide. At the start of the forming process the blank is held under pressure between the draw die and the blank holder. The slide of the press pushes the blank holder downwards over the draw die – against the upward-acting force of the draw cushion. The part is formed via the downward movement of the die over the stationary draw punch. The press slide must apply both the pressing and the blank holder forces.

Thus, using a single-action tool, the part does not have to be rotated after the drawing process. Furthermore, today the use of hydraulically controlled draw cushions, even with deep drawing processes up to depths of 250 mm, produces a workpiece quality comparable to that of double-action presses (cf. Sect. 3.1.4).

An energy-saving and cost-effective alternative in stamping is *counter drawing or reverse drawing (Fig. 4.2.4 and cf. Fig. 3.1.10)*. In this operation, once again, a single-action press with a draw cushion is used – normally a hydraulic one. The top die is attached to the slide. The lower die is mounted on the press bed with the blank holder. The punch is located in an opening in the center of the press bed on the draw cushion.

During deformation, the blankholding force is transferred via the slide from above and the draw punch force acts from below through the

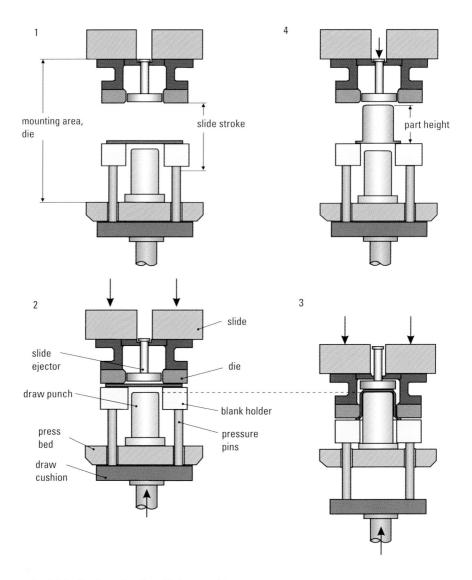

▲ **Fig. 4.2.3** Single-action die with draw cushion

active draw cushion. The punch forms the part by means of its upward movement while the blank holder rests on the die.

The active counter drawing combines the advantages of static blank-holding and the low power usage of double-action dies with the advantage of a single-action die, in which it is not necessary to rotate the part. Admittedly, it is necessary to have a special modification of the dies for performing this operation, since the forces operate in the opposite direction to those of a normal single-action deep drawing press with draw cushion. Thus, these dies cannot be installed in a normal single-action press.

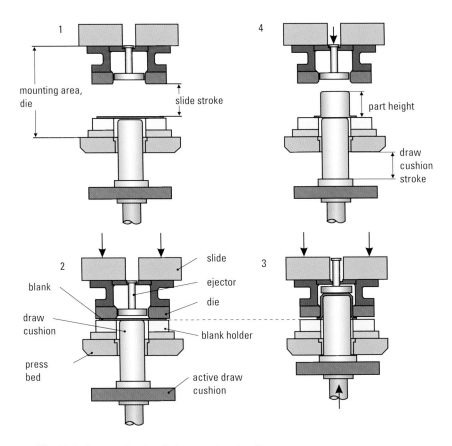

▲ **Fig. 4.2.4** Counter drawing die/reverse drawing die

Calculation of the blank size

Before starting drawing operations the size and form of the blank must be determined for the desired final part geometry and die layout. This should be shown using the example of a simple rotationally symmetrical body. In order to calculate the blank diameter, it is necessary to devide the entire axisymmetric part into various individual axisymmetric components, in accordance with Table 4.2.1 and then calculate the surface areas of these components. The total surface area as a sum of the individual areas enables the calculation of the diameter of blank D. This is shown in Table 4.2.1 for commonly used drawn shapes, starting from the desired inner diameter d. As the material will be somewhat stretched in the drawing process, there is more surplus material on the upper edge of the draw part, which cannot be precisely calculated. With high parts this can lead to distorted edges, because of the non-uniform deformation properties of the blank material (anisotropy). Therefore, in general the drawn parts must be trimmed accordingly on the edge, when produced via deep drawing. The selection of the blank size for non-symmetric and irregular parts is often carried out on a trial and error basis, as it is not possible to use simple formulas. Based on practical experience, the blank geometry is determined with experiments. Initially a sufficiently large blank size is selected for the drawing operation. After observing the actual material demand and flow, the blank size is reduced to satisfy the material requirements. More recently, computer programs are being increasingly used for the determination of the blank size (cf. Sect. 4.1.2).

Table 4.2.1: Formulas for the circular blank diameter D

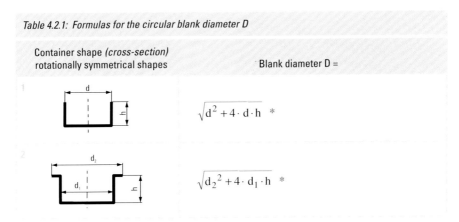

Container shape (cross-section) rotationally symmetrical shapes	Blank diameter D =
1	$\sqrt{d^2 + 4 \cdot d \cdot h}$ *
2	$\sqrt{d_2{}^2 + 4 \cdot d_1 \cdot h}$ *

* *Containers with small (bottom) radii r < 10 mm*

Container shape *(cross-section)*
rotationally symmetrical shapes

Blank diameter D =

$$\sqrt{d_2^{\ 2} + 4 \cdot \left(d_1 \cdot h_1 + d_2 \cdot h_2\right)} \quad *$$

$$\sqrt{d_3^{\ 2} + 4 \cdot \left(d_1 \cdot h_1 + d_2 \cdot h_2\right)} \quad *$$

$$\sqrt{d_1^{\ 2} + 4 \cdot d_1 \cdot h + 2 \cdot f \cdot \left(d_1 + d_2\right)} \quad *$$

$$\sqrt{d_2^{\ 2} + 4 \cdot \left(d_1 \cdot h_1 + d_2 \cdot h_2\right) + 2 \cdot f \cdot \left(d_2 + d_3\right)} \quad *$$

$$\sqrt{2 \cdot d^2} = 1.414 \cdot d$$

$$\sqrt{d_1^{\ 2} + d_2^{\ 2}}$$

$$1.414 \cdot \sqrt{d_1^{\ 2} + f \cdot \left(d_1 + d_2\right)}$$

$$1.414 \cdot \sqrt{d^2 + 2 \cdot d \cdot h}$$

* *Containers with small (bottom) radii r < 10 mm*

Container shape *(cross-section)* rotationally symmetrical shapes	Blank diameter D =
11	$\sqrt{d_1^2 + d_2^2 + 4 \cdot d_1 \cdot h}$
12 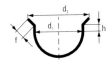	$1.414 \cdot \sqrt{d_1^2 + 2 \cdot d_1 \cdot h + f \cdot (d_1 + d_2)}$
13	$\sqrt{d^2 + 4 \cdot h^2}$
14	$\sqrt{d_2^2 + 4 \cdot h^2}$
15	$\sqrt{d_2^2 + 4 \cdot (h_1^2 + d_1 \cdot h_2)}$
16	$\sqrt{d^2 + 4 \cdot (h_1^2 + d \cdot h_2)}$
17 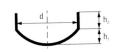	$\sqrt{d_1^2 + 4 \cdot h^2 + 2 \cdot f \cdot (d_1 + d_2)}$
18 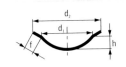	$\sqrt{d_1^2 + 4 \cdot [h_1^2 + d_1 \cdot h_2 + 0.5 \cdot f \cdot (d_1 + d_2)]}$
19	$\sqrt{d_1^2 + 2 \cdot s \cdot (d_1 + d_2)}$ *

* *Containers with small (bottom) radii r < 10 mm*

Container shape *(cross-section)*
rotationally symmetrical shapes   Blank diameter D =

$$\sqrt{d_1^2 + 2 \cdot s \cdot (d_1 + d_2) + d_3^2 - d_2^2} \quad *$$

$$\sqrt{d_1^2 + 2 \cdot \left[s \cdot (d_1 + d_2) + 2 \cdot d_2 \cdot h \right]} \quad *$$

$$\sqrt{d_1^2 + 6.28 \cdot r \cdot d_1 + 8 \cdot r^2} \quad \text{or}$$

$$\sqrt{d_2^2 + 2.28 \cdot r \cdot d_2 - 0.56 \cdot r^2}$$

$$\sqrt{d_1^2 + 6.28 \cdot r \cdot d_1 + 8 \cdot r^2 + d_3^2 - d_2^2} \quad \text{or}$$

$$\sqrt{d_3^2 + 2.28 \cdot r \cdot d_2 - 0.56 \cdot r^2}$$

$$\sqrt{d_1^2 + 6.28 \cdot r \cdot d_1 + 8 \cdot r^2 + 4 \cdot d_2 \cdot h + d_3^2 - d_2^2} \quad \text{or}$$

$$\sqrt{d_3^2 + 4 \cdot d_2 \cdot (0.57 \cdot r + h) - 0.56 \cdot r^2}$$

$$\sqrt{d_1^2 + 6.28 \cdot r \cdot d_1 + 8 \cdot r^2 + 2 \cdot f \cdot (d_2 + d_3)} \quad \text{or}$$

$$\sqrt{d_2^2 + 2.28 \cdot r \cdot d_2 + 2 \cdot f \cdot (d_2 + d_3) - 0.56 \cdot r^2}$$

$$\sqrt{d_1^2 + 6.28 \cdot r \cdot d_1 + 8 \cdot r^2 + 4 \cdot d_2 \cdot h + 2 \cdot f \cdot (d_2 + d_3)} \quad \text{or}$$

$$\sqrt{d_2^2 + 4 \cdot d_2 \cdot (0.57 \cdot r + h + 0.5 \cdot f) + 2 \cdot d_3 \cdot f - 0.56 \cdot r^2}$$

$$\sqrt{d_1^2 + 4 \left(1.57 \cdot r \cdot d_1 + 2 \cdot r^2 + d_2 \cdot h \right)} \quad \text{or}$$

$$\sqrt{d_2^2 + 4 \cdot d_2 \cdot (0.57 \cdot r + h) - 0.56 \cdot r^2}$$

* *Containers with small (bottom) radii r < 10 mm*

Example:
A cup with a conical outline and flange is to be produced. The base diameter is 60 mm, the upper diameter 100 mm, the outside flange diameter 120 mm and the can body measurement 75 mm. The required diameter of the blank piece is calculated according to formula 20 from Table 4.2.1:

$$d_1 = 60\,\text{mm}, \ d_2 = 100\,\text{mm}, \ d_3 = 120\,\text{mm}, \ s = 75\,\text{mm}$$

$$D = \sqrt{d_1^2 + 2 \cdot s \cdot (d_1 + d_2) + d_3^2 - d_2^2} = \sqrt{60^2 + 2 \cdot 75 \cdot (60 + 100) + 120^2 - 100^2}$$

$$D = \sqrt{32{,}000} = 178.9\,\text{mm}$$

Therefore, for deep drawing this cup, a circular blank of 179 mm diameter is required.

Draw ratio

The draw ratio β is an important numerical value for cylindrical draw parts in determining the required number of drawing steps. The draw ratio is the ratio of the diameter of the initial blank form to the diameter of the drawn part. For the first drawing process, that is to say for only one drawing operation, the draw ratio β [–] results in the following equation:

$$\beta = \frac{\text{Blank} - \varnothing\,D}{\text{Punch} - \varnothing\,d},$$

$$\beta = \frac{D}{d} \qquad \text{with one drawing,}$$

$$\beta_1 = \frac{D}{d_1} \qquad \text{for the first drawing,}$$

$$\beta_2 = \frac{d_1}{d_2} \qquad \text{or} \qquad \beta_3 = \frac{d_2}{d_3},$$

and so on. In subsequent drawing steps, d_1 represents the diameter of the draw punch in the first drawing operation and d_2 the diameter of the second, etc. The total draw ratio β_{tot} is calculated as the ratio of the diameter of the blank to the final diameter of the drawn part:

$$\beta_{tot} = \frac{Initial - \varnothing\ D}{Final - \varnothing\ d}$$

The maximum draw ratio depends on the properties of the blank material used. As a rough estimate, the first drawing can be calculated as having a maximum ratio of $\beta = 2$. To achieve $\beta > 2$ several drawings are required and it must be noted that β, because of work hardening, can only achieve a level of 1.3 in the next drawing step. If the part is annealed before the next drawing operation, a β of 1.7 can be assumed.

With several drawing steps, the total draw ratio becomes a product of the individual draw ratios:

$$\beta_{tot} = \beta_1 \cdot \beta_2 \cdot \ldots \cdot \beta_n$$

Example:
From a 1 mm thick blank of sheet metal material number 1.0338 (St 14) a hollow cylindrical part with a diameter of 45 mm and a can body height of 90 mm is to be drawn. First, the initial diameter of the circular blank D is calculated as per Formula 1 from Table 4.2.1:

$$D = \sqrt{d^2 + 4 \cdot d \cdot h} = \sqrt{45^2 + 4 \cdot 45 \cdot 90} = \sqrt{18,225} = 135\ mm$$

Thus, the total draw ratio β_{tot} is established:

$$\beta_{tot} = \frac{D}{d} = \frac{135}{45} = 3$$

Because β_{tot} is greater than the remaining limiting draw ratio of 2, several drawing operations are needed whereby β_1 may have a maximum value of 2, and all following β values are less than 1.3.

$$\beta_{tot} = \beta_1 \cdot \beta_2 \cdot \ldots \cdot \beta_n = 3$$
$$3 = \beta_1 \cdot \beta_2 \cdot \ldots \cdot \beta_n < 2 \cdot 1.3 \cdot 1.3$$

The required draw ratio can be attained with three operations. In order to have a robust and reproducable production sequence, the maximum permitted draw ratios should not be fully utilised. In the event that $\beta_{tot} = 1.95 \cdot 1.25 \cdot 1.25 = 3$ is selected. This calls for punch diameters of:

→

$$d_1 = \frac{D}{\beta_1} = \frac{135 \text{ mm}}{1.95} = 69.2 \text{ mm} \Rightarrow d_1 = 69 \text{ mm}$$

$$d_2 = \frac{d_1}{\beta_2} = \frac{69 \text{ mm}}{1.25} = 55.2 \text{ mm} \Rightarrow d_2 = 55 \text{ mm}$$

$$d_3 = 45 \text{ mm (before mentioned) with } \beta_3 = \frac{55 \text{ mm}}{45 \text{ mm}} = 1.22$$

The maximum draw ratio β_{max}, does depend however on other factors as well (cf. Sect. 4.2.3):

$$\beta_{max} = f (D, d, \mu, r - \text{value, s...})$$

When the friction between the drawn part and the punch is low, then failures will occur in the base of the part. If the friction between the part and the punch is high, the base of the drawn part will be increasingly stressed with increasing friction in the can body so that the failure zone will be moved to the body of the drawn can. In order to ensure a safe production process, it is preferable to select a draw ratio that is rather modest and less than the maximum possible value.

Drawing force
For drawing round parts in a single drawing, the maximum drawing force F_U can be calculated per the following equation:

$$F_U = \pi \cdot (d_1 + s) \cdot s \cdot R_m \cdot 1.2 \cdot \frac{\beta - 1}{\beta_{max} - 1} \quad [N]$$

where:

d_1 = punch diameter [mm]
s = sheet metal thickness [mm]
R_m = material tensile strength [N/mm²]
β = actual draw ratio [–]
β_{max} = maximum draw ratio [–]

Example:
Material: 1.0338 (St 14) with tensile strength $R_m = 350$ N/mm^2
Sheet metal thickness: s = 0.8 mm
Punch-$\varnothing$: d = 240 mm
Draw ratio: $\beta = 1.75$ (D = 420 mm)
max. permitted draw ratio: $\beta_{1max} = 2.0$
Drawing force calculated as:

$$F_U = \pi \cdot (240 + 0.8) \text{ mm} \cdot 0.8 \text{ mm} \cdot 350 \text{ N/mm}^2 \cdot 1.2 \cdot \frac{1.75 - 1}{2 - 1} = 190{,}637 \text{ N} = 191 \text{ kN}$$

The calculated drawing force is equal to the drawing slide force on double-action drawing presses. With single-action presses, the drawing slide force increases by the amount of force applied to the blank holder, since the blank holder force counteracts the movement of the drawing slide during the complete drawing cycle *(Fig. 4.2.3)*. If several operations are needed to produce a pressed part, it is essential to calculate the drawing force with the largest diameter punch. The forces required for the drawing of elliptical, square, rectangular and similarly shaped draw parts, where the corner radii are not too small, can be calculated in a similar way to those for deep drawn cylindrical shapes:

$$d = 1.13 \cdot \sqrt{A_{St}} \text{ [mm]} \text{ and } D = 1.13 \cdot \sqrt{A_Z} \text{ [mm]},$$

where A_{St} [mm^2] is the cross section of the punch and A_Z [mm^2] is the area of the blank.

Pressing force for sheet metal with beads
Beads are stretched out of the thickness of the sheet metal blank *(cf. Fig. 2.1.19)*, that is to say an increase in the surface area takes place without changing the external dimensions of the part. Therefore, it is necessary to assure that prior to the drawing process there are sufficient "strain reserves" as the sheet metal would otherwise be fractured.

The pressing force for forming stiffening beads or ribs is calculated as follows:

$$F_U = l \cdot s \cdot p \text{ [N]},$$

where l [mm] is the rib length, s [mm] the sheet thickness and p is the pressure. The pressure p in the area of the stiffener in N/mm^2 can be cal-

culated graphically from the drawing of a force parallelogram or math-
ematically for one side of the profile starting from the tensile strength
R_m [N/mm²] *(Fig. 4.2.5)*:

$$p = R_m \cdot \cos\alpha \ \left[N/mm^2 \right]$$

Thus, for a profile wall with $\alpha = 0°$, $p = R_m$ or $p = 2 \cdot R_m \cdot \cos\alpha$ are for
the symmetrical profile. The pressing force is therefore solely dependent
on the taper of the sides and the strength of the material. The total force
for pressing the bead is the sum of all the single forces:

$$F_U = s \cdot \left(l_1 \cdot p_1 + l_2 \cdot p_2 + \ldots + l_n \cdot p_n \right) \ [N]$$

Example:
A 1 mm thick sheet metal of material number 1.0333 (USt 13) is to be reinforced
with five symmetrical support ribs of l = 200 mm in length *(Fig. 4.2.5)*. The
height of the ribs h measures 3 mm, the upper rib width b = 6 mm, the angle
$\alpha = 20°$ and the tensile strength $R_m = 370$ N/mm².

The required pressing force is calculated as:

$$F_U = l \cdot s \cdot p_{tot} = 200 \ mm \cdot 1 \ mm \cdot 5 \cdot \left(2 \cdot 370 \ N/mm^2 \cdot \cos 20° \right)$$

$$\Rightarrow F_U = 695.4 \ kN$$

Width b and height h of the support pieces are not taken into the calculation,
only the length l of the profile and the angle α of the ribs. As a comparison:
With an angle of 0° the pressing force increases to the maximum value of
740 kN, as the angle increases the required force becomes smaller, for example
523.3 kN at 45°.

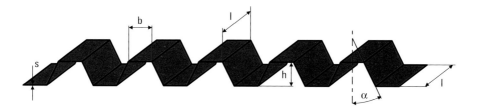

▲ **Fig. 4.2.5** Profile with beads: s (sheet metal thickness), l (length of ribs), α (angle of ribs)

Blank holder force

The magnitude of blank holder force is not only an important factor in avoiding wrinkling of the material but also vital because of its influence on the straining of the upper surface. As the blank holder force is increased, the restraining effect on the material located between the die and the blank holder becomes stronger so that the strain on the upper surface layer is also increased.

The blank holder force required for deep drawing F_B [N] is calculated from D [mm], d' [mm] (die diameter), r [mm] (die corner radius) and p [N/mm²] (specific blank holder pressure):

$$F_B = \frac{\pi}{4} \cdot \left[D^2 - (d' + 2 \cdot r)^2 \right] \cdot p \ [N]$$

Often the die corner radius and the die/punch clearance can be neglected, in comparison with the punch diameter d [mm], the approximate calculated value of the blank holder force is:

$$F_B \approx \frac{\pi}{4} \cdot \left(D^2 - d^2 \right) \cdot p \ [N]$$

The required level of the specific blank holder pressure depends on the material and punch diameters as well as on the thickness of the sheet metal. It can be estimated by using the following values:

- *steel:* p = 2.5 N/mm²
- *copper alloy:* p = 2.0 ... 2.4 N/mm²
- *aluminium alloy:* p = 1.2 ... 1.5 N/mm²

Example:
A sheet metal blank from material number 1.0338 (St 14) with R_m = 350 N/mm² tensile strength is to be formed with a punch having a diameter of d = 240 mm and a draw ratio of β = 1.75 (circular blank diameter D = 420 mm). Thus the blankholder force is estimated to be:

$$F_B = \frac{\pi}{4} \cdot \left(D^2 - d^2 \right) \cdot p = \frac{\pi}{4} \cdot \left[(420 \text{ mm})^2 - (240 \text{ mm})^2 \right] \cdot 2.5 \text{ N/mm}^2 = 63{,}303 \text{ N} = 63 \text{ kN}$$

Draw beads

With very shallow, arched parts, for which the drawing force will be relatively low, a very high blank holder pressure is required so that friction between the blank holder and the die acts to restrain the flow of the sheet metal. Thus, the tension stresses are increased and the material is formed around the punch without forming any wrinkles while it is stretched to achieve the required strength level. In such cases, in order to lower blank holder forces, draw beads are provided in the tooling. Beads are incorporated into the die as per *Fig. 4.2.6* or into the blank holder as per *Fig. 4.2.7*, while a corresponding recess is allowed for in the matching side of the die. The sheet metal is pressed around the bead by the action of the blankholder and thus must flow over the bead as part of the drawing process. The force required for this deformation is added to the drawing force. Draw beads assure that the material flows in a more uniform way, especially in deep-drawing non-circular work parts. They have the task of distributing the longitudinal stress in the drawn part over the whole perimeter. For this reason, they are positioned in locations which are lightly stressed during the drawing process such as, for example, on the longer sides of rectangular parts. If a single bead is inadequate to achieve the required braking effect, several beads can be located one next to the other.

In single-action drawing with a draw cushion it is possible to control the blank holder force during the forming stroke or in different areas of the blank holder in order to carry out the forming operation (cf. Sect. 3.1.4). With double-action mechanical presses, the blankholding force can be varied by altering the pressure in the pressure points of the blank holder slide (cf. Sect. 3.2.8).

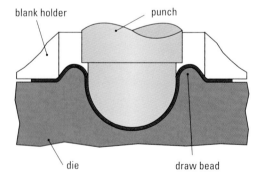

blank holder   punch

die   draw bead

◀ **Fig. 4.2.6**
Draw bead in a die

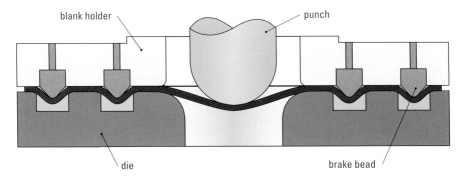

▲ **Fig. 4.2.7** Beads restraining material flow in the blank holder

Draw energy

For drawing operations on double-action presses, the required draw energy W_d [Nm] is the product of the pressing force F_U [N], the drawing stroke h [m] and a correction factor x [–]:

$$W_d = x \cdot F_U \cdot h \ \ [Nm]$$

The correction factor, which is dependent on the material and the draw ratio $\beta = D/d$, takes into account the actual drawing force characteristics; this factor fluctuates between 0.5 and 0.8. The higher level is for soft materials, which can be drawn with the maximum draw ratio β_{max} and for flanged shells, which are only partially but not completely drawn through. The lower level is for a small draw ratio, that is to say for short drawing depths or for harder grades of sheet metal. When drawing normal materials, calculations can be based on x ≈ 0.65 up to 0.75.

A sufficiently accurate rough calculation formula for estimating the draw energy of a double-action press is:

$$W_d = \frac{2}{3} \cdot F_U \cdot h \ \ [Nm]$$

With a single-action press with a draw cushion, the slide force is increased by the blank holder force F_B [N]:

$$W_e = \left(x \cdot F_U + F_B\right) \cdot h \ \ [Nm]$$

Example:

For a drawn cup from 0.8 mm thick sheet metal (material number 1.0338 – R_m = 350 N/mm^2) with a diameter of 60 mm and a height of 40 mm (circular blank diameter 115 mm, β = 1.9) the following pressing work for drawing with a double-action press is required:

$$W_d = \frac{2}{3} \cdot F_U \cdot h = \frac{2}{3} \cdot \pi \cdot (d_1 + s) \cdot s \cdot R_m \cdot 1.2 \cdot \frac{\beta - 1}{\beta_{max} - 1} \cdot h$$

$$W_d = \frac{2}{3} \cdot \pi \cdot (60 \text{ mm} + 0.8 \text{ mm}) \cdot 0.8 \text{ mm} \cdot 350 \text{ N/mm}^2 \cdot 1.2 \cdot \frac{1.9 - 1}{2.0 - 1} \cdot 40 \text{ mm}$$

$$W_d = \frac{2}{3} \cdot 18,385.9 \text{ N} \cdot 0.04 \text{ m} = 490.3 \text{ Nm} = 490.3 \text{ J}$$

Deep drawing without a blank holder is only carried out if the part does not tend to wrinkle, i.e. with thick sheet metal or small amount of deformation (with small draw ratio β).

Drawing defects

In order to eliminate the possibility of parts being rejected, it is necessary to investigate the scrapped parts very thoroughly and to identify thereby the causes of any defects and to take preventive measures. Some of the most common defects are illustrated in Table 4.2.2.

■ 4.2.2 Materials for sheet metal forming

Unalloyed and alloyed steels as well as copper alloys, copper sheet, and light alloys are all among the commonly used materials in sheet metal forming technology. Among alloyed steels, a difference is drawn between phosphor-alloyed steels, types of steel with additional hardening after heat treatment (bake-hardening steels), ferritic chromium steels, austenitic chromium nickel steels and cold-rolled sheet with higher tensile yield strength for cold forming (micro-alloyed steels). Table 4.2.3 lists popular materials with the following specifications:

– material number,
– tensile strength R_m [N/mm^2],

Table 4.2.2: Deep drawing defects

Defect	Possible Cause	Prevention, remedy
drawing grooves	tool wear	replace draw punch/female die, improve lubrication
tear marks on the base of the drawn part	drawing ratio β set too high / drawing gap too narrow / curvature on draw punch or female die too small / blank holder force too great / drawing speed too great	use a sheet metal with greater deep drawing capability or employ multi-stage drawing process / extend drawing gap / check sheet metal thickness / increase die radii / reduce blank holder force / reduce forming speed
vertical tear marks on upper edge	insufficient material in the defective area / drawing gap too wide or drawing edge curvature too great	replace the female die
on rectangular parts, the corners are higher than the side walls	too much material at the corners	change the blank shape / recess material in the affected areas / trim the edge afterwards
vertical creases in the upper area of the body (vessel wall), also in conjunction with vertical cracks	blank holder force too low / drawing gap too great / curvature at the female die too great	increase blank holder force / renew female die
crease formation at the flange	blank holder pressure too low	increase blank holder pressure
other non-specific defects / faulty or irregularly formed drawn parts	incorrect blank shape / asymmetrical material location / unsuitable sheet metal / incorrect lubricant	correct blank shape / check die and stops / use suitable sheet metals and lubricants

- lower yield strength R_{eL} [N/mm^2] for sheet metals with an accentuated yield strength or 0.2 % elongation limit $R_{p0,2}$ [N/mm^2] for those without accentuated yield strength,
- ultimate elongation A [%]: either relative to an initial length in the tensile test specimen L_0 of 80 mm (A_{80}) or to $L_0 = 5 \cdot d_0$ (A_5), with an initial tensile test specimen diameter of d_0,

- field of application.

Table 4.2.3: Summary of the more commonly used sheet metal materials

Unalloyed steels

	Material No.	R_m [N/mm²]	$R_{p0,2}$ or R_{eL} [N/mm²]	A_{80} [%]	Fields of application, examples
St 12	1 0330	270...410	max. 280	28	automotive engineering
USt 13	1 0333	270...370	max. 250	32	automotive engineering
St 14	1 0338	270...350	max. 225	38	automotive engineering

Phosphor-alloyed steels

	Material No.	R_m [N/mm²]	$R_{p0,2}$ or R_{eL} [N/mm²]	A_{80} [%]	Fields of application, examples
ZStE 220 P	1 0397	340...420	220...280	30	automotive engineering
ZStE 260 P	1 0417	380...460	260...320	28	car body work: doors, hoods, roofs
ZStE 300 P	1 0448	420...500	300...360	26	automotive engineering

Bake-hardening steels

	Material No.	R_m [N/mm²]	$R_{p0,2}$ or R_{eL} [N/mm²]	A_{80} [%]	Fields of application, examples
ZStE 180 BH	1 0395	300...380	180...240	32	automotive engineering
ZStE 220 BH	1 0396	320...400	220...280	30	automotive engineering
ZStE 260 BH	1 0400	360...440	260...320	28	automotive engineering
ZStE 300 BH	1 0444	400...480	300...360	26	automotive engineering

Ferritic chromium steels

	Material No.	R_m [N/mm²]	$R_{p0,2}$ or R_{eL} [N/mm²]	A_{80} [%]	Fields of application, examples
X 6 Cr 17	1 4016	450...600	250...270	20	automotive engineering: bumpers, hub caps, food industry, household, appliances: deep-drawing parts with high corrosion resistance properties: sink unit panelling cutlery
X 6 CrMo 17 1	1 4113	480...630	260...280	20	hub caps, bumpers, car window frames radiator grille surrounds
X 6 CrTi 12	1 4512	390...560	200...220	20	automotive engineering: shock absorbers, exhaust system components
X 6 CrNb 17	1 4511	450...600	250...260	20	dairy, brewery, food industry, soap industry and dyeing plants: parts (containers) requiring welding and which are exposed to weak acid solutions

Austenitic CrNi steels

	Material No.	R_m [N/mm^2]	$R_{p0,2}$ or R_{eL} [N/mm^2]	A_{80} [%]	Fields of application, examples
X 5 CrNi 18 10	1 4301	500...700	min. 195	s<3 mm A_{80}: lengthwise: 35; transversely: 40 / 3<s<75 mm A_5: 40	household items, food industry: wear-resistant, deep-drawable, weldable appliances, devices, containers
X 5 CrNi 18 12	1 4303	490...690	min. 185	s<3 mm A_{80}: lengthwise: 35; transversely: 40 / 3<s<75 mm A_5: 40	chemical industry, paper and textile industry: cold extruded parts, screws, nuts

Micro-alloyed steels

	Material No.	R_m [N/mm^2]	$R_{p0,2}$ or R_{eL} [N/mm^2]	A_{80} [%]	Fields of application, examples
ZStE 260	1 0480	350...450	260...340	24	automotive engineering: car body parts
ZStE 300	1 0489	380...480	300...380	22	automotive engineering
ZStE 340	1 0548	410...530	340...440	20	automotive engineering
ZStE 380	1 0550	460...600	380...500	18	automotive engineering
ZStE 420	1 0556	480...620	420...540	16	automotive engineering

Copper and copper alloys

	Material No.	R_m [N/mm^2]	$R_{p0,2}$ or R_{eL} [N/mm^2]	A_{80} [%]	Fields of application, examples
SF-Cu F 22	2.0090.10	220...260	max. 140	42	semi-finished products with very good welding and brazing properties
CuZn 28 F 27	2.0261.10	270...350	max. 160	50	all types of deep-drawing parts, all types of instruments and sleeves
CuZn 33 F 28	2.0280.10	280...360	max. 170	50	wire mesh, radiator strips, tubular rivets
CuZn 36 F 30	2.0335.10	300...370	max. 180	48	metal and wood screws, printing cylinders, radiator strips, zip fasteners, cup springs, hollow goods, ball-point pen cartridges
CuZn 37 Pb 0,5 F 29	2.0332.10	290...370	max. 200	50	deep-drawing parts
CuZn 40 F 34	2.0360.10	min. 340	max. 240	43	good hot and cold forming properties: hot extruded parts, suitable for bending, riveting, upsetting and flanging

Table 4.2.3 (continuation)

Aluminium and aluminium alloys

	Material No.	R_m [N/mm^2]	$R_{p0,2}$ or R_{eL} [N/mm^2]	A_{80} [%]	Fields of application, examples
Al 99,5 W7	3.0255.10	65...95	max. 55	40	equipment construction, electrical installations, electrical construction, high-frequency engineering, aerials, coaxial cables screening; aluminium foils in capacitors, as packaging material, for heat insulation; household equipment
AlRMg 1 W 10	3.3319.10	100...140	35...60	23	architecture, automotive engineering, metal goods: sheet metals, strips, profiles, rods, wires, pipes, forgings, extremely good cold forming properties, weldable, highly seawater resistant
Al 99,9 Mg 1 W 10	3.3318.10	100...140	35...60	23	dito
Al 99,85 Mg 1 W 10	3.3317.10	100...140	35...60	23	dito
AlMg 2,5 G 25	3.3523.29	250...290	min. 180	7	car body work, pull-tab can lids
AlMg 3 W 19	3.3535.10	190...230	min. 80	20	food industry, equipment construction, automotive engineering and shipbuilding, architecture: semi-finished products, roofing, packaging, screws, rivets
AlMn 1 F 12	3.0515.24	120...160	min. 90	7	equipment construction, refrigeration, technology, heat exchangers, roofing, panelling
AlCuMg 1 W	3.1325.10	max. 215	max. 140	13	mining, automotive engineering, aircraft construction, mechanical engineering: sheet metals, profiles, rods, pipes, forgings, screws, rivets
AlMg 2 Mn 0,8 F 22	3.3527.24	220...260	max. 165	9	automotive engineering, shipbuilding, equipment construction: parts resistant to stress at high temperatures
AlMgSi 0,8 F 28	3.2316.71	min. 275	min. 200	12	mechanical engineering, automotive engineering
AlMgSi 1 W	3.2315.10	max. 150	max. 85	18	architecture, mining, automotive engineering, shipbuilding, mechanical engineering, food and textile industries: all types of semi-finished products, screws

Titanium and titanium alloys

	Material No.	R_m [N/mm^2]	$R_{p0,2}$ or R_{eL} [N/mm^2]	A_{80} [%]	Fields of application, examples
Ti 99,8	3.7025.10	290...410	min. 180	30	chemical equipment construction, electro-deposition, aircraft and spacecraft construction, surgery and orthopaedics: protheses, bone screws, implants and splints
Ti 99,7	3.7035.10	390...540	min. 250	22	dito
TiAl 6 V 4 F 89	3.7165.10	min. 890	min. 820	6	mechanical engineering, aircraft and spacecraft construction, electrical engineering, optics, precision mechanical engineering, medicine technology, fittings
TiAl 5 Sn 2 F 79	3.7115.10	min. 790	min. 760	6	mechanical engineering, aircraft and spacecraft construction, fittings

■ 4.2.3 Friction, wear and lubrication during sheet metal forming

As particularly important areas of tribology, friction, wear and lubrication have a significant influence on process engineering related to sheet metal forming. The practice has shown that by a purposeful application of appropriate die materials, together with process-optimized lubricants, wear of forming and blanking dies can be significantly reduced.

Friction
Friction plays a major role in sheet metal forming. Friction is the result of internal deformation of the deformed material and of externally applied loads. In its different manifestations, friction results in complex wear mechanisms at the contact surfaces between the die and the workpiece. During forming and shear cutting operations, sliding friction conditions apply where lubrication is used. In describing friction in sheet metal forming, as a rule, the coefficient of friction μ [–] is used and it is expressed as the friction force divided by the normal force, in the form:

$$\mu = F_R / F_N \quad [-] \quad \text{or}$$

the frictional shear stress divided by normal contact stress

$$\mu = \tau_R / \sigma_N \quad [-]$$

At lower sliding velocities, dry or maximum friction conditions apply; that is to say the dies/the workpiece are primarily separated by the reaction or adsorption layers – so-called boundary layers – of a molecular order of magnitude, whereby solid body friction is also possible in submicroscopic ranges. Here, the workpiece slides on lubricating films of molecular magnitude, i. e. on these boundary layers. In the absence of sufficient lubrication, the boundary layers can break up locally and weld joints can develop. Coefficients of friction in the range of $\mu = 0.10$ to 0.15 are indication of dry friction conditions.

With an increase in sliding velocity, portions of the interface are supported by hydro-dynamic fluid lubrication. The thickness of the lubricating film increases, but the lubrication gap h [µm] still remains smaller than the total of the surface roughness R_t [µm] of both, the die and the workpiece, i. e.:

$$h < R_{t\,die} + R_{t\,workpiece}$$

The frictional forces F_R [N] drop sharply and come close asymptotically to a limiting value, which is the transition point to pure hydro-dynamic friction.

The influence of sliding velocity, normal stress and lubricant viscosity is at its maximum in the area of boundary or mixed friction. Here, the die/workpiece pair has the topography of the task of holding the lubricant for as long as possible during the sliding action. As a rule, all forming processes involving boundary friction operate with frictional coefficients of $\mu = 0.05$ to 0.10. With further increases in the sliding velocity, complete hydro-dynamic friction can be achieved in which the lubricant film h becomes greater than the sum of the surface roughness R_t of both workpiece and die. This condition, however, cannot be achieved in practical metal forming operations:

$$h > R_{t\,die} + R_{t\,workpiece}$$

High local plastic deformations occur during friction in sheet forming, i. e. with the alteration of the surface of the workpiece a transformation in the topography takes place. Surface transformation takes place depending upon the state of stress, strength of the sheet metal material and of the grain size and alignment. This surface transformation which can alter considerably the amount of lubricant that is retained at the die/material interface. These mechanisms receive minimum consideration in the laws of friction, as does the influence of frictional temperature and pressure on lubricant viscosity during the sliding operation.

The coefficient of friction can be considered neither as a material variable nor as a constant for the die/workpiece pair. However, if the coefficient of friction is selected for making predictions, its dependence on various process parameters must be considered. In sheet metal forming, the following such parameters are of major importance:

- the material pair and the metallic surfaces which are in contact with each other,
- the die geometry,
- the temperature-dependent properties of the lubricant as well as its volume and the location of lubricant application,
- the actual normal contact stress, which is a function of the load bearing surface portions of both surfaces,
- the magnitude of the sliding velocity and the frictional path of the contacting surfaces and
- the surface characteristics of the workpiece material as well as the changes of these characteristics that take place during the process.

Wear

In general, the assessment of the wear behavior of dies is difficult because of the presence of several superimposed wear mechanisms. In order to describe the various forms of damage, wear mechanisms should be divided into five basic fundamental phenomena *(Fig. 4.2.8):*

Deformation

Under external loads, a micro-geometrical adaptation of the surface pair takes place, initiated by the local deformation at the contacting peaks of the roughness of the workpiece surface.

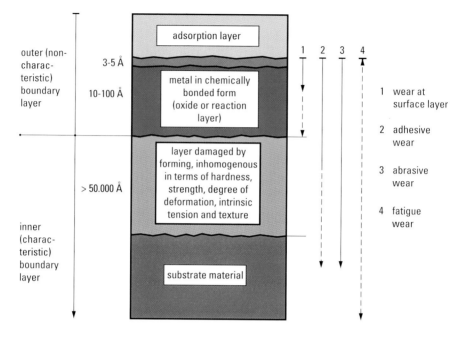

▲ **Fig. 4.2.8** Depth effect of wear mechanisms on metallic surfaces

Wear at surface layer

During the application of load chemical reactions between the die and the workpiece may take place. These are due to the influences of frictional heating, lubricant and the surrounding medium which lead to the built-up of an interface layer at the die/material interface. The shear strength of the interface layers is less than that of the metals, so that they are removed under frictional shear stress. Consequently, the coefficients of friction occurring between the die/workpiece surfaces are low when an interface layer is present. Thus, it is possible to maintain the functional capability of the contact surfaces even during short term overloading through wear of the interface layers.

Adhesion

Adhesion – also-called cold welding – occurs as a result of nuclear adhesive forces between the frictional partners. The emergence of adhesive forces pre-supposes the following:

- plastic deformation at the peaks of the roughness of contacting surfaces,
- surface expansion brought about by the forming process, which removes the adhesion presenting boundary layers and brings bright metal surfaces in contact with one another,
- similar material structures of the frictional partners.

The greater is the difference between the metals and alloys of the frictional partners, the lower is the tendency to develop cold welding. This tendency is at its lowest between metallic and non-metallic materials of corresponding hardness.

Abrasion

Abrasive wear includes all parting or separation processes that take place at the interface of frictional partners, where material is released in submicroscopic particles through a material removal process. The difference in the hardness between the two surfaces of frictional partners is an important factor for assessing abrasive friction. The wear resistance of a die increases with the hardness of its surface. Due to the relative movement of the two surfaces of the frictional partners, a part of the energy used is converted into heat. With most materials, the hardness drops as the temperature increases and the resistance to wear is reduced. Abrasive wear represents a generally inevitable long-term wear effect on the die.

Surface fatigue

Fatigue wear is considered to be the separation of micro- and macroscopic material particles. This separation is caused through fatigue cracking, progressive cracking and residual fracture due to mechanical, thermal or chemical stress-related conditions on the loaded surfaces. As a rule, surface fatigue plays a minor role in sheet forming dies. However, fatigue is common in cutting tools, as an increasing normal tension stress is applied onto the front surface in addition to a variable frictional shear stress that exists over the whole external surface area.

Lubrication

Lubricants are subject to different temperatures and compressive stresses during their use. Important variables influencing the correct choice of suitable lubricants are viscosity, density and compressibility, whereby viscosity demonstrates the greatest dependence on the two parameters: pressure and temperature. Particularly at contact points with high surface pressure, attention must be paid not only to the temperature, but also to the influence of pressure. Lubricants can be classified as follows:

- water soluble,
- non water soluble,
- solid lubricants,
- foils and varnishes.

Liquid lubricants are predominantly oil based. These are generally blends without a consistent composition, as the portions of paraffin, aromatic and naphtha content varies according to the origin of the oils. Animal-based, vegetable and synthetic oils also belong in this category. Paste-type lubricants are stabilized blends of mineral or synthetic oils, greases and waxes as well as soap. Powdered or needle shaped hard waxes and hard soaps are used as solid lubricants. The properties of lubricants can be altered by additives and adjusted to suit the respective application. Additives serve to improve load bearing capacities, bonding strength, and viscosity-temperature-pressure characteristics, and also assist in preventing corrosion. The additives create either physical adsorption layers or chemical reaction layers.

By means of so called "friction modifiers" (for example animal and vegetable fats, fat soap etc. also belong in this category), the lubricant bonds itself to the metal surface without causing any chemical reaction. This type of physically-acting additives are, however, inherently temperature-dependent. The reduction in cohesive and adhesive strength under increasing temperatures leads to an increase in the coefficient of friction. The effect of these additives ranges from a total absence of physical adsorption up to a stable chemical bond (formation of metallic soap). At high levels of interface pressures, anti-wear additives are used. They form a wear-reducing protective layer. A sub-division of this group are the extreme pressure (EP) additives, which form reaction layers under higher temperatures. With a combination of selective additives, in this range lubricants can be optimized to suit just about any particular application.

Based on the discussion given above, the following is a summary of the performance requirements of lubricants:

- development of a cohesive pressure and temperature-resistant lubrication film, which separates the surfaces of the workpiece and the die,
- high adhesive and shearing strength as well as good wetting capability,
- no involuntary physical or chemical reactions on the surface of the frictional partners,
- easy and complete removal of the lubricant from the finished part,
- no unhealthy or environmentally harmful substances.

4.2.4 Hydro-mechanical deep drawing

Hydro-mechanical deep drawing involves the use of a pressure medium, generally oil/water emulsion, to perform the forming process *(cf. Fig. 2.1.13)*. Hydro-mechanical deep drawing is carried out using double-action hydraulic presses. In addition, a bed cushion is utilised for hydro-mechanical reverse drawing.

Die layout
The lower die is a so-called pressure medium reservoir or water container. It is designed as a pressure chamber and serves as a holder for the part-specific female die *(Fig. 4.2.9)*. The water container is connected to the pressure regulator in the press. In order to fasten the female die on the water container a clamp or a shrink ring is used. This ring has grooves, which carry away emerging leakage fluid through the overflow. The female die has a circular groove at the edge of the drawing radius in order to locate a string like polyurethane seal. This is overlapped at both edges by an angle of approx. 45° when inserted in the sealing groove. This sealing groove can be eliminated in the case of rotationally-symmetrical hollow components, provided that the female die is designed accordingly.

The upper part of the die consists of the blank holder and the draw punch. The blank holder is normally a casting into which the part-specific blank holder insert is attached. This contains a circular splash ring designed to contain leaking fluid. This portion of the die serves also to bridge a larger blank holder opening in the press. The draw punch is located inside the blank holder. This is, in general, provided

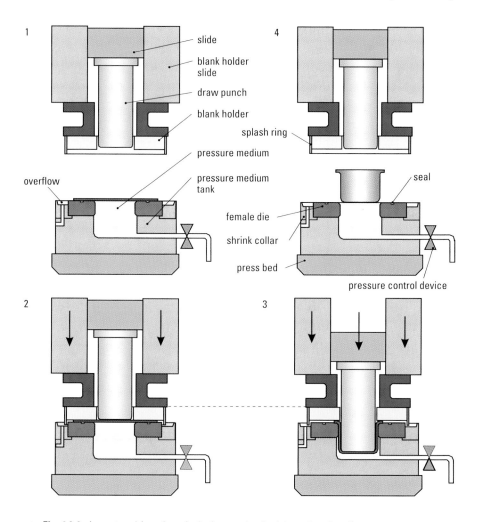

▲ **Fig. 4.2.9** Layout and function of a hydro-mechanical deep drawing die

with an extension piece. The punch extension is guided in the blank holder. In production dies, an internal mechanical stop should be provided to limit the draw depth.

Functional sequence

The press is opened and the water container is filled in the home position *(Fig. 4.2.9)*. After the insertion of the blank, the press closes and the blank holder grips the blank. The blank holder pressure, set at the press, seals the pressure chamber and the actual forming process is initiated. The medium pressure builds up as a result of penetration of the draw punch into the water container. During deformation, the sheet metal is pressed against the draw punch. During the forming phase, the control system which is linked to the pressure chamber controls the application of the hydraulic pressure in function of the draw depth.

After reaching the mechanically limited draw depth, the pressure in the chamber is released and the press travels back to its home position.

Special features of the Hydro-Mec forming process

The reaction pressure activated within the water container by the penetration of the draw punch has a multi-lateral effect during the forming process and forces the metal to be shaped against the punch *(Fig. 4.2.10)*. The dry friction between the punch and the sheet metal is thereby substantially increased. As a result, the drawing forces increase to levels considerably higher than those reached in conventional deep drawing. At the same time, the pressure of the fluid in the exposed part of the blank between the female die and the punch causes the material to bulge upwards. This deformation creates radial tensile stresses and tangential compressive stresses.

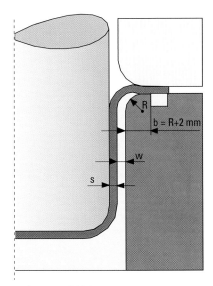

s: sheet metal thickness, w: water gap,
R: drawing radius, b: web width (at least 7 mm)

Fig. 4.2.10
Formation of the drawing gap during hydro-mechanical deep drawing

Hydro-mechanical deep drawing allows to achieve a significantly larger draw ratio than that achievable in conventional drawing operations. Whilst the limiting drawing ratio using conventional technology is $\beta \approx 2.0$, deep drawing ratio limits of up to 2.7 can be achieved in hydro-mechanical deep drawing. Since no intermediate drawing and annealing operations are required, very cost-effective forming is possible. The tool costs can also be reduced because of the reduction in the number of necessary forming operations. Another notable benefit is the quality of the surfaces of the drawn parts, as the sheet metal has not been drawn over a rigid drawing edge but over a fluid bead. Costs for finish processing operations such as polishing or grinding are substantially reduced or often completely eliminated.

Due to the pressing of the blank against the punch which reduces the amount of spring back, dimensionally accurate part production is possible. This is particularly important in reflector production, as not only measurement tolerances but also optical qualities are tested. The thickness of the sheet metal for hydro-mechanically deep drawn parts remains consistent within narrow limits. This applies particularly to the small reduction in thickness on the base radii, so that thinner blank can frequently be used for forming the desired part.

Press force is higher in hydro-mechanical drawing than in other forming methods using rigid tools, due to the reaction pressure in the water container. Here, the slide force F_{St} [kN] is the sum of the conventional forming force F_U [kN] and the reaction force F_{Re} [kN], which acts on the punch surface through the pressure medium *(Fig. 4.2.11)*.

Depending upon the particular blank materials being processed, the following pressures occur in the water container:

- aluminium: 50 –   200 bar
- steel: 200 –   600 bar
- stainless steel: 300 – 1,000 bar

4.2.5 Active hydro-mechanical drawing

Depending on their design, large-panel components for the automobile industry such as hoods, roofs and doors possess only minimal buckling strength in the center region of the parts. The reason for this is the low

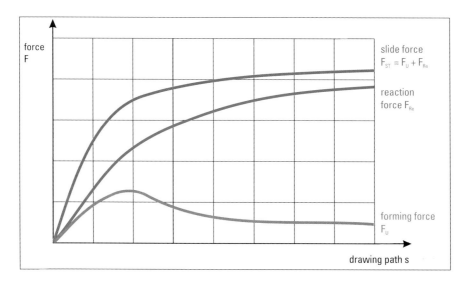

▲ **Fig. 4.2.11** Force versus displacement during hydro-mechanical deep drawing

level of actual deformation in this area. The small amount of deformation means that insufficient strain hardening takes place and the panels are liable to give way under only a low application of pressure, resulting in low dent resistance. This low component stability has a negative effect on crash resistance of vehicles and resistance to impact from hailstones. In addition, undesirable vibrations can occur at certain driving speeds.

As a result, higher strength materials or reinforcing elements and beadings are used. In some cases, this can considerably raise the unit cost of the part and increase the component weight. "Active hydro-mechanical drawing" (Active Hydro-Mec) is a deformation process that allows to overcome these problems and reduce the tool costs.

Range of parts
The Active Hydro-Mec method can be used to produce a wide range of parts, including practically all large panels and, particularly, flat external body parts of a vehicle such as hoods, trunklids, roofs, doors and fenders.

The smaller the degree of deformation required in the first operation when producing an outer panel (conventionally: deep drawing of the contoured blank), the more suited is the operation for using the Active Hydro-Mec method.

Operating principle

The Active Hydro-Mec method involves forming with an operating medium (oil-water emulsion), in which a large-size blank is clamped all around with a watertight seal between the female die and the blank holder – in the form of a water container *(Fig. 4.2.12/2)*. At this stage, a defined, part-specific gap exists between the clamped blank and the punch. As soon as the blankholding force has built up, the emulsion is actively introduced into the water container by means of a pressure intensifier and a defined pressure is built up (depending on the part between 20 and 30 bar). This pressure causes a controlled bulging of the blank which is pre-stretched over what is to become the entire surface of the component until it comes to rest against the punch in the center of the punch surface. This controlled expansion of the sheet metal results in work hardening of the workpiece which leads to a substantial improvement in buckling strength of the formed part.

After the pre-stretching process, the whole system – blank holder, blank and female die – is pressed upwards against the stationary punch. This process corresponds to the counter-drawing principle *(Fig. 4.2.4)* in which the emulsion is displaced. During the process, the pressure level can be kept constant or can be regulated to increase or decrease progressively or degressively. After completion of the drawing process, the emulsion pressure is actively built up (to a maximum of 600 bar) in order to ensure optimum application and formation of the blank against the contour of the punch.

Press

The press developed for the Active Hydro-Mec-process is a double-action straight-sided press design with Hydro-Mec pressure intensifier and a quick die changing system. *Figure 4.2.13* provides a schematic view of the press.

The press force of max. 50,000 kN is applied from below by eight cylinders located in the press bed. Six blank holder cylinders in the press crown apply a blankholding pressure of max. 16,000 kN during the pre-stretching process. Six powered mechanical stops can be adjusted for each individual part to determine the limit of the draw depth.

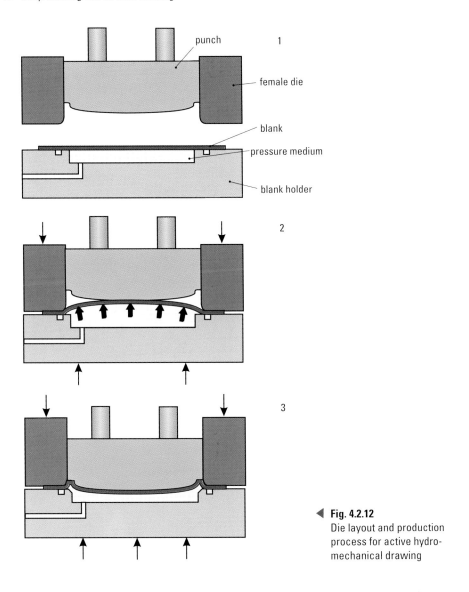

punch 1

female die

blank

pressure medium

blank holder

2

3

◀ **Fig. 4.2.12**
Die layout and production
process for active hydro-
mechanical drawing

The Hydro-Mec equipment consists mainly of a pressure intensifier (transmission ratio emulsion: oil = 2 : 1; max. emulsion pressure
600 bar, max. volumetric flow 20 l/s), and an emulsion tank. A 200 kW
motor and a pump in the main press drive system are used to drive the
pressure intensifier.

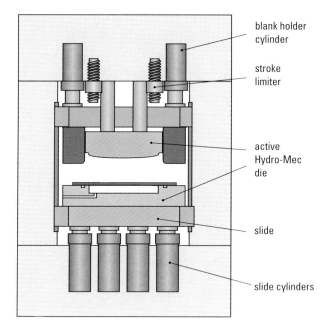

blank holder
cylinder

stroke
limiter

active
Hydro-Mec
die

slide

slide cylinders

▲ **Fig. 4.2.13** An Active Hydro-Mec press

The quick die change system consists of quick-action clamping elements and a die ejection system, and permits to achieve minimum set-up times. The blank holder base plate, female die and punch are clamped using fast hydraulic clamping elements. The ejection system transports the base plate from the machine together with the die. The die – comprising the blank holder, female die and punch – is lifted from the base plate and exchanged. The Hydro-Mec terminals are located in the base plate. These are automatically coupled by means of a docking system during inward travel into the press.

Advantages of the Active Hydro-Mec process
The Active Hydro-Mec process is characterized by a number of special features which distinguish it from conventional production. Quality benefits include the improved component stability and buckling resistance made possible by the pre-stretching process described above. Similar to conventional hydro-mechanical drawing processes, the sheet metal does not rub against the punch or female die but is pressed

against the punch by the emulsion. At the same time the process ensures optimum surface quality of the outer skin. Due to the lower sheet metal thicknesses used and the elimination of reinforcements, the overall assembly weight can be reduced. Another benefit in process engineering terms is the ability to process different materials (steel, aluminium, high-strength steels) and sheet thicknesses using a single set of dies.

In addition, the Active Hydro-Mec process represents a cost saving over conventional production methods. Tooling costs, in particular, can be reduced by up to 50%, as only one half of the die needs to be machined to have the part geometry. Furthermore, dies achieve a far longer service life due to the lower die wear involved in the hydro-mechanical drawing method. Depending on the part contour and the design of the die, it is often possible to save a complete processing step.

■ 4.3 Coil lines

The method of sheet metal feed plays an important role when considering the overall cost effectiveness of forming and blanking lines. When producing small sheet metal parts, the sheet metal coil is fed directly by a coil feed line into the press.

The function of the coil line is to unwind coils of sheet material, pass it through the straightener and to feed it by means of a roll feed precisely to the defined position of the processing line *(cf. Fig. 4.6.4)*.

The main components of a conventional coil line are the decoiler, straightening machine, coil loop and roll feed *(Fig. 4.3.1)*. The straightening machine draws the coil stock from the decoiler and flattens the material (cf. Sect. 4.8.3). Using sufficiently large dimensioned roller assemblies – in order to avoid the material from becoming bent again – the material is transported from the straightening machine continuously to the coil loop. The electronically controlled roll feed draws the stock intermittently from the loop and feeds it, correctly positioned, into the press. The maximum feed output of a high-performance coil line for small and medium-sized parts *(cf. Fig. 4.4.4)* and a blanking press (cf. Sects. 4.4.5, 4.4.7, 4.4.8) is indicated in *Fig. 4.3.2*. The stroking rate of the press is specified depending on the feed length and the crank angle which is available for the feed depending on the press and the tool in use.

Compact design coil feeding lines do not require the pit for the coil loop *(Fig. 4.3.3)*. They consist of a decoiler and a feed/straightening machine. The decoiler unwinds the stock downwards, opposite the direction of processing, so forming a horizontally aligned coil loop. The

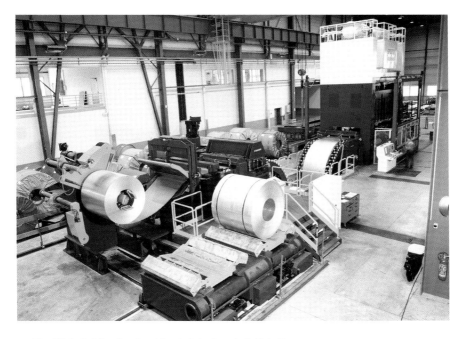

▲ **Fig. 4.3.1** Coil line for direct feed of stock material into the press

feed/straightening machine draws the stock on an intermittent basis off this loop, straightens it and feeds it, correctly positioned, into the press.

Whether the conventional coil feed line or the compact design is used depends on the material to be processed. Coil stock which is thin or has a sensitive surface can be very successfully processed using conventional methods. However, in cases where generally thick coil stock is being fed to the press, and if the material is suitable for interrupted straightening, the compact design is generally preferred.

When working with progressive dies and compound dies, the feed/straightening machine can be released briefly during immersion of the pilot pins for feed correction *(cf. Fig. 4.7.25)*. This applies equally to roll feed. For roll feed, the release period is around 0.03 s and for the feed/straightening machine around 0.05 s.

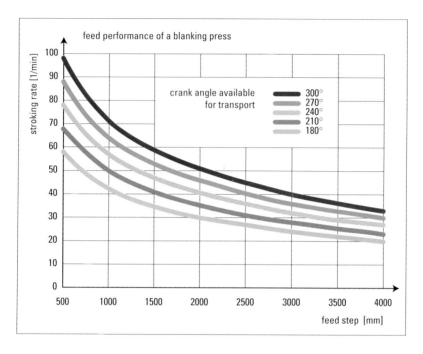

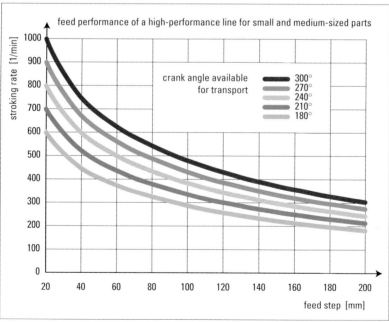

▲ **Fig. 4.3.2** Feed performance diagrams of a blanking press and of a high-performance line for small and medium-sized parts

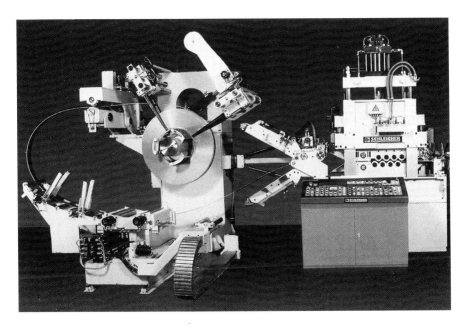

▲ **Fig. 4.3.3** Compact design coil line

■ 4.4 Sheet metal forming lines

■ 4.4.1 Universal presses

Universal presses are characterized by their high degree of flexibility. They are suitable both for blanking and forming operations *(Fig. 4.4.1)*. In combination with individual, progressive blanking, compound and transfer dies, this opens up a wide range of application possibilities for the production of small to medium-sized parts (cf. Sect. 4.1.1). To increase the number of formed parts and so enhance the cost-effectiveness of the production process, universal presses are equipped with coil lines *(cf. Fig. 4.3.1)*, scrap disposal systems and stacking units for finished products *(Fig. 4.4.41)*. Fully automatic die change plays an important role in ensuring economical production (cf. Sect. 3.4).

Comparison: mechanically and hydraulically driven universal presses
Generally speaking, universal presses are driven either mechanically or hydraulically. In the case of mechanical universal presses, the movement of the eccentric shaft is transmitted to the slide by means of the connecting rod. Accordingly, the sinusoidal curve of the stroke is prescribed by the geometry of the eccentric or crank shaft *(cf. Fig. 3.2.3)*.

For a mechanical and a hydraulic press with a forming stroke of approx. 100 mm, the stroke versus time, the time required for one work cycle (cycle time) and the time available for part handling (transportation time), are shown in *Fig. 4.4.2*. It is seen that the handling time available is 2.4 s, the cycle time of the mechanical press is 3.6 s, so that the corresponding stroking rate is around 17 strokes per minute.

◀ **Fig. 4.4.1**
Hydraulic universal press
(nominal press force 5,000 kN)

In the case of hydraulic universal presses, the slide velocity is adjusted to the process *(Fig. 4.4.3)*. Compared to mechanical presses, the impact speed of the slide on the workpiece can be drastically reduced. The desired gentle impact of the top die leads to lower tool stresses and smoother material flow, but increases the cycle time. In hydraulic presses, the pressure has to be released at the bottom dead center (BDC) without causing an impact, and the hydraulic system has to be reversed. Therefore, the workpiece contact time increases *(cf. Fig. 3.3.1)*. As a result, with a handling time of 2.7 s, the cycle time is 1.4 s longer for the hydraulic press compared to its mechanical counterpart. In our example, this corresponds to 12 strokes per minute. The difference in output between hydraulic and mechanical presses can be smaller than this but also greater depending on the specific production processes involved.

The energy consumption of the two press types also differs *(Fig. 4.4.2)*. The slide velocity of hydraulic presses is controlled in line with the motor output. In contrast, in mechanical presses energy is extracted from the flywheel, causing a drop in flywheel speed. If a

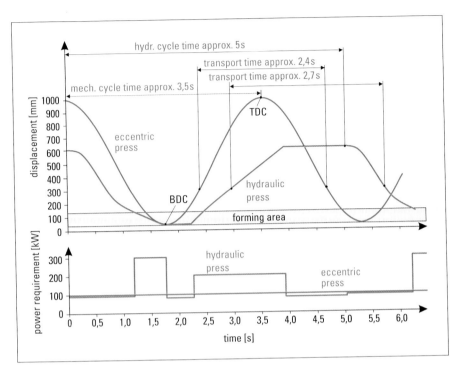

▲ Fig. 4.4.2 Slide displacement and power requirement versus time in mechanical and hydraulic presses with cycle and transport times

greater amount of energy is required for a work cycle, for example when working with high drawing depths, the drop in flywheel speed must not be more than 20 % of the idle running speed. A constant supply of energy for the drive motor is drawn from the electrical net. The energy capacity of the drive motor is selected such that the motor can bring the flywheel up to its idle speed after each working stroke. If the energy requirement of both press types is compared, it becomes apparent that hourly power requirement for mechanical presses can be some 30 % lower than that of hydraulic presses despite their lower output. However, this drawback can be compensated by changing the forming technology used, for example through counter-drawing (*cf. Fig. 3.1.10*).

High-performance transfer presses
The high-performance transfer press is derived from a mechanical universal press with particularly high output, developed for high-precision

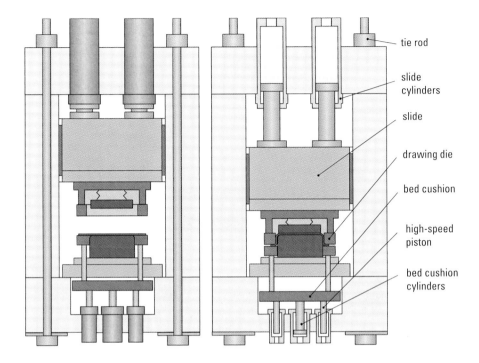

tie rod

slide
cylinders

slide

drawing die

bed cushion

high-speed
piston

bed cushion
cylinders

▲ **Fig. 4.4.3** Components and design of a hydraulic universal press

parts *(Fig. 4.4.4)*. This type of press is used for example for the production of ball and needle bearing sleeves. Depending on the press size – the rated press force of standard presses ranges from 1,250 to 4,000 kN – up to 300 strokes per minute can be achieved using a special tri-axis transfer system. The desired quality and output of the produced parts place stringent demands on the press configuration, in particular on the frame, drive system and slide gibs.

The required standard of drive system precision is achieved using an eccentric shaft mounted in roller bearings. The shaft is driven by a compact drive system with planetary gear. The roller bearings ensure only minimal play in the power delivery system of the press. A further benefit of the eccentric shaft mounted in roller bearings is that no additional release devices are required in case of press jamming due to an operating error. The eccentric shaft drives not only the main slide but also the transfer system, the draw cushion and a secondary slide. The

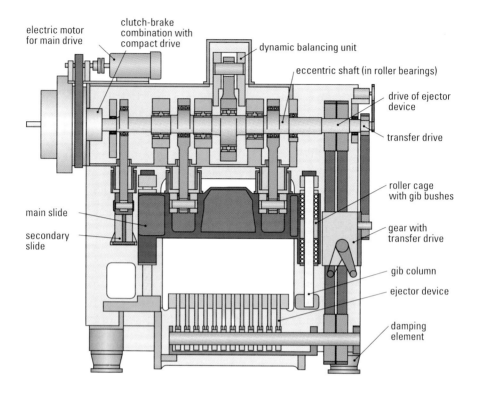

electric motor
for main drive

clutch-brake
combination with
compact drive

dynamic balancing unit

eccentric shaft (in roller bearings)

drive of ejector
device

transfer drive

roller cage
with gib bushes

main slide

secondary
slide

gear with
transfer drive

gib column

ejector device

damping
element

▲ **Fig. 4.4.4** Components and design of a high-performance transfer press

latter is also mounted in roller bearings and is used to cut the blank
from the coil stock. A zigzag feed device is used to improve material utilization *(cf. Fig. 4.5.7* and *4.5.8)*. In order to prevent any unwanted
effects of the blanking impact on the drawing operation carried out by
the main slide, the drive of the secondary slide by the eccentric shaft is
performed with a phase shift. A counterbalance mounted eccentrically
in the center of the drive shaft ensures smooth running of the high-performance transfer press. For this reason, simple press installation is possible by means of damping elements.

   Precise slide guidance is provided by four hardened gib columns with
roller cages and gib bushes on the forming level and at the top of the
slide. This system ensures optimum static and dynamic rigidity on the
horizontal plane, so enhancing the service life of dies *(cf. Fig. 3.1.5)*. The
rigid press body is a single-part welded structure, ensuring minimal tilt-

ing of the slide even in the event of eccentric loading. This is an important advantage which is particularly important during coining work.

Hybrid presses
Mechanical hydraulic presses, also known as hybrid presses, represent a new development in the field of press engineering *(Fig. 4.4.5)*. This type of press combines the benefits of mechanical and hydraulic presses. The hybrid drive system allows gentle impact of the top die on the workpiece, and also optimum control of the force exerted during the forming process. In contrast to hydraulic presses, hybrid presses offer higher output due to their mechanical basic drive system. Hybrid presses are configured for both deep drawing and also complex stamping, blanking and coining work, combining good flexibility and equipment availability with high output.

As in mechanical universal presses, the top slide is driven by the electric motor via the compact drive system, eccentric shaft and connecting rod (pitman aim) with eccentric bushings for adjustable stroke

▲ **Fig. 4.4.5** Hybrid press (nominal press force 3,150 kN)

(Fig. 4.4.6). The hydraulic bottom slide is used to alter the motion characteristics and for tilt angle compensation. The impact speed on the
workpiece can be reduced by penetration of the bottom slide into the
top slide by up to 200 mm/s *(Fig. 4.4.7).* If the top slide passes through
the bottom dead center (BDC) position, the bottom slide can be moved
up again, for example for coining. During this process, the bottom slide
and top die remain in the BDC position while the top slide is already
moving upwards.

In order to compensate for the lateral offset between the top and
bottom die, in particular when dies are used without guide posts or
gibs, an *automatic centering device* has been developed. This type of
device is commonly used, for example, in the cutlery industry. The bed
plate position is detected by position sensing systems and corrected
by comparison with the workpiece geometry. Correction is performed
by six horizontally arranged hydraulic cylinders, which are able to displace and rotate the ball bearing-mounted clamping plate on the horizontal plane within a range of hundredths of a millimeter to 1.5 mm

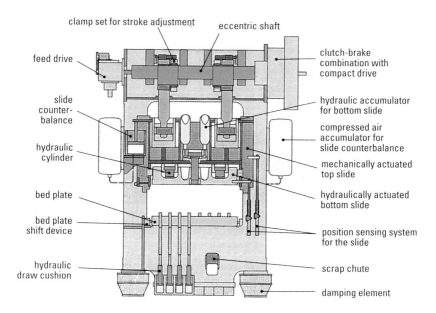

▲ **Fig. 4.4.6** Components and design of a hybrid press

(Fig. 4.4.8). Once correction is completed, the bed plate is precisely clamped in its new position by releasing the hydraulics.

Universal presses with modified top drive system
If an operating sequence involves a wide variety of work processes such as drawing, blanking, bending, stamping or reducing, it is rarely possible to achieve ideal working conditions for the respective forming process when using conventional press drive systems.

The modified knuckle-joint system used in mechanical presses offers considerable benefits *(Fig. 4.4.9)* when it is necessary to satisfy the widely differing demands made on the optimum forming velocity for universal operating sequences. By actuating the press drive and slide in this special way, it is possible to alter the movement and velocity profile of the slide *(Fig. 4.4.10).* A velocity reduction is achieved over a relatively large slide stroke in the working area, while the idle strokes for forward and return travel are conducted at increased speed (compared to eccentric and knuckle-joint drive system) *(cf. Fig. 3.2.3).*

This drive system provides power transmission from the drive to the slide, allowing to achieve high press forces and nominal force-displacement curves with a relatively small drive torque. As a result, the gearbox with clutch and brake can be produced in a compact and cost effective way.

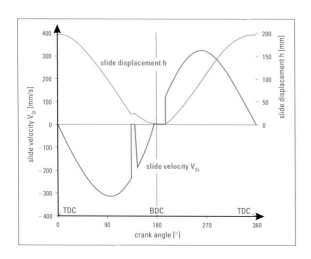

Fig. 4.4.7
Slide displacement and velocity versus crank angle of a hybrid press

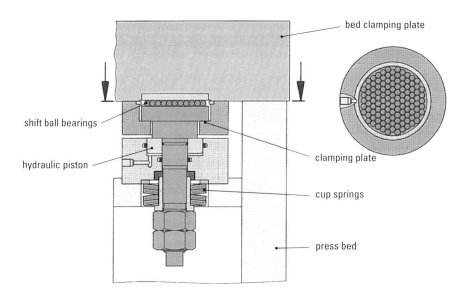

bed clamping plate

shift ball bearings

hydraulic piston

clamping plate

cup springs

press bed

▲ **Fig. 4.4.8** Clamping and release device for an adjustable bed plate

◀ **Fig. 4.4.9**
Universal press with modified top drive
(nominal press force 8,000 kN)

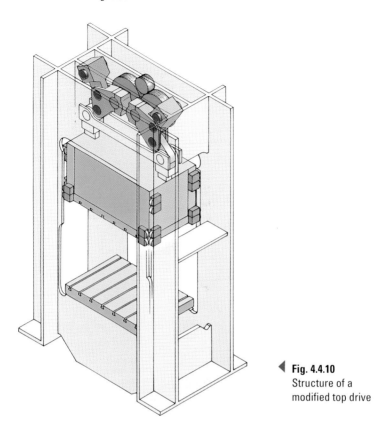

◀ **Fig. 4.4.10**
Structure of a
modified top drive

The two links are located at the outsides of the slide and allow for good compensation of the eccentric forces that may occur. Direct application of force in the upright area of the press crown helps to reduce deflection and bending and makes a major contribution to high system rigidity. The lateral forces, caused as a result of the force exerted by connecting rod, are absorbed by a crossbar, so that they cannot adversely affect the slide gibs. This system permits extremely precise slide guidance under load.

The reduced forming speed permits a higher stroking rate during drawing and reducing processes, protects the dies and reduces noise development. In addition, the added rigidity offers a number of benefits during stamping due to precise, plane-parallel forming. A reduction of dynamic deflection effects and blanking shock, as well as lower impact speeds allow to extend service life considerably. Through this optimization of working and forming conditions improved part quality and increased tool life are achieved while increasing stroking rates.

The variety of parts which can be produced on these lines include evaporator plates, bearing elements, hinges, hardware fittings and other components supplied to the automotive industry as well as parts for refrigerators, lock elements and clutch disks.

■ 4.4.2 Production lines for the manufacture of flat radiator plates

The combined forming and coining of large-panel press parts, such as those needed to make the plates for flat radiators, calls for a press system with extreme system robustness together with a generously dimensioned die space. In view of the wide variety of radiator designs, a universally applicable die technology concept is required.

High production output levels are only achievable using fully automated production lines which permit set-up within a short period for the production of different parts. Special press systems, with a nominal press force of between 4,000 and 10,000 kN *(Fig. 4.4.11)*, have been developed for the manufacture of flat radiator plates.

In order to achieve the largest possible system robustness, a bottom knuckle-joint drive system was selected *(cf. Fig. 3.2.3)*. This principle is distinguished by the highly compact, low-deflection construction of the knuckle-joint bearing and the press bed. As blanks with large surface area require large bed surfaces, a double knuckle-joint system is provided to improve press bed support. This design corresponds in principle to the single knuckle-joint system illustrated in *Fig. 3.2.5*.

The widely spaced link joints of the double knuckle-joint ensure tilt-resistant drive of the press frame. The height position of the upper die is adjusted by two wide power driven wedges which cover the entire die width. Due to the reduction in forming speed and the optimized weight of the slide frame construction, the knuckle-joint drive system ensures a high stroking rate in conjunction with smooth, quiet press operation.

An optimized welded design of the press frame allows even wider machines, with a frame width of up to around 2 m, to be run at high speeds. A press with a nominal press force of 7,100 kN, a frame width of

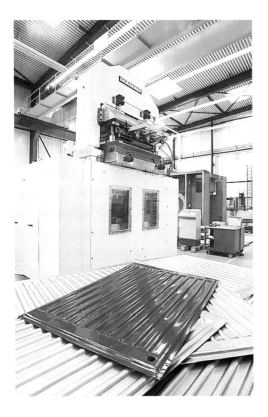

◀ **Fig. 4.4.11**
Bottom drive knuckle-joint press
for the manufacture of flat
radiator plates (nominal press
force 6,300 kN)

1,860 mm and a working stroke of 80 mm can produce 80 strokes per minute in continuous operation.

The machine is based on a modular design and can be equipped, as required, with or without a hydraulic draw cushion in the press bed and slide (cf. Sect. 3.1.4). The wedge adjustment of the slide ensures the high degree of rigidity required for drawing and coining work.

The modified knuckle-joint geometry and the connecting rod linkage prevent the knuckle-joint system from being fully extended at the bottom dead center. The bottom dead center has already been reached when the knuckle-joint system is still located some 2 to 3 mm before the maximum extended position. This design feature eliminates any possibility of press jamming, and the need to use a hydraulic overload safety system. The press force monitoring can take place exclusively with electronic measurement systems.

The slide gib is mounted at six points in spherical ball joints *(cf. Fig. 3.1.5)*. The necking action of the press frame which occurs under load is compensated by a supplementary internal slide gib in the bed. Pneumatic slide weight compensation is performed by maintenance-free bellow cylinders *(cf. Fig. 3.2.11)*. The die mounting area is designed to permit the use of all feasible die systems:

- single-action dies for individual radiator plates,
- double-action dies which can be used to stamp two adjacent plates from wide coil stock,
- sandwich dies in which two blanks are fed one above the other and two plates, i. e. top and bottom plates are pressed at the same time.

The sheet material is fed by an electrical roll feed system located on the back of the press, above the drive system. Special frame seals prevent the die lubricating emulsion from penetrating into the driving unit. Surplus emulsion is collected in a peripheral groove at the press bed or press frame and returned to the system to perform further greasing or lubrication work.

The line is controlled by a PLC control system with operator guidance and program pre-selection for different radiator types. The number of parts necessary for corresponding lengths can be put into the controller by specifying the appropriate radiator plate type.

The complete line is supplemented by a decoiler and a die change cart or a die change system with quick-action die clamping on the slide or bed plate (cf. Sect. 3.4). Scrap disposal takes place under the die by means of a scrap chute directly into a scrap container. All the drive units are accessible from the floor.

■ 4.4.3 Lines for side member manufacture

Production methods
The production of chassis side members for the truck industry is carried out in two steps: First, a blank is cut out and perforated. Next, the blank is bent to a U-shape to form a side member, while simultaneously web necks and where applicable also offsets are formed *(Fig. 4.4.12)*. Left- and right-hand members, for instance, must be configured in mirror image to each other and have reduced rib heights in the area of the dri-

ver's cab. Both operations are generally executed in a single line using two die sets.

When manufacturing side members, two different methods are used in conjunction with different types of production lines. These are distinguished from each other already in the blank feed stage.

Side members with a length of up to 7 m are generally manufactured from *plates*, which have a maximum thickness of 4 mm. The plates are fed into the press on an intermittent basis transversely to the longitudinal axis of the press. With each stroke the blanking die in the press punches a contoured blank from the plate. The blank is then bent in the next station to become a side member.

To produce side members with a length ranging from 4 to 13 m, *blanks* with a thickness of 5 to 10 mm are available. The blanks are generally cross-cut directly off the coil to the contoured blank width (cf. Sect. 4.6.1) in order to ensure optimum material utilization. The camber shaped blanks produced using this method have to be corrected in a separate unit by straightening. The straightened blanks are fed individually into the longitudinal axis of the press and processed to cre-

▲ **Fig. 4.4.12** Side members used in truck production

ate a contoured blank. This is done either by an all round blanking stroke or by local edge blanking – generally at both ends of the side member.

The advantage of processing sheet plates is that the blanks do not assume a camber shape due to the all round blanking cut used in their manufacture. Thus, the straightening process can be omitted. However, in this case a lower degree of material utilization is achieved. In the case of complex side member geometries, involving different flange and web heights, however, it is possible to save on material input by nesting the contoured blanks *(cf. Fig. 4.5.2 to 4.5.5)*. The blanking cut leaves behind scrap webs which are broken down into scrap segments.

Production line concepts
There are two basic line types used to produce side members. These differ in principle, based on the infeed and outfeed direction of the sheet metal: Whereas sheet plates below 7 m in length can still be fed into the press through the press uprights transversely to the longitudinal press axis, longer untreated blanks have to be fed parallel to the longitudinal press axis.

In lines where *transport takes place parallel to the longitudinal press axis,* the blanks are fed by means of an infeed conveyor belt inside the press uprights *(Fig. 4.4.13)*. Cross-feed units simultaneously feed the uncut blank into the die and the previously contoured blank out of the die onto the reverse outfeed conveyor. Transport systems making use of magnets are used to stack the contoured blanks emerging from the outfeed conveyor. Integrated in the line is also a blank turning station. The lines are designed in such a way that both, contoured blanks punched from uncut blanks and side members formed from contoured blanks, can be manufactured.

Important accessories for use with fully automatic lines include a blank lubricating or spraying device at the press infeed, a hole pattern camera to detect punch breakage and a device for edge milling. Fully automatic lines achieve cycle times of 3 strokes per minute. Taking into account the need for two passes and without considering the set-up time of around 15 min required for the change from blanking to bending, this results in an average line output of some 1.5 side members per minute. Die changeover also takes place in the longitudinal axis of the press.

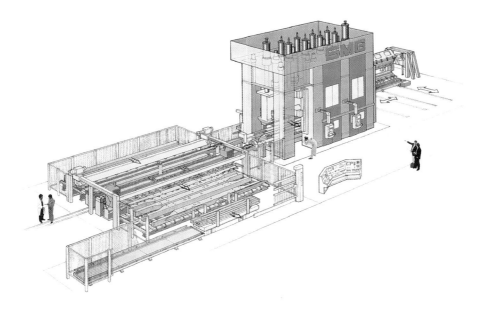

▲ **Fig. 4.4.13** Layout of a line for the manufacture of truck side members
(F_N = 50,000 kN, blanking shock damping and stroke limitation, 11 bed cushions,
25 slide ejectors, maximum off-center load: 25,000 kNm with electronic control
of parallelity)

In the case of presses with *material transport transversely to the longitudinal press axis,* the sheet plates are guided between the press uprights by means of a chain conveyor and a gripper feed device in synchronization with the press cycle. After each stroke, the remaining plate pushes the newly punched contoured blank onto the reverse outfeed conveyor belt. When an operation has been completed, the waste strip of the old plate must be disposed of together with the blanking waste from the new plate.

The equipment used for destacking the bales of plates or contoured blanks and the other elements of the line are the same as those used in lines transporting the material in the direction of the longitudinal press axis.

Die concept

The blanking and bending die each comprise a basic die equipped with trimming and piercing modules in accordance with the various side member lengths. Side members are structured in such a way that certain groups of holes, for example those used to fasten the front and rear suspension mounting, are identical and only change relative to the wheel base. In the area of the driver's cab at the front end of the side member, the contour and hole patterns are identical. The modules are accordingly adjusted and positioned according to the requirements of each individual side member type:

– front module for the area of the driver's cab,
– main module, for example for the rear axle area,
– intermediate module to bridge different side member lengths and
– the end module.

If at all possible, right and left-hand side members should have the same hole pattern. Only then it is possible to minimize the effort required for drawing and mounting piercing hole punches when changing set-up from right-hand to left-hand side members. The shearing edges required for blanking the side member contour do not have to be reset, as for example the left-hand contoured blank is produced from the right-hand contoured blank by simply rotating it in the turning station. However, for the bending process, bending dies, forming edges and ejector beams in the asymmetrical areas must be changed for a new set-up.

As the position of the front module does not change, if the length of the side members decreases, the position of the workpieces in the blanking and bending dies becomes increasingly off-center. If a single-sided coping cut is to be carried out at the same time, e. g. for example when trimming the front, an additional off-center load is applied. The off-center forces must be accounted for, either by a balancing module or by providing a suitable press concept. Otherwise an unacceptable amount of slide tilting action will occur.

To reduce the blanking force, the edges of the blanking die and punch are configured in a saw-tooth arrangement and the piercing hole punches are staggered on several blanking levels. The aim is to achieve uniform loading over the entire length of the workpiece at every blanking plane.

During the bending process, the web surface of the side member is supported above the bed cushion by means of a support beam. By varying the bed cushion pressure levels, it is possible to influence the flatness of the web surface, and in particular the rectangularity of the ribs. A stripping device in the die ensures controlled stripping of the angled side members from the bending punch. Parallel ejection is achieved using the support beams.

Segmented bed cushion
As a result of the variable side member length, the cushions at the end zones of the side members are only partially in use. A tilting moment occurs here which must be balanced by means of stable gibs.

If all bed cushions are activated at the same time during the upstroke of the slide for ejection, the cushions which have the smallest load reach the top dead center first. However, the side members must be ejected in parallel formation in order to avoid damage. There are two alternatives here: to eject the side member while maintaining continuous contact with the top die during the upstroke of the slide, or to control the synchronous ejection speed of all the cushions. If side members are formed with right-angled bends, different cushion strokes are required.

Control of parallelity and cylinder deactivation
The pistons of the cylinders that control parallelity are located at the outside of the slide. During the working stroke, they operate against the mean pressure of the servo valves *(cf. Fig. 3.3.5)*. The off-center loading of the dies generates a tilting moment on the slide. In the event of a deviation from parallelity, the pressure is increased in the control cylinder that is on the loading side while pressure on the opposite side is reduced. A supporting moment occurs which acts opposite to the tilting moment. In the case of side member tools, the off-center loading of the slide can be large due to the arrangement of the individual modules. In some cases, tilting of the slide can no longer be prevented by control of parallelity at a reasonable cost. In such cases, the control of parallelism is supported by the selective deactivation of individual slide cylinders *(Fig. 4.4.14)*. As a result, the center of application of the press force shifts in the direction of the forces generated in the dies. The larger lever arms of the parallelism control force help to increase the moment acting opposite to the off-center loading. However, deactivation of individual cylinders reduces the maximum available die force.

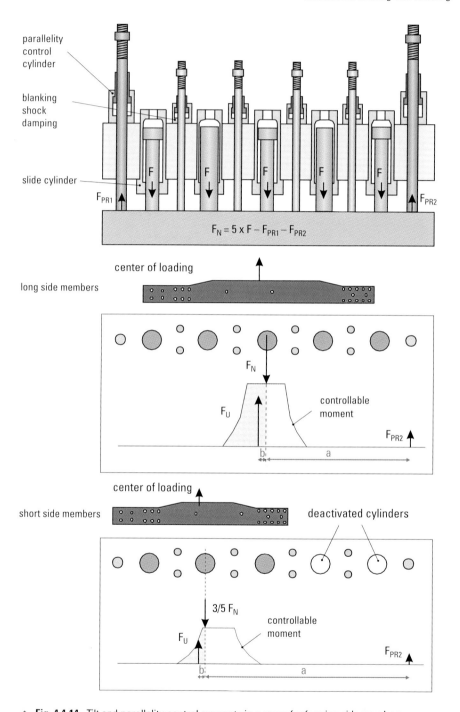

parallelity
control
cylinder

blanking
shock
damping

slide cylinder

F_{PR1} F   F   F   F   F F_{PR2}

$$F_N = 5 \times F - F_{PR1} - F_{PR2}$$

center of loading

long side members

F_N

F_U controllable
moment

F_{PR2}

b   a

center of loading

short side members   deactivated cylinders

$3/5\ F_N$ controllable
moment

F_U

F_{PR2}

b   a

▲ **Fig. 4.4.14** Tilt and parallelity control moments in a press for forming side members

■ 4.4.4 Destackers and blank turnover stations

Destackers

Destackers separate the blanks and feed them into the press in synchronization with the press cycle. Where low stroking rates are involved, the destacking and feeding operation is performed by feeders. In the case of higher stroking rates, after destacking the blank is transferred onto magnetic belts which transport them into the centering station, for example the pick up position of the press transfer.

Destackers with feeder and suction cup tooling

In the case of press lines (cf. Sect. 4.4.5) with a stroking rate of about 10 to 12 parts per minute and where suitable blank shapes are being processed, feeder systems are used. In case of a die change, the entire tooling is replaced together with the attached suction cups. During destacking, the tooling suction cups pick up the top blank from a stack. The operation is supported by fanning magnets whose magnetic field fans the top-most blanks in order to prevent the blanks from sticking to each other. The magnets can be positioned manually or automatically near the stack. Both permanent fanning magnets and electrically pulsating magnets are used.

When processing non-magnetic materials, e. g. blanks from aluminium or stainless steel, only suction cups can be used. In this case, compressed air is blown in from the side of the stack to enhance the separation of the individual blanks.

Destackers with magnetic belts

This design is equipped with one or two cylinder-actuated destacking units provided with vacuum cups or magnets *(Fig. 4.4.15)*. These are lowered between the magnet belts on the top-most blank of the respective stack, and then raise and transfer the suspended blank to the magnetic belt conveyor. Here, too, fanning magnets are used to support the removal of individual blanks. Using these systems, up to 25 relatively small and up to 15 larger parts per minute can be fed into the press.

To ensure that the press is able to operate continuously when changing from one stack to the next, either two destacking stations or systems with a reserve magazine are used. In the case of dual stacking feeders, switchover from the finished to the new stack takes place automatically.

▲ **Fig. 4.4.15** Destacker of a large-panel transfer press for maximum blank dimensions of 3,800 × 1,800 mm with a destacking bridge

Another possibility is to work with a reserve magazine. Here, the remainder of the stack is supported and destacked off the reserve forks while a new stack is added to it from underneath. Double blanks are detected electronically and automatically separated out of the system without bringing the entire line to a standstill.

In the case of non-magnetic materials, an integrated feeder with suction cups is used. Here, after destacking, the top-most blank is positioned from above onto a conveyor belt. A lower cycle rate is expected when using this type of handling system.

Another configuration used to transport medium-sized parts involves the use of magnetic belts mounted on a lifting bridge, which are lowered onto the stack of blanks. Thus, the top-most blank of the stack is picked up. This type of system is driven by means of a crank mechanism *(Fig. 4.4.16)*. The maximum achievable cycle rate with this method is about 16 parts per minute.

Prior to the forming operation, frequently the blanks must be washed, lubricated, centered and oriented. These operations are independent of the selected destacking method.

▲ **Fig. 4.4.16**   Destacker with magnetic belt for maximum blank dimensions of 2,100 × 750 mm, suitable for coated and non-coated blanks

Washing

To ensure the maximum possible part quality, the blanks – particularly those used to produce visible parts such as doors and roofs etc. – are washed and cleaned prior to forming. The washing/cleaning process is performed using cleaning emulsions or oils in conjunction with cleaning brushes. The blanks are fed towards the washing machine through a conveyor belt *(Fig. 4.4.22)*. The blank is mechanically washed in the washing/cleaning machine by a pair of roller brushes. During this process, the washing medium is continuously applied to the blank by means of nozzles. Downstream from the pair of roller brushes is a pair of squeezing rollers made of non-woven fabric. These reduce the cleaning medium uniformly and assure that there is only a minimal residue. The high throughfeed velocity based on the cycle rate is up to 150 m/min. For less complex parts, subsequent lubrication can be eliminated if the residue of the washing emulsion is sufficient for drawing, and the medium used is suitable for this purpose.

Oiling/greasing

For selective lubrication of blanks, oil spraying and greasing devices are integrated into the destacker. These can be used to apply the lubricant partially on the upper and/or lower side of the blank. Modern oiling devices are computer controlled and capable of partial greasing or oiling of approx. 200 mm length at throughfeed velocity of up to 150 m/min.

Centering/orientation

Before the blank can be automatically transported into the die, it must be aligned or centered in line with the required position. Inaccuracies may occur when positioning a blank. These may be due to the location on the stack or transport and they are corrected by means of end stops or lateral pushers. In order to obtain a high degree of flexibility concerning the dimensions and geometry of the blanks, all the positioning devices can be programmed and power adjusted. As a result, the set-up times are reduced.

Turnover devices for blank stacks

Blanks intended for further processing in forming lines are parted from the coil stock on blanking lines (cf. Sect. 4.6.2) and layered to blank stacks. In order to save tool costs, where symmetrical parts are involved, only a single blanking die is produced. Thus, it is possible, for instance, to use the blanks for the left-hand fender also for the right-hand side by simply turning over the stack. A distinction is made between the following blank stack turnover concepts:

- Friction-locked stack turning device: The stack is pressed with sufficient force so that friction holds the blanks together and then it is turned over.
- Positive-locked stack turning device: When turning in a drum or fork turning device, the stacks are mechanically supported.

Drum turning device: for blanks with complex geometries

Many blanks with irregular geometry, for example trapezoidal parts, may slip out when the stack is being turned over. Therefore, the stack must be supported on two sides by means of end stops. In addition, turning devices are often required to occupy only a minimal space and a single location to accommodate stacks both before and after turning.

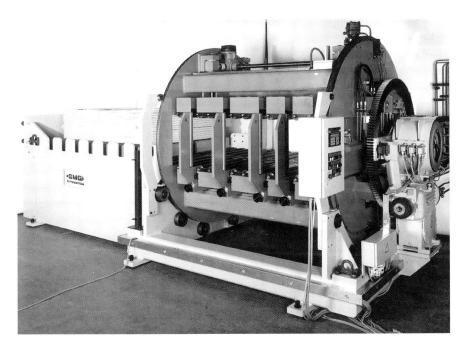

▲ **Fig. 4.4.17** Blank turning device for complex blank geometries

In the unit illustrated in *Fig. 4.4.17*, the stack of blanks is deposited on
a roller conveyor without using a pallet. The stack is transported into
the turnover device using this feeding roller conveyor and another
roller conveyor in the turnover drum itself. The blanks are held in place
by adjustable stops at two sides of the stack. A third roller conveyor
comes down onto the stack from above, and the unit rotates by 180°.
After opening the turnover device and releasing the stops, the stack is
moved out. Here, too, fast set-up is guaranteed as the stops automatically move to lean against the stack.

Fork turning device for multiple stacks
Multiple stacks of blanks which are stacked in a defined position on a
pallet and must be positioned again exactly after turning require a
highly precise turnover system *(Fig. 4.4.18)*. The stacks are initially
loaded together with the pallet on a carriage. To raise the first stack, the
carriage travels towards the turning device, allowing the lower forks to

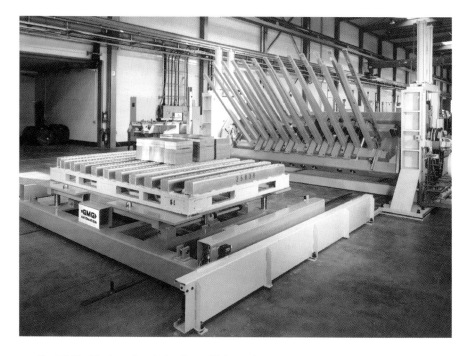

▲ **Fig. 4.4.18** Blank turning device for multiple stacks

engage between the pallet and the stack. The forks close and the stops automatically come to rest against the stack. A rotation around the horizontal axis of the turning device moves the stack of blanks to the desired position for uplifting. After the forks are opened, a second carriage with the pallet takes over the stack. Fully automatic changeover of the pallet surface and the transport directions ensures fast set-up and also allows to position several stacks of blanks on one pallet. Here, the stacks are turned over in sequence and deposited back onto the pallet in a defined position.

■ 4.4.5 Press lines

As a rule, series production of medium-sized and large sheet metal parts is performed on press lines or large-panel transfer presses. The former consist of a series of mechanical *(Fig. 4.4.19)* or hydraulic *(Fig. 4.4.20)* individual presses arranged in sequence and interlinked by means of handling devices. In many cases, press lines are being replaced by large-panel transfer presses. However, their high degree of flexibility assures that automated press lines still play a major role in the processing of complex parts which do not conform to any particular part family and which call for a high standard of quality. These include, for instance, parts requiring large draw depths, complex dies with cam action *(cf. Fig. 4.1.17)* or parts that require major repositioning between two successive operations *(cf. Fig. 4.1.11)*.

Modern press lines are expected to satisfy the stringent demands concerning output, part accuracy, die service life, equipment uptime and flexibility. These demands can be fully satisfied using mechanical driven presses with link drive systems. The use of link drive helps to reduce the impact speed during die closure to around one third of a comparable eccentric drive system *(cf. Fig. 3.2.3)*. This has major advantages to offer for the production process: With the same nominal press

▲ **Fig. 4.4.19** Mechanical press line with noise protection
(double-action lead press, nominal press force 20,000 kN,
follow-on single-action presses, nominal press force 15,000 kN)

force and force-displacement, a link-driven press can be loaded within the work range considerably higher and further away from the BDC than comparable eccentric presses, because the link-driven press has a steeper force-displacement curve (torque characteristic).

Consequently, a link-driven press can achieve higher stroking rates than an eccentric press depending on the drawing depth – the greater the drawing depth, the greater the increase. If the stroking rate is not increased, the lower impact speed results in improved drawing conditions. Even low grade sheet metals can be successfully processed. In addition, there is less wear and tear on the die and draw cushion, as well as on the clutch and brake.

Under practical conditions, the motion curve of a link-driven press with automatic part transport is configured so as to reduce the impact speed. Therefore, and also due to the use of double-helical gears in the drive system, noise is additionally reduced by up to 8 dB(A).

Number of presses

The number of presses which make up a line depends on the number of operations required to process the part in question (cf.Sect. 4.1.1). In general terms, an attempt is made to reduce the number of presses and thus also the number of necessary dies.

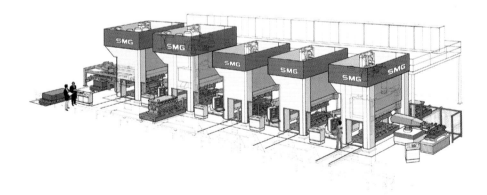

▲ **Fig. 4.4.20** Hydraulic press line automated by robots

As early as in the part design stage, through close cooperation with the component design and production departments, part simplification can lead to a situation where several different forming or blanking operations can be grouped together in a single forming station. The more operations can be combined in this way, however, the more complex the die itself becomes. This may lead to higher costs, reduction of working speed and higher susceptibility to malfunctions. Therefore, an ideal compromise has to be found when defining the number of forming processes or presses in a production line. Press lines consist of four to eight presses, although, depending on the range of produced parts, four to six presses are most commonly used.

Lead press: blank holder slide or draw cushion
While the lead press is frequently specified as a double-action press, the downstream presses are single-action, some of them incorporating a draw cushion in the press bed. In the case of double-action lead presses, separate draw and blank holder slides are available and they are driven either mechanically or hydraulically from above *(cf. Fig. 3.1.8 and 3.2.4)*. The blank holder slide contacts and holds the blank firmly at the outer contour prior to the initiation of the forming process by the draw slide *(cf. Fig. 4.2.2)*. In order to avoid wrinkle formation, the four pressure points can be separately controlled, i. e. different blank holder forces can be set at the four corners. In certain lines, it is possible to vary the blank holder forces during the drawing process using the pinch control (cf. sect. 3.2.8). These measures help to reduce the try-out period required for a new set of dies, and also improves drawing results. The double-action design of the presses allows the optimum way to achieve large draw depths. The drawback is that generally the part has to be turned over for the subsequent operation. The turnover station required after the first press limits the overall output rate of the press line *(Fig. 4.4.21)*.

New CNC-controlled hydraulic draw cushions which simulate the function of the blank holder slide have been developed. These cushions allow to form similarly deep and complex pressed parts on single-action mechanical or hydraulic lead presses (cf. Sect. 3.1.4).

In this case, the formed part does not require turning over after the drawing operation. Consequently, the output rate of a press line can be enhanced through the use of a single-action lead press with draw cush-

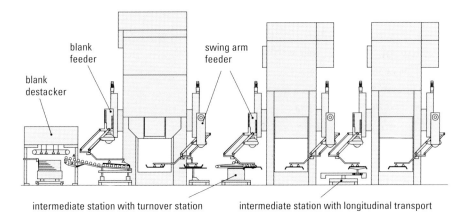

intermediate station with turnover station   intermediate station with longitudinal transport

▲ **Fig. 4.4.21**   Interlinkage of presses with swing arm feeders

ion. However, in this case the energy requirement is greater compared to a double-action press because the slide is additionally required to press against the draw cushion *(cf. Fig. 4.2.3)*.

In press lines, the design of the clutch and brake system is important because automated press lines are frequently operated in the controlled single-stroke or intermittent mode. As a result of demands for greater performance, the cycle times of the large-panel presses are continuously being increased and today can reach as many as 15 strokes per minute. Consequently, hydraulically actuated clutches and brakes are frequently used. In combination with a heat exchanger, this design alternative can sustain a higher energy rate than pneumatically actuated units, resulting in a higher stroking rate when operated intermittently, in the controlled single stroke mode. The engagement of hydraulic clutches can also be damped in order to reduce wear and tear on the drive system and to reduce noise (cf. Sect. 3.2.4).

Automation

The work sequence of a press line begins with the destacking of blanks by the destacker, cleaning, lubrication and feed of the blanks to the lead press (cf. Sect. 4.4.4). On completion of the forming operation, the parts are fed for further processing such as restriking, trimming, flanging or piercing. Stacking of the finished parts takes place at the end of the press line.

In modern press lines, workpiece transport between the individual presses as well as blank feed are automatically executed. Automatic stacking of finished parts is used wherever approximately identical stacking conditions, part shapes and sizes exist *(Fig. 4.4.41).* A number of different automation systems are used: swing arm feeders, CNC feeders or robots. The type of automation most suitable for any particular line depends on the range of parts being processed, the stroking rate and the available space.

Swing arm feeders are driven by electric motors via cams and links *(Fig. 4.4.21).* The feeder at the rear of the press removes the drawn part, which has been raised in the die by ejectors, by means of suction cups and a suction cup carrier and places it on an intermediate station. From here, the workpiece is turned over if required and loaded into the next die by the feeder of the downstream press. Swing arm feeders have fixed motion curves, making them suitable for small press distances, medium-sized parts and similar part types. They can transport between 10 and 12 medium-sized components a minute.

CNC feeders are electrically controlled and as such can be freely programmed on two axes and used for a wide variety of part types. Drive is performed by means of two spindle mechanisms or timing belt-driven carriages *(Fig. 4.4.22).* They are able to remove the part from the die

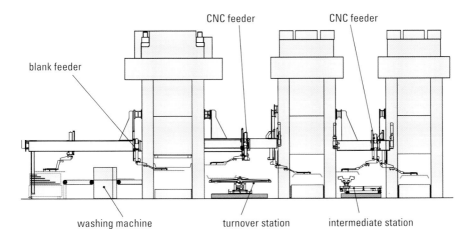

▲ **Fig. 4.4.22** Interlinkage of presses with CNC feeders and timing belt-driven carriages

without the need for an ejector. The output of a press line making use of a CNC feeder is between 8 and 10 large-sized parts per minute.

When using both, swing arm and CNC feeders, intermediate stations are required. These intermediate stations are equipped with three to five programmable axes for part transport in the longitudinal direction, for transverse and height adjustment and also additional tilting movements.

Automation using robots offers the advantage that it eliminates the need to deposit parts between the presses *(Fig. 4.4.20)*. The robot's gripper arm positions the part directly from the die of one press into the die of the next. This means that only one robot is required to service each press without the need for intermediate stations. The main drawback of robot-linked press lines is that heavy parts can only be transported slowly or not at all due to the centrifugal forces and the long transport paths. For this reason, output is largely dependent on part size and currently lies between 6 and 8 parts per minute. The freely programmable axes of the robot arm, in contrast, ensure that widely differing part shapes can be transported into many different part positions.

Die change

In addition to the interlinkage of presses in a line, automatic die changing systems also play a major role. Among die changing systems for press lines, moving bolsters with left/right-hand rails or T-tracks have proved to be the most popular *(cf. Fig. 3.4.4)*. The former require more space, but offer the shortest possible die changing times. The upper and lower die of the die set which is no longer required are moved out of the press on moving bolsters. Subsequently, the opposing bolsters move the new die sets into the press. Preparation of the dies for the next production run is performed outside the press during production. As a result of die changing automation it is possible to achieve changeover times of between 10 and 30 min depending on the workpiece transport system.

Nowadays, press lines are often enclosed in order to enhance safety and noise protection. Noise emissions can be reduced by around 15 dB(A) if the space between the individual presses and the peripheral equipment is also enclosed *(Fig. 4.4.19)*.

■ 4.4.6 Transfer presses for small and medium sized parts

Mechanical and hydraulic transfer presses are used for the manufacture of ready-to-assemble components for the automotive industry, its suppliers, the electrical and household appliance sectors, and other branches of industry *(Fig. 4.4.23)*. These presses guarantee high output rates because all forming operations required to manufacture a workpiece are included in one and the same line. Furthermore, all these operations are performed nearly simultaneously. A part is completed with each stroke of the slide, irrespective of the number of die stations in use.

Transfer presses are either supplied from stacks of blanks from the blanking line *(Fig. 4.4.15)* or they process stock directly off the coil *(cf. Fig. 4.3.1)*. In the former case, the press is equipped with a destacker; in the second it is equipped with its own coil line including decoiler, straightener and feed system. By locating the shearing station for blank-

▲ **Fig. 4.4.23** Mechanical transfer press for the production of wheel hubs
nominal press force: 42,000 kN; number of stations: 11; feed pitch: 700 mm;
slides: 3; strokes per minute: 10-28

ing outside of the press upright, it is possible to choose between lateral or longitudinal material feed to the press. Lateral coil feed offers the benefit that the scrap web can be rolled into a coil again. The drawback here is the greater space requirement of a lateral coil feed system. Generally speaking, however, the space requirement of a transfer press is considerably less than that of a press line; the lower space requirement also serves to reduce ancillary costs.

Forming and blanking operations can be economically performed both on mechanical and hydraulic transfer presses: blanking, initial draw and subsequent drawing processes, contour pressing, calibration, flat pressing, flanging, trimming, hole punching, etc. On the one hand, even parts with extreme draw depths and complex or asymmetrical shapes can be manufactured on transfer presses without any problems. Horizontal wedge-driven cutting and forming tools are also used *(cf. Fig. 4.1.17)*. On the other hand, dies can often be simplified by making most effective use of the existing number of stations. The number of die stations within the die sets used does not necessarily need to be identical. Idle stations can easily be bypassed. Transfer presses with a large number of die stations permit the production of two different parts simultaneously provided the die sets are able to form the part on fewer stations. In these cases, the blanks are fed synchronously to two different points in the press.

The total nominal force and energy costs of transfer presses are generally lower than it is the case for individual interlinked presses. Operating and maintenance costs are reduced to a single unit, and favorable conditions are created for reducing production costs due to the elimination of transport distances and intermediate stations.

Mechanical transfer presses
Mechanical transfer presses are built with two, three or four uprights. Depending on the range of parts to be manufactured and the number of stations, the presses are equipped with one, two or three independently driven slides *(Fig. 4.4.23)*. Each slide is designed individually to satisfy the requirements for size, nominal force and stroke. In particular for larger presses with different press force requirements at various forming stations, this type of multiple upright construction offers considerable advantages. It is customary in transfer presses to integrate a draw cushion in the bed *(cf. Fig. 3.1.11)*. A hydraulic overload system

protects against overload or in extreme cases against breakage of the slide, die or press as a result of excessive forces *(cf. Fig. 3.2.10)*. In the case of transfer presses with high loads at individual stations, a total press overload system is also provided.

To eject parts that may stick to upper dies, it is possible to equip each station individually with pneumatic, hydraulic or mechanical ejectors *(cf. Fig. 3.2.12)*. Their ejection force can be adjusted by means of pressure reducing valves or pressure axes. In conjunction with mechanical hold-back devices, the ejectors can also be used for thin-walled parts.

In mechanical transfer presses, parts are transported from one station to the next by a gripper rail system whose two or three-dimensional movement is generally coupled mechanically to the slide motion *(Fig. 4.4.24)*.

In two-dimensional systems, the gripper rails execute only the opening and closing as well as forward and return feed movements. If the workpieces have to be raised at certain stations of the forming sequence, this is executed by special grippers. A three-dimensional transport system is used in cases where workpieces require a lift motion at most sta-

▲ **Fig. 4.4.24** Tool area of a transfer press for producing oil filter housings

tions. The entire gripper rail system is then equipped with a lift stroke. This principle, which is described in Sect. 4.4.7 "tri-axis transfer presses", is generally in the production of larger panels.

All three axes of the gripper rail drive systems are positively driven by double cams. Configuration of the motion curves in accordance with mathematical principles prevents jerky accelerations – in other words, no centrifugal effects are created which would jeopardize reliable workpiece transport. The distance between the gripper rails perpendicular to the direction of transport can be adjusted infinitely or on a step-by-step basis. In the case of parts with small dimensions, this eliminates the need for long grippers and corresponding motion of large masses.

Hydraulic transfer presses

In the case of hydraulic transfer presses, an electrically driven gripper rail system transports the workpieces from the destacker to the storage location behind the press *(Fig. 4.4.25)*.

The manufacture of parts using the counter drawing method is generally used only in hydraulic presses *(Fig. 4.4.26, cf. Fig. 3.1.10 and 4.2.4)*. The drawing process is performed, generally in two stages, by active draw cushions in the press bed followed by trimming and flanging of the edge. The slide closes and displaces the blank holder while the two rear dies form the edge. The slide reaches its bottom dead center when it makes contact with mechanical stops. Only after this stage is reached, the parts are drawn using the punches by switching over the drive pumps from slide to bed cushion operation. The blanks are held at every die station using the pre-selected force level acting from above. Once the forming operation has been completed, the slide opens while the bed cushions return to their starting positions. The advantage of this forming method is that no tilting moment acts on the slide during the forming process, as the two halves of the die set are resting on each other during the entire drawing operation. Thus, it is possible to achieve a high standard of part quality and also a favorable energy balance (cf. Sect. 3.3).

If the parts are produced in the transfer press using a single-acting system with a draw cushion, large off-center loads can occur if drawing forces come into effect in the first stations, e. g. in the front-most dies, while the rear die stations have not yet been subjected to loading. This situation leads to tilting of the slide, which can be compensated with the aid of a hydraulic system controlling the slide parallelity *(cf. Fig. 3.3.5)*.

Fig. 4.4.25
Hydraulic transfer
press: nominal press
force: 42,000 kN;
number of stations:
11; feed pitch:
700 mm; slides: 3;
strokes per
minute: 10-28

Within the distance between the die centers – which corresponds at the same time to the transfer step – the draw cushions and bed ejectors are located underneath the bolster. Each drawing die set requires its own draw cushion whose blankholding force can be individually adjusted for the drawing process, carried out at each individual station (cf. Sect. 3.1.4). During the upstroke motion of the slide, the bed ejectors raise the workpieces to the transfer plane to allow the gripper rails to engage the part and transport it forward.

Hydraulic ejectors in the slide help to prevent the formed parts from sticking in the upper dies. The force and the speed of these ejectors must be configured in such a way that the parts remain in a defined position on the lower dies *(cf. Fig. 3.2.12)*. Optionally, it is also possible to equip the tools with low-cost spring loaded slide ejectors. However, in this case, the ejection of the parts is not well controlled.

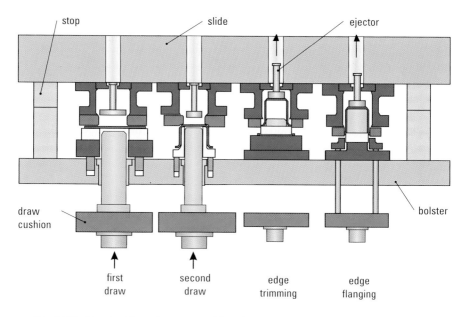

stop   slide   ejector

draw
cushion

bolster

first
draw

second
draw

edge
trimming

edge
flanging

▲ **Fig. 4.4.26** Transfer die set for a press with active counter drawing and fixed stops

■ 4.4.7 Large-panel tri-axis transfer presses

During the sixties and seventies, developments in the field of press line technology led to continuous performance improvement, culminating in the synchronized press line with fully automatic part transport (cf. Sect. 4.4.5). As a result of developments in large-panel transfer presses, it was possible to improve production times per part and to further reduce capital investment, production and storage costs. In conjunction with the integrated gripper rail transfer system, this press concept incorporates all the processing stages required in a press line *(Fig. 4.4.27)*.

Large-panel transfer presses with a distance between die centers of approx. 2,200 mm and die widths of up to 3,000 mm can process large workpieces measuring up to 2,500 × 1,500 mm. The forming stations are moved so close together that several stations can be accommodated under a single slide. This reduces the space requirement by 50 and 70%. The required power is reduced by 40 to 50% and the investment by 20 to 40%. Other benefits are gained as a result of reduced manpower requirement, increased safety against accidents and reduction of noise at lower cost.

In the initial layouts of press lines, the large-panel transfer press was combined with a double-action draw press *(cf. Fig. 3.1.8 and Fig. 4.4.28a)*, meaning that the drawing technique customary on double-acting presses was retained. The additionally required turnover device and synchronization of the presses, however, reduced the output to approx. 12 parts per minute. In addition, due to the use of the turnover device (cf. Sect. 3.1.3), the time required for die changing was considerably longer. Investment, operation and space costs still substantially exceeded those of a single-action large-panel transfer press.

▲ **Fig. 4.4.27** Large-panel tri-axis transfer press
nominal press force: 38,000 kN; slides: 3; work stations: 6;
feed pitch: 2,000 mm; strokes per minute: 8-18

Through the use of recently developed, controlled draw cushions in the press bed, today it is possible to execute all the necessary production steps on a large-panel transfer press (cf. Sect. 3.1.4). As workpiece transport from the first to the last processing station now only takes place using a single transfer system, higher transport speeds are possible. Depending on the transfer step, these lines are capable of an output of between 15 and 25 parts per minute. This corresponds to an increase in output of some 50% over an automated press line.

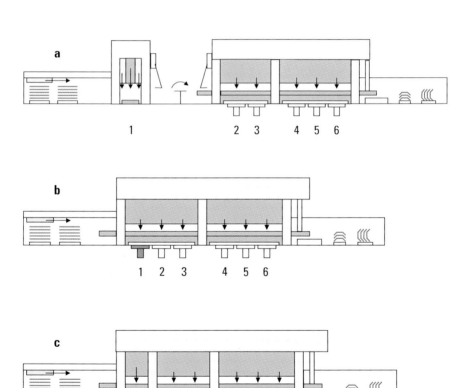

▲ **Fig. 4.4.28** Large-panel transfer press systems
a double-action press combined with a transfer press
b press with three uprights
c press with four uprights

Press layout

The design of a large-panel transfer press depends on the number of stations and the distribution of forces. The four to seven die sets required for the production of sheet metal parts are distributed over one, two or three slides. The transfer step depends on the size of the parts. The slides are guided in eight tracks *(cf. Fig. 3.1.5)*. In two- and three-upright presses, two, three or six die sets are each distributed to one slide *(Fig. 4.4.28c)*. Here, tilting of the slide can occur as a result of off-center loads, and must be partially compensated by appropriate countermeasures in the die set. In the case of extremely high drawing forces in the first work station, a four-upright press with a separate slide for the first station is recommended *(Fig. 4.4.28c)*.

Modern large-panel transfer presses are equipped with link drive systems in order to reduce the closing speed of the dies *(cf. Fig. 3.2.3)*. The press crown is split into two parts, the press bed into two or three parts. The press body is clamped by hydraulically pre-tensioned tie rods *(cf. Fig. 3.1.1)*.

Automation

Blank feed, workpiece transport, draw cushion control, part removal and in some cases also the stacking of finished parts as well as electrical control of the entire transfer press are fully automatic *(Fig. 4.4.11)*.

When changing from one stack to the next and when ejecting double blanks, continuous operation of the destacker is essential. Supplementary attachments for washing and lubricating the blank in the destacker are also frequently integrated. The first blank is separated off the stack by fanning magnets and lifted by suction units. It is then transferred to the centering station by means of magnetic belts and roller conveyors (cf. Sect. 4.4.4). The feeder to the first station lifts the blank from the centering station and transfers it to the first press station. Subsequent workpiece transport is performed by the tri-axis gripper rails. The gripper rails index the workpieces from one station to the next *(Fig. 4.4.29)*. Depending on the blank geometry, the rails are equipped with pneumatically actuated active grippers or shovels which support the workpiece during transport. In order to achieve a higher output, the gripper rail mechanism is manufactured for optimum weight savings as a box-type construction using hollow profiles. The entire transfer system is electronically monitored to check for part presence, position and other functions.

If there is an upright located after a die station, the grippers deposit the part on an intermediate station. The three movements executed by the gripper rails are performed in the longitudinal direction to transport the parts from one station to the next, in the transverse direction in the form of a closing movement for gripping the workpieces and in the vertical direction to lift out the drawn parts. The path covered by the gripper rails in the longitudinal direction is equal to the distance between die centers.

The longitudinal movement of the gripper rails is actuated directly from the press drive via intermediate gears, cams and cam levers. The closing boxes for lifting and lowering, and for opening and closing the gripper rails, are driven via cam levers and thrust rods.

Thus, part transfer is mechanically synchronized with the press drive system and it is therefore precisely reproducible. It is particularly beneficial for the entire transfer drive system to be located above floor level, as this reduces distances and also changes in the direction of motions. Lower masses, less strain and less play are the result thus permitting a

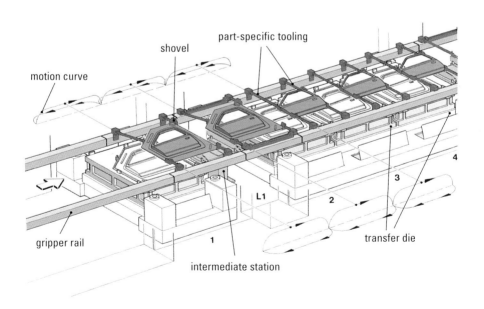

▲ **Fig. 4.4.29** Transportation of parts using mechanically driven gripper rail transfer

higher stroking rate and reliable part transport. A special computer program optimizes the motion curve of the transfer movements regarding mass and acceleration so that the shortest possible cycle times and extremely smooth movements are ensured for all stroking rates *(Fig. 4.4.30)*.

Motion curves are established for each transfer system and they are used to check the clearance of the gripper rail system relative to the die movement. A safety clearance of at least 20 mm must be ensured between the tool and the curve in each case. All elements which may impede this clearance, such as cam drives or gib elements must be taken into consideration (cf. Sect. 4.1.6).

Electrically driven transfer
Instead of mechanical transfer systems, electrically driven systems can also be used for large-panel presses *(Fig. 4.4.31)*. The path, acceleration and speed values of electrical transfers can be freely programmed, which allows the system to be adjusted to different sets of dies.

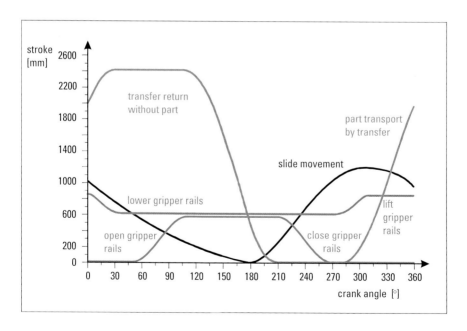

▲ **Fig. 4.4.30** Motion diagram of a tri-axis transfer system

Workpiece-specific axis data, lifting strokes, closing distances and where applicable also transfer steps are entered into the press control system. Electrical transfers make particular sense wherever just-in-time production of small batch sizes is called for *(Fig. 4.4.32)*. Here, the transport system can be adjusted to a different number of stations, in some cases different distances between die centers and other parameters by accessing the values for each part from the data memory. However, for reasons of safety, this high degree of flexibility means compromising on cycle speeds in comparison with mechanically coupled transfer systems.

In the case of gripper rails for high stroking rates with high values of acceleration and speed, extremely light composite materials and aluminium alloys are used.

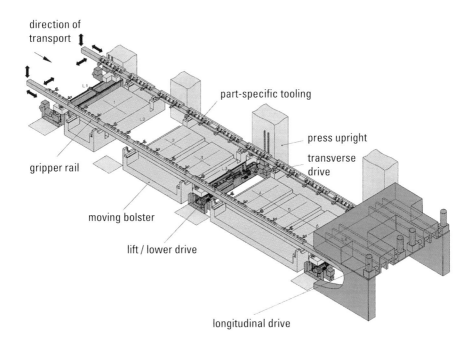

direction of transport

part-specific tooling

press upright

transverse drive

gripper rail

moving bolster

lift / lower drive

longitudinal drive

▲ **Fig. 4.4.31** Layout of an electrical transfer system

Draw cushion

When producing car body parts on large-panel transfer presses, the blank holder function must be performed by the draw cushion in the press bed in order to permit the part to be further processed in the same position *(cf. Fig. 3.1.11)*. Provided the geometry of the drawn part is suitable and with adequate pre-acceleration of the draw cushion, good results can be obtained using pneumatic draw cushions. However, the blank holder force control is even more favorable when using microprocessor-controlled hydraulic draw cushions *(Fig. 4.4.33* and *cf. Fig. 3.1.12)*. Even with large drawing depths up to 250 mm, the use of this type of draw cushion produces workpiece quality comparable to that achieved on double-action presses. However, this system necessitates adjusting the die design to the press *(cf. Fig. 4.1.16)* and transfer system, so that marginally higher die costs must be taken into account. Retrofitting of existing die sets for use in this type of large-panel transfer press must be reviewed individually in each case as regards costs and feasibility. A particularly econom-

▲ **Fig. 4.4.32** Electrical transfer in a large-panel tri-axis transfer press

ical solution is to integrate a large-panel transfer press into the production process when a model change is made and when new die sets have to be manufactured.

Die change

In large-panel transfer presses, dies are also exchanged using moving bolsters which move automatically in and out on the left and right of the press, in some cases simultaneously *(cf. Fig. 3.4.4)*. The die change area is safeguarded by light barriers and partially monitored by video cameras.

The die changeover sequence is depicted in a die change diagram. Initially, the part-specific contoured nests of the intermediate stations are automatically transferred to the moving bolster. The part of the gripper rails between the uprights to which the part-specific grippers are attached, must be exchanged as part of the die change process.

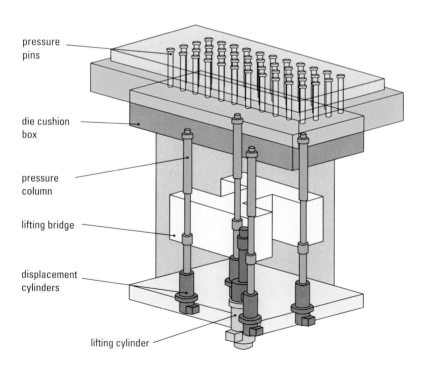

pressure pins

die cushion box

pressure column

lifting bridge

displacement cylinders

lifting cylinder

▲ **Fig. 4.4.33** Hydraulic four-point draw cushion with pressure pins

These components are separated from the part of the rail which remains in the press by means of an automatic coupling device, and deposited on the moving bolster.

After releasing the power connections, the moving bolster with the die and part-specific accessories travels to one side out of the press, while the sliding bolster with the next die enters from the other side *(Fig. 4.4.27)*. After automatic coupling of the gripper rails and power connections, the die data such as pressure levels and positions are accessed from the data memory and automatically set. When the moving bolster with a new die set is set up outside the press, all the part-specific devices such as grippers and contoured nests are exchanged and prepared for the next production process.

A characteristic feature of large-panel transfer presses is their high productivity with an outstanding degree of flexibility made possible by short set-up times. The automatic die change process possible in large-panel transfer presses can be completed in less than 5 min. Thus, a major precondition is satisfied for the reduction of capital costs and of the need for intermediate storage of finished parts between the stamping plant and the body assembly plant. This means that, depending on batch size, it may be possible to perform several die changes per production shift (cf. Sect. 4.9.2).

Depending on the press operating period, on a large-panel transfer press between 5 and 30 different die sets are used with batch sizes of 3,000 to 10,000 panels. To ensure economic utilization of the press, parts with nearly identical form and size and a similar degree of forming complexity should be grouped to create families of parts *(cf. Fig. 4.9.2 and 4.9.3)*.

■ 4.4.8 Crossbar transfer presses

Where large panels from around 2.5 × 1.5 m in size are processed, the low inherent stability of the parts during transportation calls for special measures. Conventional gripper rails which only grip or hold the parts around their outer periphery *(Fig. 4.4.29)* are not suitable. Particularly large unstable parts such as roofs, full body side panels or floor assemblies are accordingly produced using crossbar transfer presses. The application of this type of press also covers the dual production of medium-sized parts, in which two parts are formed from one blank or two adjacent blanks *(Fig. 4.4.34* and *cf. Fig. 4.9.2).* Double parts produced from a single blank are separated as required in a subsequent operation. The production of double parts could refer to any suitable parts, for example passenger car doors. This kind of production doubles the output per press stroke and increases considerably the economy of the system, compared to gripper rail transfer systems. Moreover, dual production helps to better utilize the press load capacity when forming a single large and unstable part does not fully require the available press load.

Large-panel crossbar transfer presses can be structured in different ways. Ideally, each die should be assigned to a separate slide to ensure optimum peripheral conditions for the die and for the forming process. The design is based on a modular structure comprising individual machines in which all the drive systems are connected to the main press drive system by means of central longitudinal drive shaft and intermediate couplings *(cf. Fig. 3.2.9).* This ensures synchronous running of all the stations: The transport system can be operated by a continuous transfer with only a minimal safety clearance to the top die.

To transport large unstable parts in transfer presses, a two-axis transfer system equipped with crossbars and suction cups is used *(Fig. 4.4.34).* Unlike the off-center positioning of parts on feeder or robot arms as used on press lines *(Fig. 4.4.35),* here the parts are held directly by suction cup carriers located above the part. Due to the symmetrical arrangement of the suction cups relative to the center of gravity of the parts, higher accelerating forces acting on the part are permissible. In comparison with feeder mechanisms, stroke rates can be increased from 13 to 15 parts per minute with transport steps between 2,000 and 2,600 mm.

The crossbars are fastened on carriages which execute the longitudinal feed step from one die to the next. The carriages run on the two lift beams of the transfer which move in the vertical direction to raise and lower the part. No supplementary aids, such as lifters or ejectors in the die, are required for part transport.

The time-motion diagram differs markedly from that of the tri-axis transfer *(Fig. 4.4.36).* In contrast to a tri-axis transfer, the crossbar transfer cannot return during the forming process, as the crossbars are located between the dies. When the dies are closed the crossbars are located in a parked position outside the die area. The die spacing must in any

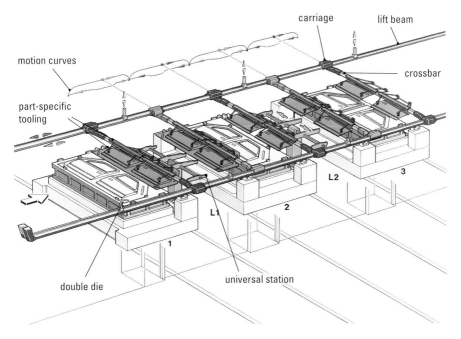

▲ **Fig. 4.4.34** Part transport with crossbar transfer system in the production of doors using a double die

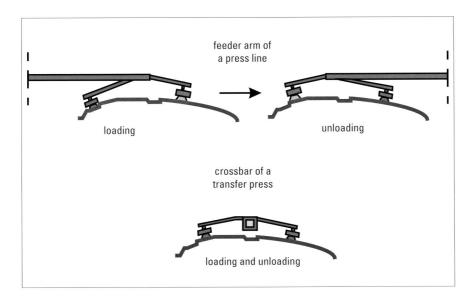

▲ **Fig. 4.4.35** Positioning of parts on a feeder or robotic arm of a press line and on a crossbar

case be sufficiently great to accommodate the crossbars. With a suitable number and arrangement of the suction cups on the crossbars, it is possible to transport large one-piece body side panels for passenger cars or even minivan-type vehicles. The blank dimensions for this type of side panel can be up to 4,200 × 1,900 mm.

In order to achieve smooth transfer in conjunction with maximum output, the movement sequences are computer-optimized *(Fig. 4.4.36)*. Composite carbon fiber materials are used for the crossbars in order to improve rigidity, damping and to reduce mass. This permits more reliable part transport to be achieved even at high stroking rates.

With the use of universal stations between the uprights, it is possible to achieve an optimum degree of production flexibility *(Fig. 4.4.37* and *4.4.34)*. Program control of up to five axes allows the transport of parts into the ideal position for subsequent operations, thus reducing the expense of the dies and improving the reliability of die functions. Furthermore, the universal stations allow the crossbars to assume an asymmetrical parking position during die closure. This means a marked reduction in the length of the part transfer and consequently an

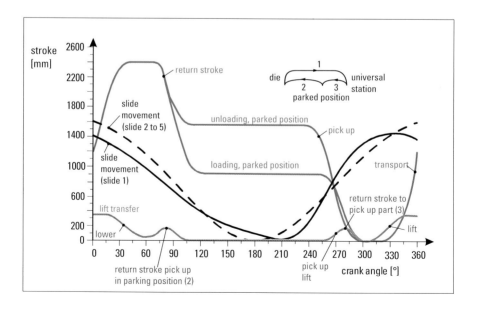

▲ **Fig. 4.4.36** Motion diagram of a crossbar transfer press

increase in the stroking rate of 10 to 15% without generating high forces during part acceleration. The motion paths and acceleration values are calculated once the part family has been defined.

The press concept in which every station is equipped with a separate slide *(Fig. 4.4.38)* offers certain advantages over slides which accept several dies. For example each slide and therefore each die station can be individually adjusted concerning height, press force and response of the overload system. The slide is generally subjected to on-center loads, as the individual die stations do not influence each other. Forces are absorbed by the four connecting rods with minimal tilt of the slide. The slides are guided in eight-way gibs which reduce tilting of the slide *(cf. Fig. 3.1.5)*. The deflection in the direction of transport is reduced in both the press beds and the slides.

These features ensure – particularly in the case of large dies – an advantageous effect on part quality and on the service life of dies. Moreover, it results in reduced die try-out periods and run-in times for new die sets until production of the first acceptable part.

Separate slides also offer the benefit that the spacing between individual dies – still with the advantage of reduced part transfer – increases.

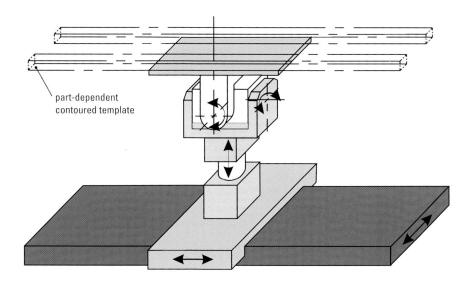

part-dependent
contoured template

▲ **Fig. 4.4.37** Universal station with three to five programmable axes

As a result, quick die clamps, tooling at the transfer and universal stations and the dies themselves are more easily accessible, allowing for more generously dimensioned scrap chutes. Malfunctions are less frequent and easier to locate and remedy. In general, separate slides help to increase the utilization rate and economy of the press.

In addition to crossbar presses with separate slides described here, this press type can alternatively be designed to accept several dies under one slide. This represents a solution at lower capital investment similar to the solution used in a tri-axis transfer press *(Fig. 4.4.27).*

Basically this system transfers the parts directly from one die to the next die without employing universal stations for part repositioning in between. Consequently, the part transport needs to cover the distance between die centers with the inclusion of the crossbar parking space which results in about 20% more horizontal transportation and consequently reduced stroking rate.

Die changing equipment plays a significant role in reducing press set-up times. Transfer presses are equipped with moving bolsters

▲ **Fig. 4.4.38** Crossbar transfer press with moving bolsters outside the press
nominal press force: 52,000 kN; slides: 5; work stages: 5; feed pitch: 2,000 mm

(cf. Fig. 3.4.3) for automatic die change. For each die station there are two moving bolsters. One die set is used for production inside the press, the other set is located outside the press being prepared for the next production run *(Fig. 4.4.38* and *cf. Fig. 3.4.4).* This allows the die and part-specific tooling to be prepared while the press is still producing.

While the lift beams stay in the press, the crossbars and the tooling are separated by automatic couplings from the lift beams for die change. They are then positioned on the moving bolster of the die and moved out of the press at 90° to the direction of part flow. Every die set is equipped with its own crossbar and suction cup tooling. When setting-up the moving bolsters outside the press, the crossbars assigned to the moving bolster are equipped with the tooling required for the next part.

The steps performed when changing dies and resetting all relevant parameters of the press are executed automatically. These steps include: deposit and unclamping of top dies, release of crossbar couplings, raising of lift beams to clear traversing of moving bolsters, exit of moving bolsters; entry, lowering and centering of new bolsters, clamping of upper dies, replacing the next crossbars with tooling and contoured nests of the universal stations.

All the die change parameters are automatically set by a programmable logic control. Accordingly, all the adjustable parameters of the peripheral press systems such as the destacker or part stacking are recalled and automatically set (cf. Sect. 4.4.11).

A complete die changeover requires about 10 min, as various resetting processes run simultaneously. The individual phases of the die changing process are locally and centrally displayed on screens for monitoring purposes.

When manufacturing large parts such as one-piece body sides, in particular, individual and direct control of the blankholding forces is very important to achieve good part quality. Just as in tri-axis transfer presses, power is transmitted via pressure cylinders to the blank holder frame. Generally, systems with four displacement cylinders are used. The fundamental advantage of the hydraulic displacement system, compared to pneumatic draw cushions, is that the draw cushion force can be optionally controlled at each individual hydraulic cylinder during the drawing process by means of microprocessor-controlled servo valves. The pressure can be adjusted between 25 and 100% of the rated pressure *(cf. Fig. 3.1.12).*

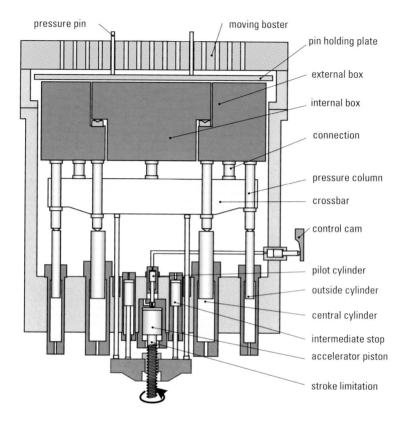

▲ **Fig. 4.4.39** Hydraulic draw cushion with eight displacement cylinders and a split pressure pad

Double parts can be most successfully produced *(Fig. 4.4.39)* on eight-cylinder draw cushions with a split pressure pad. This allows individual adjustment of four cylinders per part.

■ 4.4.9 Presses for plastics

Metal components are being replaced by plastics in a number of applications, including the automotive engineering industry. The materials used here include FRTP (glass fiber-reinforced thermoplastics) and SMC (sheet molding compound). The manufacture of these parts imposes particularly stringent requirements on the pressing process *(Fig. 4.4.40).* The major difference between pressing glass fiber-reinforced plastics and sheet metal forming lies in the material behavior, the wall thickness of the produced components and the forming speed.

In sheet metal forming, the thickness of the part depends on the starting blank. The material thickness of a plastic part, in contrast, is determined by the parallelism and the cavity of the tool. Furthermore, plastic is inserted in the press in a malleable condition. It is shaped, when using the SMC process, by the application of heat and when using the FRTP process by the dissipation of heat to form a stable com-

▲ **Fig. 4.4.40** Hydraulic plastic press for SMC and FRTP parts
nominal press force: 18,000 kN; maximum stroke: 1,300 mm

ponent. This imposes additional demands on the functions of the press, in particular on the speed of the slide. This calls for the use of special control systems (cf. Sect. 3.3.3).

When pressing thermoplastics in refrigerated dies, forming must be performed quickly, and the cooling process must take place at a complete standstill and with optimum parallelism of the two die halves.

In the case of thermoset plastics, a layered mat is inserted in the die. The compound material – glass fiber and resin – must be pressed according to a precisely specified speed pattern with precise parallelism of the two die halves, in order to avoid fiber orientation. The material hardens as heat is applied. In order to achieve a uniform part thickness, the slide must be kept < 0.1 mm/m parallel. These process parameters are essential for the production of high-grade SMC parts and can only be implemented using a hydraulic press with multiple axis control. The technological requirements

- fast closure, in order to minimize heat transfer between the workpiece material and the die before forming,
- pressing at the specified speed pattern to ensure uniform front flow behavior,
- pressing with the specified force pattern to ensure application of the special pressure level required for precise forming,
- holding the pressure at a prescribed level while the workpiece hardens at the precisely parallel position of the two die halves,
- controlled opening motion to avoid damage to the workpiece

can only be achieved using a hydraulic press. The drive system is buffered by pressure accumulators in order to assure that the required different oil volumes are available (cf. Table 3.3.1). The slide operates in a closed control loop, monitored by sensors and controlled by servo valves. The parallelism of the slide is maintained within extremely close tolerances by four cylinders acting on the slide corners *(cf. Fig. 3.3.5).*

■ 4.4.10 Stacking units for finished parts

Finished parts leaving the press are palletized either manually or with the aid of an automatic stacking unit to permit their storage or transportation. It is often uneconomical to convert the unit for widely varying parts, as the stacking unit would have to be reset at the same speed as the forming line, i. e. in 10 min or less. This would necessitate a fully automatic resetting process. In addition, the stacking position in transport containers often varies considerably for different parts. Accordingly, stacking units for finished parts only make sense where very similar parts are processed. The unit illustrated in *Fig. 4.4.41*, for instance, stacks only similar passenger car doors. There are two different types of system for stacking finished parts: using an overhead feeder system or using industrial robots.

The destacking feeder of the press places the sheet metal parts on the part lifting and locating device, which positions them on synchronized intermediate deposit stations or belt conveyors. The parts travel through a test and inspection station, at the end of which they are transferred to another part lifting and locating device, aligned and moved into the stacking position.

In systems with overhead feeder, depending on part size and the press cycle rate, one or two CNC feeders engage two, four or eight parts, orient them as required and stack them in the transport containers.

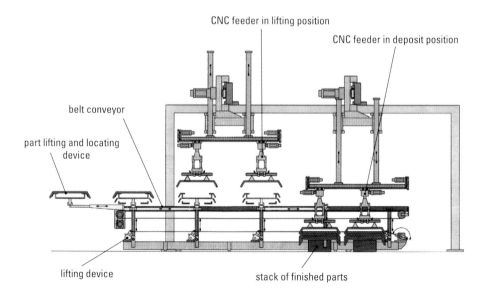

▲ **Fig. 4.4.41** Stacking unit for passenger car doors

Transport containers with stack latches permit individual removal by industrial robots. The unit is also equipped with a detection system for transport containers and with an electrical and mechanical control system for the stack latches. This ensures that the transport containers belonging to the various finished parts are made ready and that the stack latches are adjusted to the geometry of the formed sheet metal parts.

The transport containers arranged on both sides of the conveyor belt guarantee continuous stacking even during container changeover. A supplementary conveyor belt can be used to transport parts out of the press in the event of a malfunction in the stacking unit or for inspection purposes without interrupting the press operation.

In addition to the solution using overhead feeders, industrial robots with six axes can be used for removal and stacking of finished parts. This type of automation concept makes particular sense when the parts are placed to be ready in the assembly position by the handling device, necessitating an additional rotating or tilting movement. This type of concept is found not only in presses but also for flanging and welding operations used in plant construction applications (cf. Sect. 4.6.5).

■ 4.4.11 Control systems for large-panel transfer presses

Thanks to the progress made in microelectronic engineering now electronic control systems, industrial PCs, high-speed communication systems and digital drive systems are available and can be used in the development of economical press control systems offering maximum control capability for the operating team.

Control systems able to comply with the specific requirements of large-panel presses must fulfil the following criteria:

– clearly arranged, easily understood press control structure,
– use of standard modules and identical parts to reduce component variety,
– clearly arranged operating and visual display system without excess amounts of information,
– operator support wherever operators are located,

- efficient troubleshooting, back-up in correcting malfunctions and repairing failed components,
- use of proven standard components from manufacturers offering worldwide service,
- use of standard communication systems,
- non-standard solutions only in functional areas where this is un-avoidable,
- compliance with customer regulations and specifications,
- coordination with and approval from customer regarding the control concept.

As an example of control engineering applied in large-panel presses, the following is a description of *control concepts for large-panel transfer presses with crossbar transfer system* (cf. Sect. 4.4.8). The modular mechanical structure and the active universal stations are of particular relevance for the structure of the control system.

The consistently modular decentralized control structure directly reflects the modular mechanical structure of the press itself *(Fig. 4.4.42)*.

The blankloader, each press module and the associated universal station are each equipped with their own control unit, the so-called *local control unit*. It consists of the conventional part, the PLC system, the decentral input/output system and task-specific individual electronic control systems, and each has its own local visualization unit.

Central functions such as mode selection, clutch/brake, main drive, transfer system with loading feeder, operating units and tool data management are the responsibility of a *central control unit* in conjunction with a central visualization unit.

All the local control units of the press are identical in their internal structure, only the first control unit differs from the others. Here, the functions for the bed cushion are added and the functions relating to the universal stations do not apply. The PLC programs of all local control units are identical: Different functional requirements are detected by the program by means of hardware identifiers at the PLC inputs.

The logical communication structure is strictly star-shaped, i.e. neither data nor signals are exchanged between the local control units. These structural characteristics result in standardized, easily manageable control units with a transparent flow of information and a number of appreciable benefits for the user:

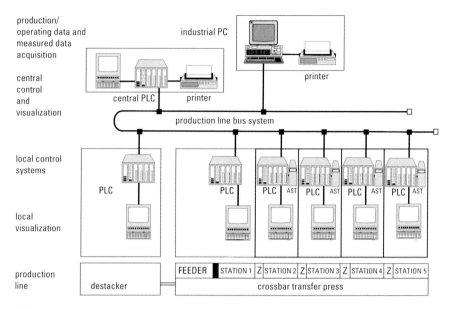

PLC: programmable, logic controller   AST: axis control   VIS: visualization   Z: universal station (5 NC axes)

▲ **Fig. 4.4.42**   Control structure of a crossbar transfer press

- reduced familiarization periods for the unit operators and maintenance staff,
- simplified troubleshooting and maintenance,
- simple data link to higher-level systems.

The underlying characteristics of the electrical equipment and thus also the press control system were described in Sect. 3.5. The explanation provided here deals with the specific properties and functional characteristics of the control system for the above-mentioned press type.

Operating modes of the press
The user has the following press operating modes available:
- *Set-up:* This mode is used for executing manual operations, for example on initial press run-up, when restarting after a malfunction or when running in new dies.
- *Single cycle:* One press cycle is executed at a low stroking rate.

– *Automatic:* This is the mode used for part production. The press executes cyclical production strokes at the set stroking rate, e. g. 15 strokes per minute.
– *Die change:* In this mode, the line is reset for the next job. The entire die set and all the die-specific tooling are exchanged. In case of fully automatic die change, all functions are controlled by the PLC without any intervention by operating staff. The sequence can be monitored at the visualization system using the synoptic die change display.

Conventional control system
The components and main functions of the conventional press control system were described in Sect. 3.5. A major component of conventional control systems is *safety control*. Today, this still generally makes use of conventional technology using contacts, although machines supplied to automobile manufacturers in the US are frequently operated using redundant PLC systems.

For considerations of operating safety, the aim is to also allow transfer presses to operate even in the set-up mode only with closed protective gear; this permits considerable simplification of the safety control system while at the same time enhancing working safety for personnel.

Electronic control system
The electronic control system comprises the programmable logic controllers (PLC), the input/output system, the axis control systems and the communication system (press bus system). Every PLC is assigned to an input/output system. The components of the *input/output system*, the input/output modules, are located in the terminal boxes located at the individual structural elements of the machine.

Communication between the PLC and the relevant input/output modules is executed by a high-speed, easily installed *field bus system* with a deterministic (constant, pre-calculable) time response. To control the NC (Numerical Control) axes, the universal stations and other functional units, *axis control systems* are used in conjunction with highly-dynamic servo drives and absolute position measuring systems. The axis movement cycles must be executed in synchronization with the transfer movement. The *line communication system* links the programmable logic controllers with each other and with the central units of the visualization system.

The *switch cabinets* which accommodate the components of the electrical control system are arranged wherever possible on the press crown. Switch cabinets for the various units such as the hydraulic unit, the lubrication system and the hydraulic draw cushion are mounted in the press pit. Overall, this produces an efficiently functioning, decentralized, modular-structure press control system.

Operating and visualization system

In complex production lines which are spread over a wide area such as large-panel transfer presses, the operating system is of major significance in determining productivity levels. A detailed, clearly arranged representation of the line status, comprehensive operator prompting for set-up functions, and an efficient troubleshooting system are important prerequisites for safe and efficient operation, as well as for reducing press downtime.

Control units are positioned at all the major operating stations of the press line *(Fig. 4.4.43)*. Local control panels are mainly located on the press operating side and also in reduced form at the rear of the press.

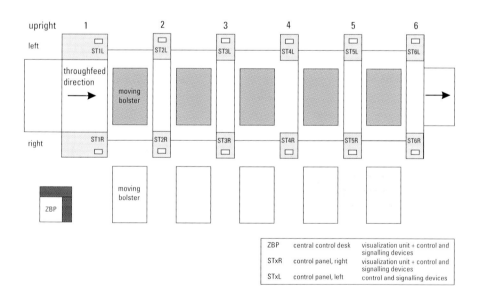

▲ **Fig. 4.4.43** Operating points on a crossbar transfer press

Local operating units are equipped with all the necessary functions related to the operating station in question.

Central control functions such as preselection of operating modes, main drive system ON or start of die change, are executed from the central control desk. In machines delivered to US automobile manufacturers, pendent control units are frequently mounted along the length of the press which allow central control functions to be made available locally to operating staff.

Alongside the classical operating and display elements such as push-buttons and pilot lamps, modern operating systems for complex lines frequently make use of color graphic visualization systems. Depending on the level of complexity of the visualization system, industrial PCs or compact, intelligent operating terminals with graphic capability are used in local control panels.

The visualization system is used for user prompting and guidance during start-up, operation and monitoring of the equipment, and also for troubleshooting and diagnostic functions. *Figure 4.4.44* indicates

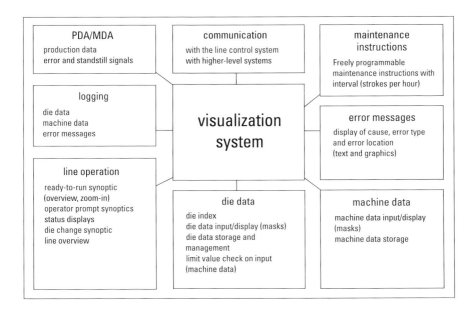

▲ **Fig. 4.4.44** Functions of the visualization system used on large-panel presses

the functions of a visualization system used in large-panel presses. Using modern operating techniques such as pull-down menus and windows technology, the press operator is able to access the required functions quickly and simply. It is also possible to jump across between functional areas and from one image to the next. The image frame is structured identically for all the masks, and the function key assignment and operating procedures are standardized throughout. No graphic input device such as a mouse or trackball is required. This makes working with the user interface of the visualization system extremely easy to learn. With a minimum of practice, the system can be quickly and safely operated.

The *ready-to-run synoptic display (Fig. 4.4.45)*, frequently also referred to as the press synoptic, serves to prompt the user one step at a time until ready-to-run status is reached. Green squares represent fulfilled and red squares unfulfilled conditions. Alongside ready-to-run statuses of the functional areas of the press, ready-to-run statuses of other line

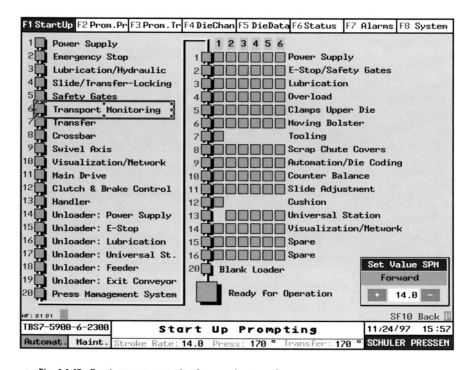

▲ **Fig. 4.4.45** Ready-to-run synoptic of a crossbar transfer press

modules are also indicated; in this case of the blank loader. With the aid of the so-called zoom function, it is possible to access a detailed synoptic for each of the listed conditions. This can be actuated manually by selecting with the cursor, or in the "Automatic" synoptic mode. Here, a detailed synoptic is displayed automatically with the first unfulfilled condition. As synoptic displays are so clear and easy to understand, they are also used for the die changing sequence and for operator prompting. The *die change synoptic* displays all the sequential steps and their current execution status.

Operator prompt synoptics provide the user with an indication of all the conditions for initiating individual functions in the "set-up" and "die change" modes *(Fig. 4.4.46)*. The visualization system makes this type of synoptic available for each set-up function.

The *malfunction diagnostics* system detects malfunctions occurring anywhere throughout the press. The visualization system indicates the existence of a malfunction in each image by highlighting the menu

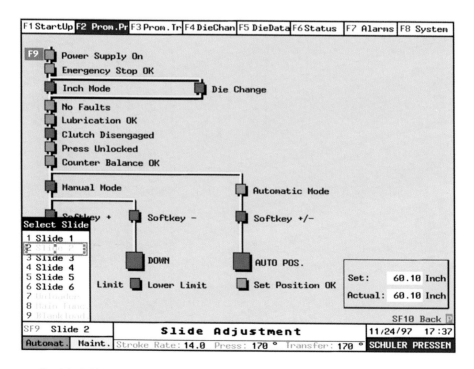

▲ **Fig. 4.4.46** User prompting for slide adjustments in a crossbar transfer press

point "Error" in the menu line. Error messages contain an error number, plain language, reference to the affected assembly (operating resources identification) and reference to the point in the PLC program. This information, in conjunction with the status displays, supports the detection and localization of the source and cause of the malfunction.

Die-dependent data related to all dies used in the line can be entered and edited at the visualization system. They are stored and managed in the visualization system. During die change, this data is automatically transmitted to the relevant control units.

Alongside the visualization system, which is imperative for line operation and monitoring, PC-based systems are also frequently used for the acquisition and processing of machine and operational data as well as measured values.

Machine and production data acquisition and evaluation
The machine and production data acquisition system (MDA/PDA) – or PMS (Press Management System) for short – logs error and standstill data as well as production data accumulated in the press line. As such, it plays an important role in detecting and analysing weak points and in registering and monitoring press line productivity.

The hardware basis of the PMS is an industrial PC which is directly linked to the line's control system. The central PLC is responsible for making available the necessary data, and so functions as the communication partner for the PMS. Table 4.4.1 indicates the spheres of operation of a PMS for large-panel presses.

The field of machine data acquisition is subdivided according to the acquisition and evaluation of errors and standstill periods.

According to the definition used here, errors are constituted by machine statuses which are the direct result of faulty or damaged line elements or components and which generally necessitate repair work. Errors are automatically classified by the PMS. In addition, a freely configured event-related comment is possible for each error message.

Errors do not, however, always cause a standstill; Standstills, in turn, do not necessarily result in errors. As a result, alongside the detection of errors, separate acquisition of all press standstill statuses is required.

Standstill, in accordance with the definition used here, consists of all times during which no parts are produced (cf. Sect. 4.9.2). Depending on the line in question, a difference is made between approx. 30 differ-

Table 4.4.1: Functional areas of machine and production data acquisition systems in large-panel presses

Machine/production data acquisition and evaluation		
automatic acquisition	errors	standstill – during production – during die change	job data, production data
manual acquisition	commentaries (plain language)	commentaries (plain language), classification	number of reject parts, number of rework parts
evaluations	chronological list	chronological list	chronological list, individual printout (job data table)
	sorting according to: error frequency, total duration, error category frequency and total duration	*sorting according to:* standstill frequency, total duration, error category frequency and total duration	*sorting according to:* die number, workpiece number, batch size
	selection according to: period, shift, job number, die number, error number, error categories	*selection according to:* period, shift, job number, die number, standstill number, standstill categories	*selection according to:* period, shift, job number, die number

ent causes of standstill. The conditions of the ready-to-run status synoptic form the basis for monitoring standstills.

An important function in monitoring standstill periods is the determination of the actual cause of the standstill. The control system frequently determines only the so-called secondary causes, in particular in the case of manual operator intervention. The actual cause of the problem, termed the primary cause, can be determined by the user by means of event-related modification of the standard standstill classification and by the addition of event-related comments.

The field of *production data acquisition* is subdivided into job data acquisition and, based on this, production data determination. Job data refers to the individual production orders. Its acquisition stops on completion of the order, and the relevant data is stored *(Fig. 4.4.47)*.

Job data is automatically acquired, supplemented by optional manual inputs in the press control system. At the end of a shift, and at the end of an order, the current data table is transmitted to the PMS. The *production data* is ascertained on the basis of this information for an evaluation period which can be freely defined by the operator.

The production status illustrated in *Fig. 4.4.48* offers a survey of the most important data from all the shifts of a production day. In addition, the pre-processed data can be exported in a standard data format in order to give the user the chance to execute his or her own evaluation.

Data acquisition and processing for draw cushions

The acquisition and processing system for measured values for the draw cushion, known as the Cushion Monitoring System (CMS), permits the observation of the drawing process. In addition, this provides support for repair personnel in terms of malfunction diagnostics and also when resuming production after a component failure. All the relevant set values, control signals and process variables (actual values) for pre-acceleration and for the actual drawing process are monitored and stored, for example

SCHULER PRESS MANAGEMENT SYSTEM

F1-File F2-Handle F3-View F4-Evaluation F7-System F8-Info

Selection		Sort		Chronologically
Time Period from - to		Die Code	Part Description	Job
27.09.96 00:00:00	27.09.96 23:59:59	0	All	All

Die Code	7
Part Description	5532 DACH AUSSEN
Theor. Strokerate	9

Job Number	532 9/27/96		
No.of Parts scheduled	3000	Set Strokerate	10

Begin of Job	27.09.96 05:25:00	End of Job	27.09.96 12:24:00
Duration of Job	418:36	No. of Parts produced	1310
Production Time	143:11	OK-Parts	1310
Die Change Time	34:43	Rejected Parts	0
Idle Time	57:47		
Fault Time	182:55		

SPMS 4.1	Job Data	25.11.97 09:05:31	SCHULER PRESSEN

Single view of job data NF

▲ **Fig. 4.4.47** Job data

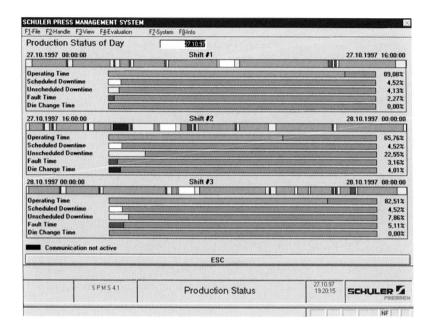

▲ Fig. 4.4.48 Overview of one day's production data

- set drawing pressure values
- actual drawing pressure values
- actual system pressure values
- press angle
- slide position and
- draw cushion position.

The CMS comprises a unit for measured data acquisition and an industrial PC with software capability for acquisition, evaluation and display of the above signals and data. The system permits the graphic display of the progress of pre-acceleration, the drawing process and upstroke both in the form of an overview (*Fig. 4.4.49* and cf. Sect. 3.1.4) and detail displays. The measurement can be depicted in real time for every individual stroke. Zoom functions and the optional representation of measured values relative to the time or the press crank angle permit detailed analysis of measured values. The measurements are stored in the PC and can be accessed at a later date if required. In addition, the CMS provides diagnostic images for troubleshooting and settings during start-up of the press line.

Outlook

Further developments in the field of control engineering in large-panel presses will be determined by technical and economic factors and also by organizational requirements. Initially, the innnovation in automation components is relevant.

Nowadays, low-cost small control systems with communication capability, fast field bus systems and decentral input/output modules are available. This opens up the possibility for further decentralization of control systems, and consequently for a substantial reduction not only of costs and system complexity but also of input for installation, start-up and troubleshooting.

Analogue technology, in particular in controllers for electrical and hydraulic drive systems, is being replaced by digital technology. This permits the parameterization of controllers by means of data transmis-

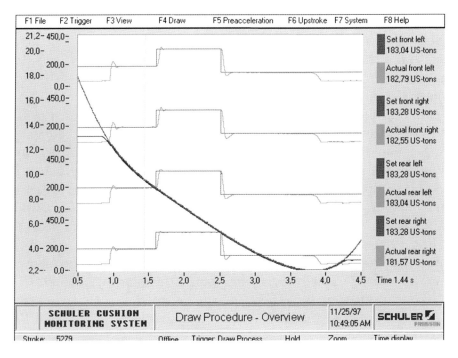

▲ **Fig. 4.4.49** Overview of process signals for the drawing process using a four-point draw cushion: slide displacement (red), draw depth (blue), set pressure value (red right-angled curve), actual pressure value (green right-angle curve), cursor (yellow)

sion following a change of requirements or after exchanging defective components. The benefits here include shorter repair times and thus an increase of press up-time.

Highly-dynamic electrical feed systems in conjunction with efficient control and regulation systems for movement synchronization are increasingly replacing mechanically coupled drive systems. This development not only allows mechanical mechanisms to be simplified and flexibility to be further increased, but in certain cases also enhances productivity.

The changes taking place in the field of work organization in many stamping plants are imposing new demands, in particular on the line control center. Production lines are being managed increasingly by independently acting teams of employees on their own responsibility. These teams require information on site related to production and machine data, to production quality, waiting production jobs and maintenance work to be carried out. Only with this information at their fingertips are team members in a position to optimize the productivity of their line and to assume control functions on the production level. As a result, machine and production data acquisition systems such as those described above will continue to be more extensively used and the functional scope of their application extended to cover ever wider fields.

Multi media-based production line information systems which permit fast, selective access to stored line documentation and additional information on the elimination of faults and routine maintenance will be integrated in the future into the press control center. Using these systems, it will also be possible for operating personnel to enter their own findings into the information data base and so develop their own experience data base.

In general terms, future developments in the field of control engineering will concentrate on reducing complexity and on the development of more efficient systems for repair and maintenance support to reduce the costs of press control and increase production line uptime.

■ 4.5 Blanking processes

In blanking operations, a difference is made between open and closed contours *(cf. Fig. 2.1.29)*. Open contour blanking techniques are used mainly for blanking strips made of sheet plates. This technique is similar to the action of a pair of scissors *(cf. Fig. 2.1.31)*. The blanking elements used for open blanking can be either longitudinal or circular knives. In closed blanking processes, the sheared contour is closed, for example when piercing. Here, the process is applied using blanking dies: The relative movement of the blanking punch to the female blanking die seperates the metal *(Fig. 4.5.1)*. The punch makes contact with the sheet metal, initially causing elastic deformation. The plastic deformation stage then follows, leaving the sheet metal with a permanent camber. The upper edge of the sheet metal then bends and draws in, followed by a shearing action which leaves a visible, smooth area on the cut surface. If the shearing strength is exceeded, cracks are formed. These generally run from the edges of the female blanking die and lead

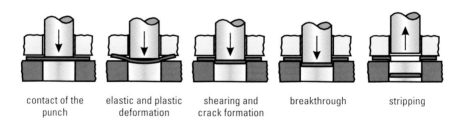

| contact of the punch | elastic and plastic deformation | shearing and crack formation | breakthrough | stripping |

▲ **Fig. 4.5.1** Phases of the blanking process

to complete breakthrough of the metal as the movement of the punch progresses. A force-time diagram of a typical blanking process is provided in *Fig. 4.6.7*. A rough fracture zone then forms on the cut surface underneath the shearing zone *(cf. Fig. 4.7.8, left)*: The sheet metal material springs back after the blanking process, causing it to clamp onto the lateral surface of the punch. The sheet metal has to be separated from the punch by means of a stripper during the return stroke.

Due to the wide variety of applications in stamping plants, only processes used in the production of single parts by closed dies will be described here.

Positioning of blanks in the strip and material savings
Guidelines on the economical layout of blanked, punched and drawn parts:

Close cooperation between component developers and die designers can help to substantially reduce material waste.

Workpieces should be designed with the smallest possible surface area; their shape should be such that they can be lined up or nested in each other in the sheet metal strip so as to ensure minimum waste. The arrangement of parts illustrated in *Fig. 4.5.2 b,* for instance, is more favorable than that in *4.5.2 a*. The best possible material utilization is achieved where the surface shapes are completely interlocking, i.e. can be blanked of the strip without leaving any scrap whatsoever *(Fig. 4.5.3)*. The drawback of scrap-free blanking is that burr occur on both sides of the workpiece and the design of the blanking tools needed in this case is more complicated.

In cases where it is impossible to design workpiece shapes so that optimum in-line arrangement is possible in the strip, it may be possible to

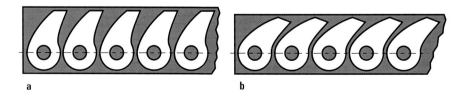

a   b

▲ **Fig. 4.5.2** Arrangement of parts in the strip: **b** offers greater material savings than **a**

▲ **Fig. 4.5.3** Examples of favorable part arrangement in the sheet metal strip

utilize the remaining scrap for other workpieces *(Fig. 4.5.4)*. If scrap skeletons or webs have become deformed during the initial blanking process, they must go through a straightening machine before further processing (cf. Sect. 4.8.3).

T, L and U-shaped workpieces are ideally produced in alternatively reversed manner (top to tail) *(Fig. 4.5.3 e, f* and *Fig. 4.5.5)*. It is important to remember when blanking top to tail that the straight strip bends within the strip plane if subjected to unilateral stress and stretched during the blanking process, or if unilateral tensions are released. This applies in particular to soft, thick sheet material. In these cases, it may be more practical to blank using a multiple punch arrangement for the simultaneous production of several workpieces rather than using the top to tail arrangement *(Fig. 4.5.4)*.

Occasionally, it is possible to improve material utilization by blanking another usable component *b* from the main piece *a*, as demonstrated in *Fig. 4.5.6*. Figure *4.5.6c* illustrates the two parts assembled. The multiple arrangement of workpieces is often economical *(Fig. 4.5.7)*. In the interests of optimum material savings, web or skeleton widths should be kept as narrow as possible (Table 4.5.1). The designations on the blanking strip are indicated in *Fig. 4.5.8*. It is important to remem-

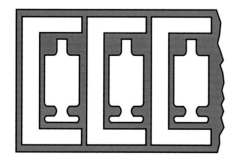

◀ **Fig. 4.5.4**
Utilization of scrap pieces

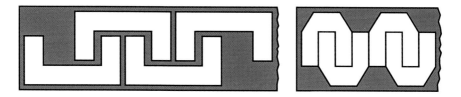

▲ **Fig. 4.5.5** Top-to-tail blanking

ber that the strip width b [mm] depends on the type of material and the strip length l_e [mm]. In any event, it is essential to assess whether the savings made in terms of material are not negated by excessively high die costs.

The material utilization of the sheet metal is:

$$\eta_A = \frac{z \cdot A}{L \cdot b_S} \ [-]$$

with the strip length L [mm] and width b_S [mm], the number of workpieces produced from the strip z [–] and the surface area of one workpiece A [mm²] (without deducting the blanked inside contours). With sizes corresponding to those in *Fig. 4.5.8*, the material utilization for round workpieces arranged in n rows in the sheet metal strip is calculated using the following equation

$$\eta_A = \frac{z \cdot A}{V \cdot b_S} = \frac{n \cdot \frac{\pi}{4} d^2}{(d+b) \cdot b_S} \ [-],$$

whereby

$$b_S = 2 \cdot a + 2 \cdot \frac{d}{2} + (n-1) \cdot a_R = 2 \cdot a + d + (n-1) \cdot 0.866 \cdot V \ [mm]$$

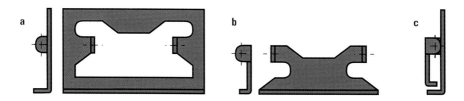

a   b   c

▲ **Fig. 4.5.6** Utilization of waste piece **b** in an assembly **c**

Table 4.5.1: Web and rim widths in accordance with VDI 3367 (7.70)

strip width b_S	web length l_e resp. rim length l_a in mm	web width b rim width a	Sheet metal thickness s in mm						
			0.1	0.5	1	1.5	2	2.5	3
to 100 mm	up to 10 and round parts	b	0.8	0.8	1	1.3	1.6	1.9	2.1
		a	1	0.9					
	11... 50	b	1.6	0.9	1.1	1.4	1.7	2	2.3
		a	1.9	1.0					
	51... 100	b	1.8	1.0	1.3	1.6	1.9	2.2	2.5
		a	2.2	1.2					
	over 100	b	2.0	1.2	1.5	1.8	2.1	2.4	2.7
		a	2.4	1.5					
	side cutter scrap i			1.5		2.2	3	3.5	4.5
over 100 mm to 200 mm	up to 10 and round parts	b	0.9	1.0	1.1	1.4	1.7	2	2.3
		a	1.2	1.1					
	11... 50	b	1.8	1.0	1.3	1.6	1.9	2.2	2.5
		a	2.2	1.2					
	51... 100	b	2.0	1.2	1.5	1.8	2.1	2.4	2.7
		a	2.4	1.5					
	über 100	b	2.2	1.4	1.7	2	2.3	2.6	2.9
		a	2.7	1.7					
	side cutter scrap i			1.5		2.5	3.5	4	5

The feed step

$$V = d + b \ [mm]$$

is calculated from the hole diameter d [mm] and the strip width b [mm]. In percentage terms, the greatest savings are made by transition from a single row to double or triple rows, while in general the greater the number of rows of holes the more favorable is the overall material utilization *(Fig. 4.5.7)*. The high percentage of savings gained when changing to two rows of holes often justifies the procurement of dou-

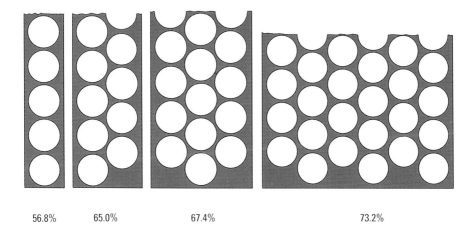

56.8%   65.0%   67.4%   73.2%

▲ **Fig. 4.5.7** Strips using single and multiple-row punch arrangement with utilization
factor η_A in %

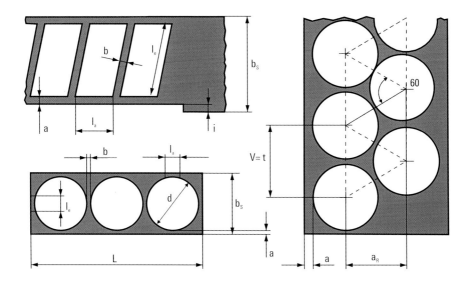

▲ **Fig. 4.5.8** Definitions on the strip: d hole diameter; a rim width; b web width;
i side cutter waste; l_e web length; l_a rim length; b_S strip width;
L strip length; V feed step; t pitch; a_R space between the rows

ble dies or, where double dies do not represent a viable investment due to small batch sizes, the use of a feed system for offset blanking.

Blanking force and blanking work

The necessary *blanking force* F_S [N] for punches with flat ground working surfaces (closed blanking contours) is calculated with the following equation:

$$F_S = A_S \cdot k_S = l_S \cdot s \cdot k_S \ [N],$$

whereby A_S is the sheared surface in mm^2 and k_S the resistance to shear, shearing strength or relative blanking force of the sheet metal expressed in N/mm^2. The sheared surface is determined from the length of the sheared contour l_S [mm] multiplied by the thickness of the sheet metal s [mm], whereby the length of the sheared contour is taken to mean the sum of all sheared edge lengths in mm. If the ratio of the punch diameter d to the sheet metal thickness s is greater than 2, the following equation is sufficient for an approximate calculation of the shearing strength k_S:

$$k_S = 0.8 \cdot R_m \ \left[\frac{N}{mm^2} \right],$$

whereby R_m [N/mm^2] is the tensile strength of the sheet metal.

The blanking force must not exceed the nominal press force within the nominal force-stroke curve, given in press specifications, as otherwise the machine will be overloaded. If the sheet metal thickness is larger than the nominal press force stroke position (i.e. distance before BDC, where nominal press force is given), then the permissible press forces are smaller (cf. Sect. 3.2.1). These can be ascertained from the load versus stroke diagram included in the operating instructions.

The force of the return stroke for stripping the workpiece off the punch is around 3 to 5 % of the blanking force when the ratio of the punch diameter d to the sheet metal thickness s is around 10 (d/s = 10). In the case of smaller d/s ratios, the return stroke forces increase substantially, amounting with d/s = 2 to around 10 to 20 % of the blanking force, while this drops further with greater d/s ratios. A greater return stroke force is required for stripping tough materials than for brittle ones. The return stroke force must be taken into account when designing the punch and dies, and in extreme cases also when dimensioning the press.

Lateral forces can also occur during blanking operations. Particular attention must be paid to these, in particular when the strip layout has a configuration of narrow webs between closely positioned blanks and in cases where the blanking process is not performed simultaneously, for example due to offset punches *(Fig. 4.5.9 g)*. In this case, the web is subjected to stress as a result of the blanking and horizontal forces. The horizontal force exerted lies approximately between 2 and 10 % of the blanking force, whereby the lower value applies to thin, brittle sheet metal and the higher one to thicker, tougher material types. The use of blunt blanking edges increases the blanking and horizontal force.

It is possible to reduce the necessary blanking force if instead of a flat punch with parallel cutting edges, a bevelled punch with an oblique shearing action is used *(Fig. 4.5.9 a, b)*. The height difference h [mm] to be selected in this case should be around 0.6 times (for brittle material) to 0.9 times the sheet metal thickness s. The bevel angle should be no greater than 5°, in order to prevent damage to the cut edge by lateral displacement of the punch. Unilateral displacement of the punch and material can be prevented by using a punch with a hollow or pointed face or with a groove *(Fig. 4.5.9 c, d)*. However, this results in deformation of the slug. If the slug is intended for use as a flat workpiece, the punch face must be flat and the female die must have a hollow or pointed configuration *(Fig. 4.5.9 e, f)*.

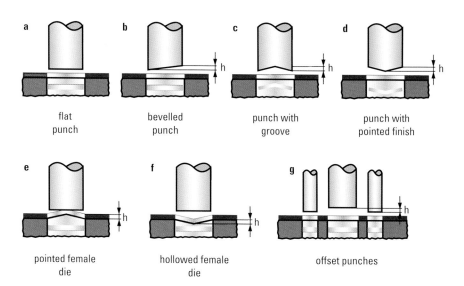

a	b	c	d
flat punch	bevelled punch	punch with groove	punch with pointed finish

e	f	g
pointed female die	hollowed female die	offset punches

▲ **Fig. 4.5.9** Punch and female die shapes (h = difference in height)

When using a bevel ground punch or female die, the blanking force is reduced by at least 30% compared to punches or female dies with a flat surface, as only a certain part of the die is engaged at any one time *(Fig. 4.5.10)*. However, overall the blanking work remains the same, as the reduced force acts over a longer stroke.

If there are several punches mounted in a die, it is also possible to reduce the blanking force by causing the punches to act in sequence *(Fig. 4.5.9 g)*. However, in order to ensure a smooth press operation and reduce the stress on the dies, when using this method, the height difference between each punch force should not be greater than the penetration depth of the die until the fracture of the sheet metal (around 0.3 to $0.4 \cdot$ s). Bevel grinding of punches or female dies and offsetting in the case of several punches also help to reduce noise during the blanking process.

Example:
Assuming a 2 mm thick, hard brass sheet Ms 63 ($k_S = 380$ N/mm^2) is to be perforated using a flat ground punch of 113 mm in diameter. The sheared surface is:

$$A_s = d \cdot \pi \cdot s = 113 \text{ mm} \cdot \pi \cdot 2 \text{ mm} = 710.0 \text{ mm}^2$$

Accordingly, the blanking force will be:

$$F_s = A_s \cdot k_s = 710 \cdot 380 = 270,000 \text{ N} = 270 \text{ kN}$$

The blanking work is determined with the following equation:

$$W_S = x \cdot F_S \cdot s \; [\text{Nm resp. kNm}]$$

with the blanking force F_S and material thickness s. The factor x [–] takes into consideration the actual progression of force when blanking and depends on the material. It lies within the approximate range of 0.4 to 0.7, whereby the lower value applies for brittle materials, a large blanking clearance and thick sheet metals, and the upper value is used primarily for tougher materials, a small blanking clearance and thin sheet metals. For approximate calculation, the following equation applies:

$$W_S = \frac{2}{3} \cdot F_S \cdot s \; [\text{Nm resp. kNm}]$$

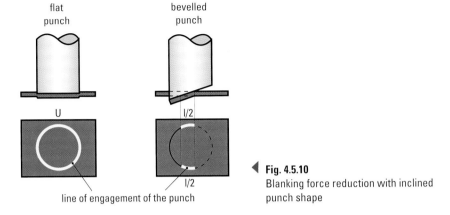

flat
punch

bevelled
punch

U

l/2

l/2

line of engagement of the punch

◀ **Fig. 4.5.10**
Blanking force reduction with inclined
punch shape

When calculating blanking energy and blanking forces, in particular
where thin sheet metals are involved, it can happen that a machine is
designed to have adequate energy for blanking but not adequate blanking force. In these cases, it is possible to reduce the blanking force by
using a bevelled or corrugated punch or a female blanking die.

Example:
Assuming that 100 mm diameter circular disks (cut contour circumference U =
314 mm) are to be produced from 2.5 mm thick sheet steel with a shearing resistance of $k_S = 100$ N/mm², with the blanking force

$$F_S = A_s \cdot k_s = 2.5 \text{ mm} \cdot 314 \text{ mm} \cdot 100 \text{ N/mm}^2 = 78.5 \text{ kN}$$

the blanking energy is calculated as follows:

$$W_s = \frac{2}{3} \cdot F_s \cdot s = \frac{2}{3} \cdot 78{,}500 \text{ N} \cdot 0.0025 \text{ m} = 130.8 \text{ N} \cdot \text{m} \approx 131 \text{ Nm}$$

To ensure the correct location of the blanking die in the press, it is
essential to know the location of the resultant blanking force (frequently also called the center of blanking forces). At this location, the
sum of the individual forces occurring around the periphery of the
blanking edges is considered to act as a single force. This force should
be applied as far as possible in the center of the slide, as this results in
on-center loading and symmetrical elastic deflection of the press. On-
center loading also ensures the best possible die life, and press overload

due to off-center loading is eliminated. When dies are provided with a locating pin for clamping in the press, this should be positioned in the location of the resultant force.

To determine the location of the resulting blanking force, the shape, scope and position of sheared contours are of importance. The position of the center of force is generally determined by *calculating the center of gravity of the blanking edges / cut contours* or where simple geometries are involved (circle, rectangle), by *ascertaining the center of gravity of the punch perimeters / cut contour perimeters*. Where asymmetrical punch configurations are involved, this calculation must be made for the x and the y direction:

For the location of the resultant force in the x direction (distance from the y axis), the following applies:

$$x_S = \frac{U_1 \cdot a_1 + U_2 \cdot a_2 + ... + U_n \cdot a_n}{U_1 + U_2 + ... + U_n} \ [mm],$$

where U [mm] represents the relevant perimeters of the individual punches and a [mm] is the distance of their centers of gravity from the selected y axis. If the line centers of gravity are taken as a basis, then

$$x_S = \frac{l_1 \cdot a_1 + l_2 \cdot a_2 + ... + l_n \cdot a_n}{l_1 + l_2 + ... + l_n} \ [mm],$$

where 1 [mm] is taken to represent the lengths of the blanking edges. The location of the resulting force in the y direction (distance from the x axis) is calculated from the corresponding equation for y, by substituting the distances of the perimeter / blanking edge lengths from the x axis for b [mm]:

$$y_S = \frac{U_1 \cdot b_1 + U_2 \cdot b_2 + ... + U_n \cdot b_n}{U_1 + U_2 + ... + U_n} \ [mm]$$

$$y_S = \frac{l_1 \cdot b_1 + l_2 \cdot b_2 + ... + l_n \cdot b_n}{l_1 + l_2 + ... + l_n} \ [mm]$$

Using these equations, it is possible to determine the position of the center of force S_R (x_S; y_S) in any selected system of coordinates. The equations can only be used for dies with simultaneously acting flat ground punches and flat female dies. It makes sense to lay the coordinate axes through as many centers of gravity as possible, so reducing the distances to zero and simplifying the calculations. Steps should be taken to avoid laying the axes between centers of gravity as otherwise attention must be paid to +/− signs (generally applicable for x: to the right positive values, to the left negative values; for y: upwards positive values, downwards negative values). The formulas required to calculate the centers of gravity of the individual perimeters or line elements can be found where applicable in books of tables. More complex part shapes can often be composed of basic figures (quadrants, rectangles, etc.). Nowadays, computer programs are frequently used to determine the surface center of gravity.

Where the two force action lines in the x and y direction interact, we obtain the center of blanking forces S_R (x_S; y_S), which should be ideally in the center of the slide, or in the case of smaller tools, at the optimum position of the centering pin.

Example:

A workpiece with outside dimensions 80 × 90 mm with a semicircular punched hole (radius 30 mm) and two round holes with a diameter of 20 mm is to be produced in two steps *(Fig. 4.5.11a)*. Two blanking operations take place in the press: piercing and blanking, as indicated in *Fig. 4.5.11b and c*. The center of force S_R is determined initially on the basis of the centers of gravity of the punch perimeters, and then using the line centers of gravity for the purpose of comparison.

$U_1 = U_2 = \pi \cdot 20 \text{ mm} = 62.83 \text{ mm}$

$U_3 = \frac{1}{2} \cdot \pi \cdot 60 \text{ mm} + 60 \text{ mm}$

$= 94.25 \text{ mm} + 60 \text{ mm} = 154.25 \text{ mm}$

$U_4 = 2 \cdot 80 \text{ mm} + 2 \cdot 90 \text{ mm} = 340 \text{ mm}$

$l_5 = l_6 = \pi \cdot 20 \text{ mm} = 62.83 \text{ mm}$

$l_7 = \frac{1}{2} \cdot \pi \cdot 60 \text{ mm} = 94.25 \text{ mm}$

$l_8 = 60 \text{ mm}, l_9 = l_{11} = 90 \text{ mm}$

$l_{10} = l_{12} = 80 \text{ mm}$

→

The distances between the centers of gravity in the respective coordinate system in mm are:

$a_1 = 0$	$b_1 = b_2 = 0$	$a_5 = 0$	$b_5 = b_6 = 20$
$a_2 = 40$	$b_4 = 25$	$a_6 = 40$	$b_7 = 50.9*$
$a_3 = 20$	$b_3 = 38.33**$	$a_7 = a_8 = 20$	$b_8 = 70$
$a_4 = 105$	$a_9 = 65$	$b_9 = b_{11} = 45$	
	$a_{10} = a_{12} = 105$	$b_{10} = 90$	
	$a_{11} = 145$	$b_{12} = 0$	

* Center of gravity of the semi – circular arc: $y_S = \dfrac{2 \cdot r}{\pi} = \dfrac{60}{\pi} = 19.10$

(drawn from the straight line)

Position in the current coordinate system: $b_7 = 70 - 19.10 = 50.9$

** $y_S = \dfrac{60 \cdot 0 + 94.25 \cdot 19.10}{60 + 94.25} = 11.67$ (drawn from the straight line)

Position in the current coordinate system: $b_3 = 50 - 11.67 = 38.33$

a) Calculation of the coordinates of the center of force on the basis of the perimeters *(Fig. 4.5.11b)*:

$$x_S = \frac{U_1 \cdot a_1 + U_2 \cdot a_2 + U_3 \cdot a_3 + U_4 \cdot a_4}{U_1 + U_2 + U_3 + U_4}$$

$$x_S = \frac{62.83 \cdot 0 + 62.83 \cdot 40 + 154.25 \cdot 20 + 340 \cdot 105}{62.83 + 62.83 + 154.25 + 340} = 66.6$$

$$y_S = \frac{U_1 \cdot b_1 + U_2 \cdot b_2 + U_3 \cdot b_3 + U_4 \cdot b_4}{U_1 + U_2 + U_3 + U_4}$$

$$y_S = \frac{62.83 \cdot 0 + 62.83 \cdot 0 + 154.25 \cdot 38.33 + 340 \cdot 25}{619.9} = 23,25$$

$\Rightarrow$ Center of gravity S_R (66.6 ; 23.25) in the coordinate system b.

b) Center of force calculated using the line centers of gravity (Fig. 4.5.11 c):

$$x_S = \frac{l_5 \cdot a_5 + l_6 \cdot a_6 + l_7 \cdot a_7 + l_8 \cdot a_8 + l_9 \cdot a_9 + l_{10} \cdot a_{10} + l_{11} \cdot a_{11} + l_{12} \cdot a_{12}}{l_5 + l_6 + l_7 + l_8 + l_9 + l_{10} + l_{11} + l_{12}}$$

$$x_S = \frac{62.83 \cdot 0 + 62.83 \cdot 40 + 94.25 \cdot 20 + 60 \cdot 20 + 90 \cdot 65 + 80 \cdot 105 + 90 \cdot 145 + 80 \cdot 105}{62.83 + 62.83 + 94.25 + 60 + 90 + 80 + 90 + 80}$$

$x_S = 66.6$ (same value as with a), as the same y–axis)

$$y_S = \frac{l_5 \cdot b_5 + l_6 \cdot b_6 + l_7 \cdot b_7 + l_8 \cdot b_8 + l_9 \cdot b_9 + l_{10} \cdot b_{10} + l_{11} \cdot b_{11} + l_{12} \cdot b_{12}}{l_5 + l_6 + l_7 + l_8 + l_9 + l_{10} + l_{11} + l_{12}}$$

$$y_S = \frac{62.83 \cdot 20 + 62.83 \cdot 20 + 94.25 \cdot 50.9 + 60 \cdot 70 + 90 \cdot 45 + 80 \cdot 90 + 90 \cdot 45 + 80 \cdot 0}{62.83 + 62.83 + 94.25 + 60 + 90 + 80 + 90 + 80}$$

$y_S = 43.25$

$\Rightarrow$ Center of gravity S_R (66.6 ; 43.25) in the coordinate system c.

This point corresponds to the point calculated using the perimeters.

Blanking clearance between punch and female die

The blanking forces ascertained on the basis of the equations provided here assume that the correct blanking clearance has been selected between the punch and the female die. The precise size of the blanking clearance u [mm] depends on the thickness of the sheet metal and its tensile and shearing strength, as well as on the blanking speed, the type of blanking plate perforation (with or without clearance angle α [°]) and the required quality of the cut surface *(Fig. 4.5.12)*. A larger blanking

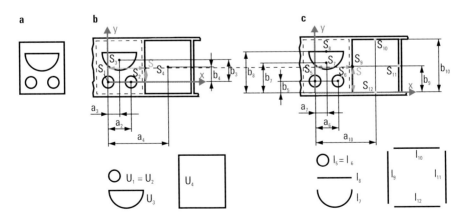

▲ **Fig. 4.5.11** Example to determine the surface center of gravity S_R:
    a part to be produced
    b determination using lengths of the perimeters
    c determination using line centers of gravity

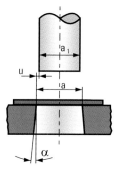

◀ **Fig. 4.5.12**
Dimensions for piercing and blanking on the punch and the
female die

clearance generally reduces the necessary force and work requirement,
and thus also tool wear; with a smaller clearance, in contrast, a qualita-
tively improved cut surface and greater part accuracy are frequently
achieved. In conventional blanking with blanking speeds in the region
of 0.1 to 0.2 m/s, the optimum blanking clearance is between 2 and 10 %
of the sheet metal thickness *(cf. Fig. 4.7.6),* whereby the lower value
applies to thinner or softer sheet metals.

When piercing sheet metal, the dimension of the pierced hole is
determined by the blanking die; the opening in the blanking die must
accordingly be selected to be twice the blanking clearance greater than
the blanking punch. When blanking external contours, by contrast, the
female die determines the size of the blanks. The punch must accordingly be configured smaller by twice the blanking clearance.

Example:

A 3 mm thick medium-hard steel sheet is to be processed in a progressive tool
(cf. Sect. 4.1.1). Initially a hole with a diameter of 10 mm is to be pierced, after
which the circular blank with a diameter of 28 mm (external contour) is to be
blanked out of the metal strip *(Fig. 4.5.13):*
1st stroke – piercing:
The piercing hole punch is given the nominal dimension 10 mm. The female
blanking die has the diameter

$$d = 10 \text{ mm} + 2 \cdot \frac{3 \text{ mm} \cdot 6\%}{100\%} = 10 \text{ mm} + 0.36 \text{ mm} = 10.36 \text{ mm}$$

2nd stroke – blanking:
The female blanking die has the workpiece dimension 28 mm diameter. The
blanking punch has the smaller diameter:

$$d = 28 \text{ mm} - 0.36 \text{ mm} = 27.64 \text{ mm}$$

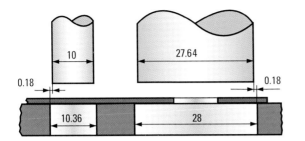

▲ **Fig. 4.5.13** Workpiece and die dimensions used in the manufacture of a pierced disk

■ 4.6 Shearing lines

■ 4.6.1 Slitting lines

The parting of coiled sheet metal stock to produce several narrow strip coils is performed using *waste-free cutting with rotating circular shearing blades* in so-called slitting lines. Materials processed in this way include hot and cold-rolled and stainless steels, NF metals, in particular aluminium, and coated sheet metals.

Alongside the decoiler and recoiler, the major components of a slitting line include the circular shear and the brake frame. The starting coil is pulled tightly off the decoiler, cut to narrow strips with the circular roller shears and coiled again to form firm rings by the brake frame and the recoiler *(Fig. 4.6.1)*. The material processed in this way has generally a thickness of 0.2 and 10 mm. The starting coils can weigh anywhere between 5 and 30 t. The minimum width of the individual coil depends on the width of the blade and the thickness of the sheet metal being processed. Circular shear blades are generally between 10 and 20 mm wide and have a diameter of 250 to 500 mm. The maximum number of cuts depends on the diameter of the shear shaft and the shaft length. Shaft deflection must not exceed a certain value.

Circular shears with a continuous shaft are equipped with circular blades, rolling rings and distance sleeves *(Fig. 4.6.2.)*. Alternatively, hydraulically fastened flanged blades can be used for greater slit coil widths. To reduce standstill periods, the shears are configured as exchangeable units that can be pre-assembled outside the running slitting line prior to production. Where electrically adjustable disk shears are used, the blade holders are positioned by a control system to the

▲ **Fig. 4.6.1** Slitting line

programmed width. Width-adjustable shears permit easy trimming of sheet metal coils of different widths.

The *brake frame* applies the necessary restraining force to the slit sheet metal, thus allowing it to be recoiled under sufficient tension. The specific tensile stress levels in the coiling lines lie between 10 and 20 N/mm². The brake frame used depends on the type of coil stock processed, the treatment or coating of the metal surface, the necessary coiling quality and the thickness and strength of the sheet metal. An underlying difference is drawn between plate-type brakes and rotating brake systems. The *plate-type brake* applies pressure through braking felt-carrying plates positioned above and below onto the surface of the sheet metal. Because of the 100 % slip between the braking felt and the surface of the coil material, this braking method can only be used on sufficiently insensitive surfaces. The braking energy generates heat in the slit coil. Where *rotating brake systems* are used, slip can be reduced to 5 %, avoiding damage of the material surface.

A *coil loop* must be provided between the circular shear and the brake frame. This is used to accommodate the length differences in coiling

frame   distance sleeve   circular blade   sheet metal strip   rolling ring

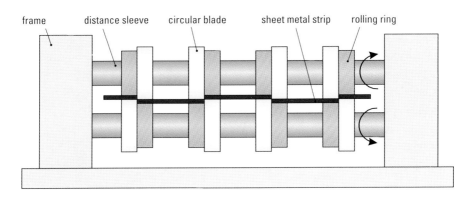

▲ **Fig. 4.6.2** Circular shears

which occur between the inner and outer slit coils with each revolution of the decoiler. The reason for these differences are the thickness variations in the starting coil material over the coil width.

The production speed achieved by slitting lines ranges from 100 to 500 m/min. The resetting of blades, as well as the removal and hooping of the slit coils with packaging tape can lead to system bottlenecks. Improved capacity utilization can be achieved by using automatic exchange of blades and by recoiling outside the line itself.

■ 4.6.2 Blanking lines

Sheet metal parts with medium and large surface areas are not produced directly off the coil, but from stacked blanks fed automatically into the forming press. These blanks are produced in *blanking lines with cut-to-length shears or blanking presses.*

Blanking lines consist of the following components:

– coil line (cf. Sect. 4.3)
– shear or blanking press
– stacking line

The blanking process used *(Fig. 4.6.3)* depends on the production volume and the degree of material utilization. Rectangular blanks can be produced using simple cut-to-length lines. Trapezoidal and parallelogram shaped blanks require a swivel mounted shear or a swivelling die in the blanking press, which move the shear or the die to the other swivel position respectively during each feed movement. The feed unit is driven either hydraulically or using program-controlled servomotors. Closed contours can only be produced on presses with blanking dies *(cf. Fig. 2.1.29)*.

Lines for cut-to-length shearing
In *cut-to-length lines*, the coil stock is cut by means of stationary, flying or rotating shears.

In lines with stationary shears, the roller feed indexes the coil stock forward one step at a time. Cutting takes place when the material is stationary *(Fig. 4.6.4)*. This represents the lowest cost solution and uses a fixed or swivel-mounted shear.

Flying shears are accelerated briefly to match the speed of the coil for shearing. After completing the cut, the shear returns to its starting position. The benefits of this design are continuous feed with a constant straightening machine speed and elimination of the pit to accommodate the coil loop, so ensuring that the material is not bent again after

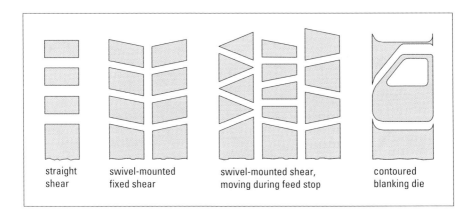

| straight shear | swivel-mounted fixed shear | swivel-mounted shear, moving during feed stop | contoured blanking die |

▲ **Fig. 4.6.3** Blank geometries, arranged according to the type of blanking line

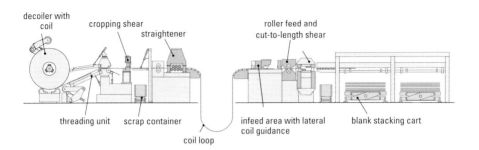

▲ Fig. 4.6.4 Cut-to-length lines: coil line, cut-to-length shear and stacking line

straightening (cf. Sect. 4.8.3). As the sheared blank also moves at the same speed as the coil, there is no need to accelerate the blank as it is the case when using stationary shearing methods. Compared to the stop and go method used with roller feed and stationary shears, the return movement involved in the flying shear method results in a slightly lower output when working with short cut lengths.

Rotating shears achieve the highest number of cutting operations. Their output lies at over 150 parts per minute, making them twice as productive as a blanking press. In the case of rotating shears, the coil material runs at a constant speed between two cutting tools moving in opposite directions. The cutting tools are arranged over the complete coil material width at right angles to the feed direction, so that the sheet metal is separated transversely to the feed direction when it makes con-tact with the cutting tools. When using this process, the surface edge speed of the cutting tools and the speed of the coil stock must match at the moment of shearing. While the cutting tools are not engaged, their rotation speed is accelerated or reduced by the drive system in order to produce the required blank lengths.

As is the case with flying shears, the feed rate and consequently also the output of rotating shears depend heavily on the blank lengths. For both shear types, it is necessary to have the supply loop for coil mater-ial and the pit required to accommodate it, as shown in *Fig. 4.6.4*. Lines with continuous feed are consequently shorter in comparison to coil lines equipped with stationary shears.

Blanking presses for contoured blanks

Contoured blanks have a complex closed contour *(cf. Fig. 2.1.29)* and can contain holes. These blanks are manufactured on blanking presses, also known as contour blanking lines *(Fig. 4.6.5)*. Blanking presses are high-speed eccentric or link drive presses similar to single-action car body presses with a maximum of 80 strokes per minute *(cf. Fig. 4.1.19 right* and *Fig. 4.6.8)*. They feature a high degree of rigidity as well as small bearing and gib clearances. The slide – generally with a maximum stroke of 300 to 450 mm – is mounted using eight guide gibs *(cf. Fig. 3.1.5)*.

Blanking presses require particular attention to *noise reduction*. Blanking noise depends on a number of factors, including the impact speed of the punch on the sheet metal, and accordingly also the slide velocity in the work area. Investigations have proven that noise emissions can be reduced by around 6 dB(A) by halving the slide velocity. In the case of *link drive systems*, the impact speed reached is only 30% that of

▲ **Fig. 4.6.5** Contour blanking line with nominal press force of 8,000 kN (six-link drive)

eccentric presses, representing a reduction in the noise level of some 12 dB(A) *(cf. Fig. 3.2.3)*. In addition, secondary measures are applied to reduce the noise emission from blanking shock in presses, for example by providing a sound enclosure. This can serve to reduce noise by as much as 25 dB(A).

As a rule, modern presses are mounted *on rocker elements* in order to reduce floor vibrations. With a natural line frequency of some 3 Hz, this type of solution achieves an isolating efficiency of around 80 %.

To allow blanking presses to also produce straight and trapezoidal blanks, swivel dies are used. These are generally configured in the form of impact dies. Vertical guidance of the upper die is integrated directly into the die set. The swivel movement is generated by hydraulic cylinders or electric motors, nowadays generally using servomotors and gear drive. The PLC control specifies the swivel angle.

Stacking lines

Both in the case of contour blanking lines and cut-to-length lines, the produced blanks are stacked onto pallets by automatic stacking units *(Fig. 4.6.4)*. *Conveyor belts* arranged crosswise or lengthwise, relative to the longitudinal axis of the press, transport the blanks either lying or hanging out of the blanking line up to the adjustable stops, where they are located and ejected onto pallets. The pallets are put on stacking carts equipped with lifting devices which are lowered continuously in order to permit precise stacking. Using two *stacking carts* in each stacking device, continuous operation is possible.

Die change

In blanking presses, die change is performed fully automatically using two moving bolsters *(cf. Fig. 3.4.3)*. The *moving bolsters* move out of the press towards the operating side and are arranged in the form of T-tracks *(cf. Fig. 3.4.4)*. Positioning takes place automatically in the machine, on the set-up station and in the exchange position. This configuration allows to achieve die change times of under 5 min.

Modern *coil lines* are also automatically reset (cf. Sect. 4.3). The control system automatically executes the sequences "fetching the coil", "depositing the coil", "coil insertion", "finish processing of the coil", "welding seam disposal" and "scrap metal disposal".

■ 4.6.3 High-speed blanking lines

Blanked and stamped parts made from thicker sheet metal such as chain link plates, cup springs, aluminium and coin blanks *(cf. Fig. 6.8.7)* as well as sheet metals used in electric motors *(Fig. 4.6.20)* are manufactured in large quantities, calling for high-performance lines with high stroking rates – so-called high-speed blanking lines *(Fig. 4.6.6)*.

In this type of press, the blanking process causes sudden changes in press forces due to the rapid nature of material breakthrough when the shear strength of the material is exceeded. These forces can result in major dynamic displacement in the dies *(Fig. 4.6.7)*. Because of possible negative effects on part quality and on service life of the dies, high-speed presses are subject to particularly high precision requirements *(Fig. 4.6.8)*.

▲ **Fig. 4.6.6** Blanking line with 1,250 kN high-speed press and die changing cart for automatic die change (long version)

The productivity of the blanking line is determined by the mechanism used for sheet metal feed, the efficiency of the press itself, workpiece removal and the equipment used for resetting. The coil material is fed by a coil line whose coil speed is automatically adjusted in line with the blanking rate of the press (cf. Sect. 4.3).

To produce coin blanks, an additional *coil thickness unit* is required *(cf. Fig. 6.8.23)*. The sheet thickness is continuously measured by a sensor and, where appropriate, a command is transmitted to the control system of the high-speed blanking line, allowing faulty parts to be discarded automatically.

A major criterion when it comes to achieving high output is the efficiency of the feed system by means of roller feed. Mean feed rates of up to 100 m/min are achieved where continuous adjustment of the feed length is possible for a ratio of 1 : 10. The adjustment process is motor powered and can also be carried out during press operation. The feed length is either manually preselected at the keyboard and digitally displayed, or where an automatic tool change system is used, it is accessed by the press control system depending on the die set in use.

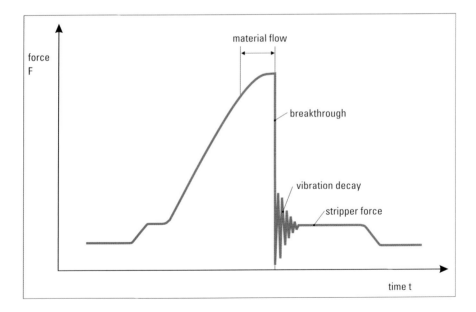

▲ **Fig. 4.6.7** Force-time curve for blanking of sheet metal

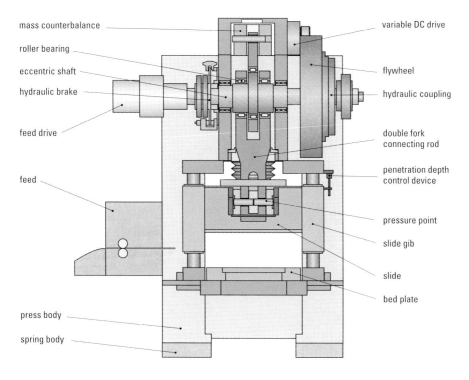

mass counterbalance

roller bearing

eccentric shaft

hydraulic brake

feed drive

feed

press body

spring body

variable DC drive

flywheel

hydraulic coupling

double fork connecting rod

penetration depth control device

pressure point

slide gib

slide

bed plate

▲ **Fig. 4.6.8** Construction of a high-speed press (short version)

Another important aspect determining the output of a high-speed blanking line is its stroking rate. High stroking rates can only be achieved where there is a *counterbalance of masses* created by the horizontal and vertical forces *(Fig. 4.6.8)*. Counterbalance systems in which the forces of the slide and upper die are completely eliminated by directly opposing masses have proved particularly beneficial. The mass compensation of presses with stroke adjustment, which are also suitable for smaller forming operations, is automatically adjusted together with the stroke. Mass counterbalance is an essential prerequisite for subcritical arrangement of the press on rocker elements, which ensures that the vibrations initiated by the stamping process are transmitted only minimally to the foundation.

The systems of *part removal*, transport and interlinking with the downstream production process increase in significance as the press and die output increases. Conveyor belts for transportation of blanked parts are frequently arranged under the press. The parts drop out of the female die onto an ejection channel, which is frequently oriented, onto the conveyor belts. There is a separator integrated in the ejection channel which segregates defective parts from the start of the coil, the end of the coil or parts of the coil which lie outside the specified tolerances.

Link-up between this device and the downstream processing stations for washing, annealing, polishing, inspection, coating or packaging of parts is increasingly implemented using conveyor belts or handling devices (cf. Sect. 4.6.4).

Stamping scrap is often collected in boxes and transported away. More satisfactory than this method is continuous disposal by means of conveyor belts positioned underneath the press. The scrap can either be fed away directly through the female die or outside the die by means of scrap chutes. If a scrap web is created, this is generally chopped by a shear at the outfeed side of the press.

In order to reduce storage of material and capital tie-up to a minimum, small batch sizes also have to be produced economically on high-speed blanking lines. Short die change and resetting times are an essential requirement here. *Resetting* for the manufacture of a different part includes set-up of the feed and removal devices, the preparation of dies, die change, conversion of the press and a final check of die and press setting data (cf. Sect. 3.4).

As the blanking process is the major factor in determining the *precision of punched parts* and also the *service life* of dies, steps must be taken to minimize the vibration occurring between the punch and the female die in the vertical and horizontal direction at the moment of sheet metal breakthrough by executing the necessary measures at the machine *(Fig. 4.6.7)*.

Vertical vibrations are created as a result of the play existing in the force flow of the press and of the elastic properties of the entire press system. They cause an increase in the penetration depth of the blanking punch in the female die, the extent of which depends on the stroking rate. The result is increased punch wear *(Fig. 4.6.9)*.

Compared to a press with friction bearings, vertical play is considerably reduced when using an eccentric shaft running in roller bearings *(Fig. 4.6.8)*. In conjunction with the hydraulic slide adjustment clamp, the total vertical play is reduced to a minimum. A high degree of verti-

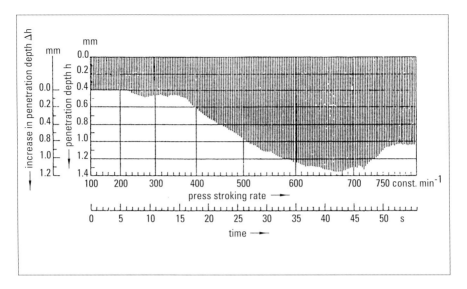

▲ **Fig. 4.6.9** Influence of the press stroking rate on the penetration depth of the blanking punch in the female die

cal rigidity is provided by the solid press body, by the slide drive system using a double fork connecting rod in short presses and two generously dimensioned connecting rods in long presses, and by the bending resistance of the slide. However, as the stroking rate-dependent increase in penetration depth has a physical cause and does not depend on the static and dynamic behavior of the press, a control device is frequently used on high-speed presses. This device detects the increase in penetration depth and automatically corrects it to the set value *(Fig. 4.6.10)*. Even when working at extremely high stroking rates, this system allows to achieve a die service life that is similar to that achieved when blanking in the lower stroking ranges.

In addition to the behavior of the press in the vertical direction, the *vibration characteristics of the slide in the horizontal direction* also influence die life. As the blanking clearance between the punch and female die amounts to only a few hundredths of a millimeter, highly stringent demands are made on accuracy of slide guidance and on the horizontal rigidity of the press body. The degree of horizontal vibration is also directly influenced by play and by the elastic behavior of the press. The horizontal play can be eliminated through the use of slide gibs with

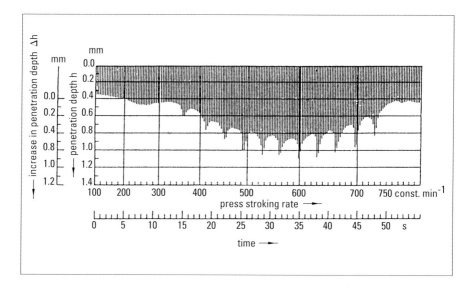

▲ **Fig. 4.6.10** Effect of penetration depth control on the increase in penetration depth

rollers, while the horizontal press rigidity is appreciably increased through special configuration of the press body.

The slide is guided without play by means of rollers at four columns *(cf. Fig. 3.1.6)*. The optimum arrangement of gibs above and on the blanking plane, in particular, serves to reduce slide vibrations compared to previous slide gib systems so that a substantial increase in die life is achieved.

In contrast to conventional presses, the configuration of the press body is now so compact that upright deflection under load is negligible, and horizontal forces generated during blanking can be absorbed evenly over the entire gib area.

■ 4.6.4 Lines for the production of electric motor laminations

Rotors and stators in electric motors and iron cores in transformers are made from individual layered pieces of sheet metal ranging from 0.5 to 1 mm in thickness, in order to reduce eddy current loss *(Fig. 4.6.11)*.

The starting material used for this type of application is silicon-alloyed iron sheet or semi-finish sheet, which is available in coil form

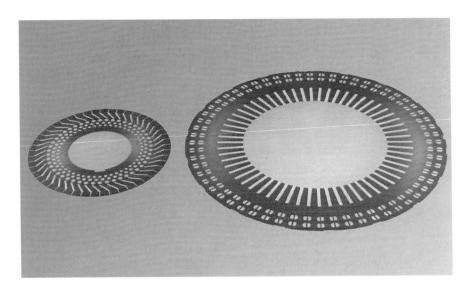

▲ **Fig. 4.6.11** Rotor and stator blank of an electric motor

in widths of up to 1,300 mm. Rotor and stator sheet laminations can accordingly be produced from a single piece only up to a maximum outside diameter of 1,300 mm. All larger diameters are composed of segments which can be used to produce stacks for electrical machines of any optional size.

Depending on the geometrical shape and production lot size, the economical production of laminations for electric motors calls for a variety of die and machine technologies. *Figure 4.6.12* indicates the fields of application for single notch, complete blanking and progressive dies. To manufacture rotor and stator laminations in larger lot sizes, up to a circular blank diameter of 600 mm, compound dies are used in high-speed blanking lines. For diameters up to 1,300 mm, complete blanking dies are used in straight-sided presses *(Fig. 4.6.13)*. Where smaller lot sizes are involved, the laminations are produced from circular blanks on notching machines using the single notch blanking method.

Notching machines
Notching machines are generally open-front presses with mechanical drive systems. The press is mounted in adjustable roller bearings on a

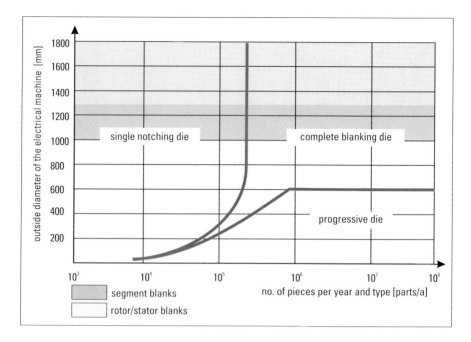

outside diameter of the electrical machine [mm]

single notching die

complete blanking die

progressive die

no. of pieces per year and type [parts/a]

segment blanks

rotor/stator blanks

▲ **Fig. 4.6.12** Die engineering methods for electric laminations

machine bed. An important element of every notching machine is the indexing device which is also located on the machine bed. This generates the required notching pitch, i.e. the number of notches per blank. The punching radius is selected by adjusting the distance between the press and the indexing device.

In *semi-automatic notching machines,* the blanking process is performed automatically, while feed and discharge are manual processes. The indexing unit is either mechanically or numerically controlled *(Fig. 4.6.14).* The mechanical control device operates with a cam tripping gear and gears that are exchangeable for obtaining different notching pitches. When using circular blank diameters between 800 and 1,000 mm, these low-cost devices achieve a higher output than numerically controlled systems. In contrast, however, numerically controlled solutions offer greater flexibility: The notching pitch can be varied as required, making this type of system ideally suited for more complex notch geometries such as magnet wheels, pole laminations, rotor rim punchings or segments *(Fig. 4.6.15 below).* With the aid of controllable dies, it is possible

▲ **Fig. 4.6.13** Complete blanking die set

to generate contours, several rows of notches in a single clamping operation, uncommon notching pitches (relative to 360°) and particularly
wide notches using double punches.

In *automatic notching machines and flexible notching systems,* the
unnotched blanks are removed fully automatically from a stack of
blanks and fed into the notching machine. Following the notching
operation, the blanks are removed, also fully automatically, and
deposited on a stack. The loading and discharge operations are performed either by a slewing ring, a linear transfer system or a robot. The
six arms of the *slewing ring of an automatic notching machine,* which
operate using a circular indexing technique, are equipped with centrally adjustable transport magnets. Although this process permits the
shortest possible indexing times, it only allows the processing of round
or polygonal blanks. In the standard version, the blanks are transported

◀ **Fig. 4.6.14**
Notching machine for circular
blanks with numerical indexing
drive system (nominal press
force 250 kN)

from one stacking mandrel to the next. With special attachments, pallets can also be used.

On automatic notching systems with one notching machine, it is
possible to carry out either

– simultaneous notching and parting operations or
– only notching operations or
– first stator notching and parting and then rotor notching in two
  passes.

However, rotor and stator laminations can be produced in a single pass
on a line with two notching machines *(Fig. 4.6.16)*. In the first case,
additional shaft hole punches are inserted in an empty station, in the
second case, the blanks are positioned in the sixth station and passed
outwards to the shaft hole notching device. Slewing arms are the most
economical and most efficient way of automating notching machines
for the production of standard laminations.

▲ **Fig. 4.6.15** Complete blanking die for the production of segment blanks

Depending on the blank geometry, *flexible notching systems* work with *transfers or robots*. In this case, transport always takes place from one pallet to the next *(Fig. 4.6.17)*. The standard number of five stations can be extended at will. As for semi-automatic notching machines with numerical control, flexible automatic notching lines are particularly suited for the manufacture of complex blanks: The same selection criteria apply concerning controllable dies. At a higher level of automation, automatic notching machines can be individually adjusted to specific production processes. One such device joins *baked enamel-coated laminations* to create a complete *lamination stack* of the type used in electric motors. The stator blanks are individually heated, inserted in a basket-like device. Pressure is applied onto the complete stack with each inserted lamination. A measurement device guarantees a consistent stack height. Once the required stack height has been reached, the basket-like device containing the laminated stack is swivelled out. The pressing process is carried out while the next stack is being formed. This curing period allows the coating on the lamina-

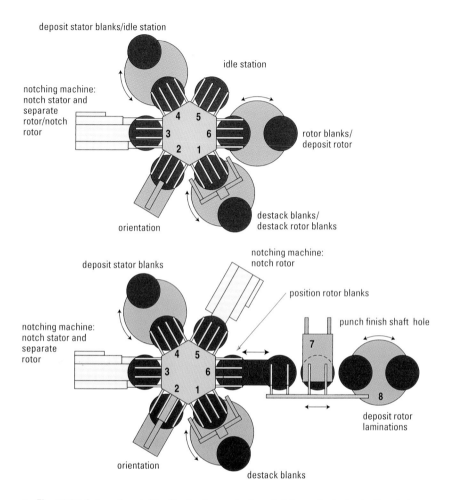

▲ **Fig. 4.6.16** Automatic notching line for the production of stator and rotor laminations in two passes with one notching machine (above) and in a single pass with two notching machines (below)

tions to harden, fusing the stack together. The stack is then ejected onto a roller conveyor.

Using a particularly flexible automation method, two individual notching machines with pallet magazines and pallet storage stations are linked by means of a pallet handling system. A host computer is responsible for joint control. This allows both lines to produce stator and rotor laminations for one size electric motors or for different independent sizes.

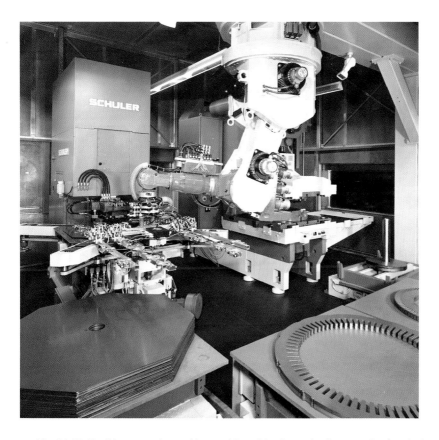

▲ **Fig. 4.6.17** Flexible automatic notching machine with robot and pallet magazine (nominal press force 250 kN)

In general, large-scale special motors are custom-produced using blanks which require up to seven different blanking operations *(Fig. 4.6.11).* The unproductive downtime involved in this type of process can account for over 50 % of the total production time as a result of the resetting and handling work involved. For this reason, these blanks are most economically produced using a *flexible production cell* which requires only a single clamping operation. A central robot is responsible for all handling functions and also for fully automatic die change. If the blanks and dies are placed ready for processing during normal working hours, the production cell is capable of operating unmanned round the clock.

Notching machines can be equipped with *supplementary attachments* for special applications. These include devices for the production of

– interrupted hole patterns,
– tapered lamination stacks and
– skewed lamination stacks.

Blanking lines with complete blanking dies

Medium-sized batches of electric motors, finished segments, raw segments and pole blanks are generally produced using complete blanking dies *(Fig. 4.6.13)*. Here, the external contours, holes and notches are blanked in a single work process, generally using straight-sided presses *(cf. Fig. 3.1.1)*. The press concept used here is similar to that of universal mechanical presses (cf. Sect. 4.4.1). It is also possible to manufacture circular blanks, which are frequently used for further processing in notching machines. The width of the individual blanks, strips or coil material can be anywhere between 300 and 1,250 mm. These presses are constructed for blanking forces ranging from 1,250 to 5,000 kN. The stroking rate can be adjusted continuously between 15 and 150 strokes per minute depending on the size of the machine and the process engineering concept used.

The *dies* are often constructed in such a way that the punch with the outside contours is located in the bottom die holder and the female die is located in the upper die holder. Several parts can be produced simultaneously. In some cases, the scrap is punched by the bottom die.

The separate stripper plate in the bottom die strips the scrap web while the ejector in the upper die ejects the finished part. This requires a plate or crossbar ejector in the press slide which is active near the top dead center.

The gib columns generally extend out of the gib bushes when the die opens. Therefore, they are usually manufactured of steel, while the gib bushes are made of bronze. The coil guiding elements are combined with the stripper plate.

As a single-unit welded construction, the press body is configured for a high degree of rigidity. Considering that high-quality blanking dies are used, the precision of the slide gib is of particular importance. The eight-track roller gib system has proven to be an ideal system here *(cf. Fig. 3.1.6)*. The slide is equipped with a total of 16 roller elements, which run, backlash-free, along hardened gib rails. Should the slide become inadvertently jammed, the press can be quickly released with the aid of hydraulic oil cushions under the connecting rods.

In earlier systems, presses equipped with complete blanking dies required frequently manual loading. Now, however, *automated* part transport is increasingly becoming a customary feature. But this is only possible where machines are equipped with coil lines (cf. Sect. 4.3) or blankloaders (cf. Sect. 4.4.4) and part discharge devices. The press dimensions are determined accordingly not only by the width of the coil stock being processed and the necessary press force and the die dimensions, but also by the greater space required to accommodate add-on units.

The processing of coil stock using a *coil line* requires the use of a decoiler and straightening device as well as an efficient feed system at the press. For small coil stock widths, a dual decoiler is generally used, while wider coils are processed by a single decoiler with a coil lifting platform. A coil loading car is also capable of accommodating two or three large coils.

The electronically controlled roller feed system has proven to be highly popular for *material feed*. It is driven by a threephase servomotor with control circuit, which allows the feed phase to be set depending on the slide stroke and die engineering concept. Today, state of the art presses are generally equipped only with one feed device at the material infeed side.

Individual components, such as unfinished segments, are inserted into the die by a mechanically or electrically driven blankloader. The roller feed for the coil stock and the *blankloader* can be combined with a vertical adjuster in such a way that one of the two is selected to be in operation at any one time.

Using a plate-like feed device, the blankloader transfers the blanks from the stack, which is raised by a lifting platform, to the press. The electrical device is driven by a threephase servomotor and chain transmission. Output is limited to some 20 blanks per minute. Mechanical transfer systems driven directly by the eccentric shaft of the press via cam indexing gears are able to transport up to 40 blanks per minute depending on the part size.

Removal of the scrap web from the tool area is performed by a power-driven magnetic roller. The scrap web created by the blanking process is finally chopped by a cropping shear positioned on the outfeed side of the press.

In the case of *automatic part discharge,* the stroke of the press slide lies between 150 and 300 mm due to the long discharge period. An adjustable slide stroke offers the advantage that progressive dies with a small stroke of for example 40 mm can be used to optimum effect *(cf. Fig. 3.2.13).* Mechanical discharge devices which are either directly driven or controlled by the press are generally used.

The parts are ejected when the press is at its top dead center and they are collected by *discharge plates,* which deposit them outside the press on a magnetic belt, as illustrated in *Fig. 4.6.18* for a finished segment. Depending on the part size, this method limits the stroking rate of the press to between 20 and 60 parts per minute. This requires a slide stroke of 150 to 300 mm. Due to the long slide stroke required, a part discharge method has been developed in which the punched parts are collected and pushed out by two *raised lugs* in the scrap web *(Fig. 4.6.19).* The dies can be operated with a substantially smaller slide stroke of 70 mm, so

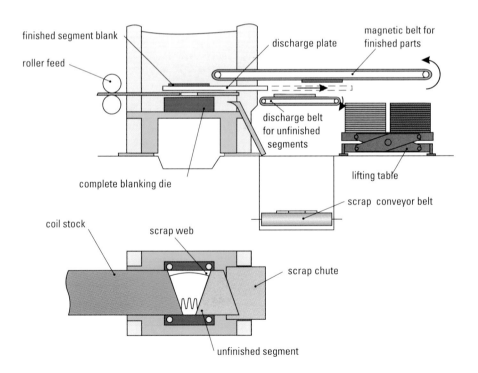

▲ **Fig. 4.6.18** Discharge plate and destacking magnetic belt for an unfinished segment

that up to 150 strokes per minute can be achieved depending on the part size. This is only possible using a special ejector in the slide, which ejects the part around 10 mm after the bottom dead center and deposits it on the scrap web. During the subsequent feed movement, the blank is engaged by the raised lugs and pushed out of the die. Behind the die is a magnetic belt which takes the part and transports it in an overhead position. The scrap web is chopped either directly behind the die or using a separate cropping shear.

In comparison with conventional discharge systems, this method permits the output to be doubled, so ensuring the economical application of complete blanking dies. However, this method can only be used when only one finished part is produced at a time and where the outside shape of the part permits discharge using raised lugs in the scrap web.

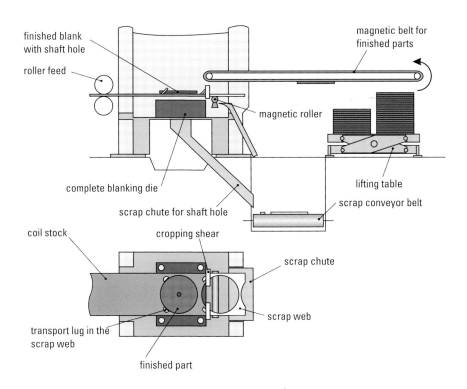

▲ **Fig. 4.6.19** Discharge of finished parts by lugs stamped into the sheet metal strip

High-speed blanking lines with progressive blanking dies

Large series of rotor, stator and magnetic laminations are produced using progressive blanking dies on high-speed blanking lines, in which coil stock feed, part discharge and frequently also die changes are all automatic (cf. Sect. 4.6.3). In progressive blanking dies, piercing and cutout are executed first for the rotor and then for the stator *(Fig. 4.6.20)*. At feed rates of up to 100 m/min over 1,000 parts per minute can be punched. Where small feed distances are involved, output is limited by the press stroking rate, while in the case of larger feed steps the output is limited by the feed system.

Small blanks are frequently discharged using stacking channels. Often, considerable time – up to three hours – is required for changeover of the stacking channels in magnetic lamination production. Therefore, automatic die change systems have been developed which permit lines to be automatically reset in less than 10 min. As a rule, these systems comprise a die change cart capable of accommodating two dies, actuators and positioning devices for the axes to be traversed, and a dialogueoriented, programmable logic controller. The dies are exchanged complete with mounted and filled stacking channels *(Fig. 4.6.21)*. After specifying the relevant die set, the complete die changing and resetting process runs fully automatically.

◀ **Fig. 4.6.20**
Progressive blanking die set for rotor and stator blanks

▲ **Fig. 4.6.21** Automatic die change using a die changeover cart

Larger rotor and stator laminations, in contrast, are still arranged in accordance with their notching pattern in the die, stacked on mandrels or in magazines up to a preselected stack height and conveyed out of the press by means of transport systems. A sequence of stacks of varying heights, for example for ventilated motors, can also be programmed.

Another method for the ordered unloading of electric laminations is by punch-bundling *(Fig. 4.6.22)*. For this purpose, additional punches are located in the die set and they punch for example lugs, cams or neps into the laminations. During the cut-out process, these raised areas press the laminations into the corresponding recess in the previously punched part *(Fig. 4.6.23)*. Once the required stack height has been reached, the following lamination is pierced where the raised area is prior to bundling by means of a piercing hole punch. This ensures that the pierced lamination is not able to link up with the previous lamination and becomes the first lamination of the new stack.

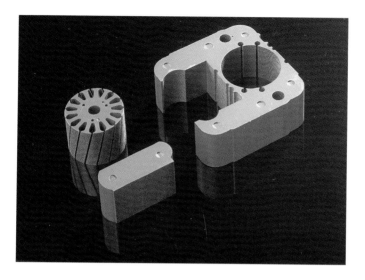

▲ **Fig. 4.6.22** Punch-bundled rotor and stator lamination stacks

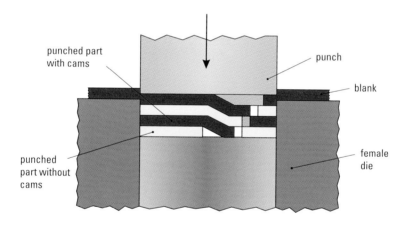

▲ **Fig. 4.6.23** Joining laminations by means of punch-bundling

■ 4.6.5 Production and processing of tailored blanks

Welded blanks comprising different materials, thicknesses or coatings
are known as tailored blanks. The most important benefits offered by
tailored blanks are weight savings and also reduction of the number of
parts required to form assemblies, for example in the case of car body

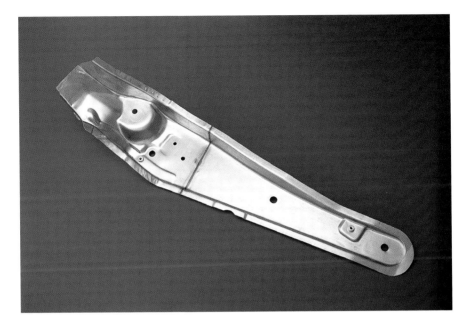

▲ **Fig. 4.6.24** Passenger car reinforcement formed from a tailored blank

reinforcing components *(Fig. 4.6.24)*. In the production of complete assemblies, this reduces lead times and the susceptibility to corrosion due to the smaller number of welding joints.

By using scrap parts from previous blanking operations, for manufacturing tailored blanks it is possible to reduce scrap in the stamping plant. Costs for input materials can also be reduced by using lower-cost sheet metal grades for low-stress applications. In addition, tailored blanks can be configured to special shapes for specific cases of application. Such shapes are generally not offered by steel manufacturers. These benefits have led to the widespread processing of tailored blanks.

Tailored blanks are manufactured on blank welding lines *(Fig. 4.6.25)*. The blanks produced on a blanking press are loaded in stack form, aligned, centered and fed to the automatic welding machine. The blanks are then welded by means of resistance, laser or induction welding methods *(Fig. 4.6.26)*.

Where resistance welding is used, a control system checks the process, assuming high output and quality. The welding unit comprises two loading and centering tables for blank feed, a traversing unit to hold the

▲ **Fig. 4.6.25** Blank welding line with continuous part handling system for the manufacture of welded side members *(cf. Fig. 4.4.12)*

overlapping blanks together, and the stationary welding rolls. Alternatively, the blanks can be stationary and the welding rolls may move.

Laser technology using CO_2 or solid lasers has proven successful as a new production technology. However, when this method is used the two sheet metal edges to be welded together must be prepared for the welding process by laser cutting or by high-precision fine shearing. Welding itself is performed with the blanks stationary and a mobile laser focusing device or with moving blanks and a stationary focusing device. The working speeds are comparable to those of a resistance welding unit. In the case of coated sheet blanks, it is actually possible to achieve higher working speeds. The benefit of laser welding is that no weld overfill occurs and even coated blanks can be processed without problems.

A third system, which is not yet used on an industrial scale, is equipped with an induction welding unit. The blanks are pressed together at the edges. The welding process itself is carried out at high speed, as the blanks are joined over the entire length of their edges within about

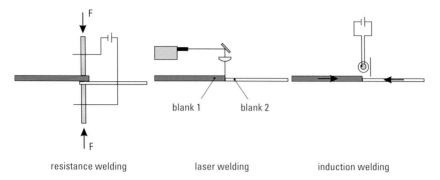

F
blank 1   blank 2
F

resistance welding   laser welding   induction welding

▲ **Fig. 4.6.26** Comparison of the principle layouts of resistance, laser and induction welding systems

2 s. The process cannot be executed on a continuous basis, as part transport must be interrupted during welding.

After the welding process, the parts pass through the monitoring station and the beading and oiling unit. The finished tailored blanks are then stacked again and made ready for further processing in the forming line.

When deep drawing, depending on the position of the welded seam, tailored blanks behave differently to conventional parts. If, for example, the sheet metal thickness differs while the blank holder force remains constant, wrinkles or tears can occur in the formed part. In these cases, special blank holders which can be adjusted according to the formed product, are used in the die set *(Fig. 4.6.27)*.

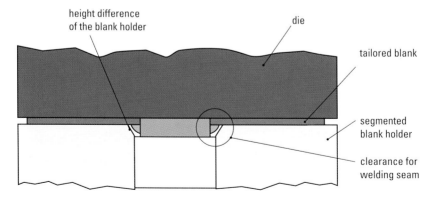

height difference of the blank holder   die   tailored blank

segmented blank holder

clearance for welding seam

▲ **Fig. 4.6.27** Modified blank holder for deep drawing of tailored blanks

■ 4.6.6 Perforating presses

Perforated sheets are in widespread use in almost every sector of industry: in the form of sieves and filters in the food industry, in mining and gravel pits, as protective covers in the electrical industry, in building machinery and machine tools, as partitions and for decorative purposes, in household appliances, office furniture, aircraft and industrial construction.

The number of punch contours and hole patterns is almost infinite. Steel, NF metals or combinations are among the materials used, and sheet thicknesses can range anywhere between 0.3 and 30 mm. Perforated sheet metal is manufactured either off the coil or in plate form. Basically, two main methods are used: Continuous or periodically interrupted hole patterns are manufactured in large series from coil or plate stock on all-across perforating presses, while strip perforating presses are used for the small-series production of optional hole patterns from plate material.

Both machine systems can be equipped with units for automatic sheet metal feed, for removal of finished parts, notching, separating and splitting, and with quick-action die changing devices, die monitoring systems and sound enclosures.

All-across perforating presses
Sheet metal with continuous or periodically repeated hole patterns are manufactured on all-across perforating presses *(Fig. 4.6.28)*. These presses process both coil and plate stock up to a thickness of around 6 mm. The sheet metal runs through the press only in the feed direction and is normally perforated in a single working stroke over the entire width. The nominal press force lies between 800 and 5,000 kN, the maximum material width between 1,000 and 1,600 mm. Depending on the press size and feed system, all-across perforating presses operate at up to 800 strokes per minute.

The press frame is configured as a monobloc *(Fig. 4.6.29)*. The slide runs in clearance-free hardened roller gibs *(cf. Fig. 3.1.6)* via two columns, and is driven by a DC or threephase control motor via a flywheel, eccentric shaft and two connecting rods. This configuration guarantees a long die life. Short switching times, short stroke lengths and high resistance to wear are achieved through the use of a quick-action clutch-brake combination *(cf. Fig. 3.2.8)*.

◀ **Fig. 4.6.28**
All-across perforating
press (nominal press force
1,000 kN)

In contrast to conventional perforating presses with a non-adjustable, rigid sheet metal stripper, in all-across perforating presses the stripper is driven by an eccentric shaft *(Fig. 4.6.30)*. Thus, the stripper and slide movement are out of phase so that the stripper plate remains at the bottom dead center *(Fig. 4.6.31)* while the slide travels upwards. The slide stroke can be increased, so extending the feed phase and increasing the feed output. Like the slide, the stripper is mounted in clearance-free hardened roller gibs at two columns, in order to ensure particularly precise punch guidance.

A major benefit of the moving stripper plate is its additional blank holder function. The stripper force and stripper stroke can be adjusted to the sheet metal thickness with the aid of the adjustable stripper linkage or a variable bed plate. The friction path between the punch and stripper plate is also reduced, as the downward movement of the plate partially coincides with that of the punch.

This system helps to reduce wear and the return stroke force at the punch, increases die service life, and ensures more gentle handling of the sheet metal. When perforating stainless steel sheets, particularly, stabilization of the sheet metal by the stripper plays a major role.

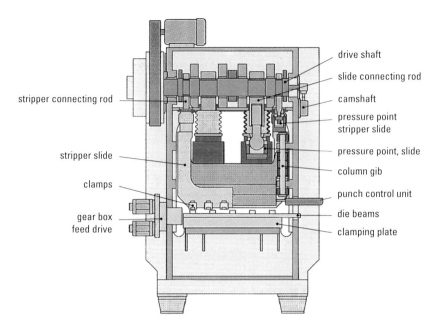

▲ **Fig. 4.6.29** Structure of an all-across perforating press

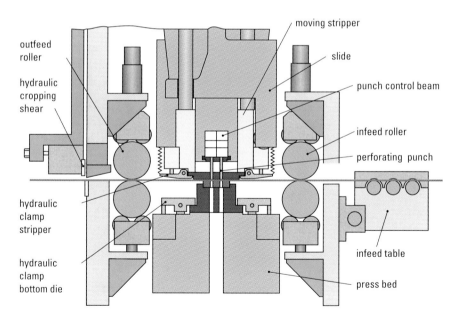

▲ **Fig. 4.6.30** Section through an all-across perforating press with die holding system and separating shear

The stripper plate is raised during the feed movement sufficiently to ensure that even corrugated coil stock can be transported through the open die. After perforation, the corrugated sheet metal can be straightened by the stripper plate.

Before an electrohydraulically powered shear, mounted at the outfeed roller upright, separates the coil into plates, it can be cut using a supplementary slitting or notching device into various widths. The slitting or notching device is mounted either at the main slide – the more economical solution – or fastened at a slide driven separately by the eccentric shaft. In this case, it must be possible to move the separating shear out of the way. If the device is mounted at the main slide, it must also be possible to remove the outfeed system to ensure improved accessibility. When changing dies, the resetting time required can be reduced by exchanging the slitting or notching device from the press outfeed side. The perforating dies can be released by means of hydraulic quick-action clamping devices and removed complete at the side of the press.

Individual or multiple rows of perforating punches in the die can be moved into place, depending on the specifications of the program, using a electrohydraulically actuated sliding beam. Even at maximum

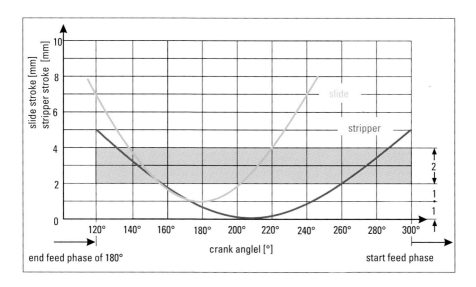

▲ **Fig. 4.6.31** Stroking curves of the slide and stripper plate with a sheet metal thickness of 2 mm

stroking rates, the control system is able to move the punch in and out of position without the need to stop the press. The feed devices arranged directly in front of and behind the tool mounting area can also be programmed. The upper and lower rollers of the feed units are connected by means of clearance-free mechanical intermediate gear drives. Both rollers are supported in several points, and the upper rollers are hydraulically pressed downward above the support in the area of the rollers. The rollers are lifted manually or by means of program control. The infeed device is removable, the outfeed device either removable or stationary.

To ensure a perfectly perforated pattern, a photoelectric camera monitoring system is used *(cf. Fig. 4.9.7)*. While the all-across perforating press is in the set-up mode, the device is programmed using the teach-in mode. If the actual hole pattern deviates from the programmed pattern during production, the press is automatically switched off.

Strip perforating presses

While perforated sheets are produced in medium and large-series on all-across perforating presses, strip perforating presses are used for the production of individual plates, particularly where large or thick sheet metal materials are used or where individual hole patterns are required *(Fig. 4.6.32)*. Strip perforating presses are available with nominal press forces ranging from 500 to 2,500 kN for sheet metal plates between 1,500 × 3,000 mm and 2,000 × 6,000 mm with a maximum thickness of 30 mm. Depending on the sheet metal thickness and feed step, it is possible to achieve between 40 and 400 strokes per minute.

The strip perforating press control system permits simple programming on screen and storage of complex hole patterns. The stored data records can be simply accessed if a particular production run has to be repeated.

The sheet metal plates are fastened on a clamping plate which can be moved horizontally in two axes. Electrical servo drive systems with clearance-free intermediate gear drives are used to power the feed system. These permit perforation to take place during forward and reverse movement of the material. As is the case with all-across perforating presses, here too a separately moved stripper plate with stripping function can be used *(Fig. 4.6.33)*.

Stationary single punches or small punch assemblies perforate the sheet metal plates. Where complicated hole patterns are involved, the

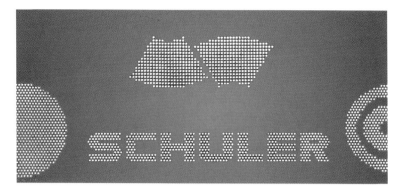

▲ **Fig. 4.6.32** Examples of hole patterns produced on a strip perforating press

punches can be driven using two additional programmable axes. Particularly complex geometrical shapes can also be programmed record by record or using the teach-in mode, whereby the patterns being programmed can be displayed on screen. Temporary deactivation of the slide movement irrespective of the clamping plate feed movement can be used to create optionally interrupted patterns.

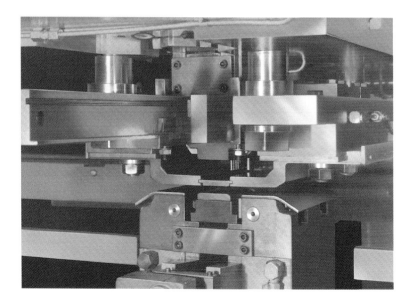

▲ **Fig. 4.6.33** Die with separately controlled stripper plate

The clamping plate is equipped with electrohydraulically actuated clamping dogs which are retracted in the area of a perforation close to the material edge, so eliminating the need to reclamp the sheet metal plate. Support rails equipped with rollers which can be moved as required on the clamping plate are positioned in front of and behind the die to prevent unwanted sagging of the sheet metal plate. This system allows to reduce the mass of inertia to a quarter of that of the old-style solid construction clamping plates. Used in conjunction with modern optimized high-performance actuators, extremely short traversing periods of for example 0.2 s with a feed length of 80 mm can be achieved.

■ 4.6.7 Control systems for blanking presses

The concept of production cells and team work is being introduced in production and assembly plants in order to improve flexibility, quality and also productivity. This involves extending the responsibility of each group or single workplace to include planning, inspection and maintenance activities. As a result, structures will become increasingly decentralized in the future. The greater degree of automation resulting from this development will inevitably lead to machines and production lines of ever greater complexity. At the same time, optimum operating and maintenance capability as well as high equipment availability must be achieved. This means that decentralized units must take increasing responsibility for the provision and updating of production and equipment-related data and information at the point of its generation. A variety of technical aids must be made available to enable the machine operator to avoid or quickly remedy any machine standstill.

Information technology
It is necessary to have an operating and information system, integrated in the machine control, that must fulfil the following criteria *(Fig. 4.6.34):*

– a simple, easily understood system of operator support when setting the operating parameters for start-up and resuming operation following a die change or machine failure,
– a tool data management system that has an overview of existing dies and related features as well as machine parameters and that permits simple access and editing,

- provision of all important data to the operator during production, to allow on-going assessment of machine status,
- support for the operator in troubleshooting and repair work in case of machine failure,
- clearly arranged indication of maintenance intervals and instructions on the execution of necessary work,
- production log to inform the operator on the type and time of previous faults and maintenance work – operators must be able to enter and access their own comments,
- a software system which is sufficiently open-ended to allow other applications such as job management, quality assurance, operational data acquisition or remote diagnostics.

Compliance with all these requirements imposes stringent demands on the control and in particular on the software system. While these requirements exceed the capability of a programmable logic controller (PLC), industrial PCs possess practically all the components necessary for compliance with these specifications. The industrial PC is not only highly efficient, but also benefits from an open-ended software system, offering graphic user interfaces and object-oriented programming through the use of powerful programming tools.

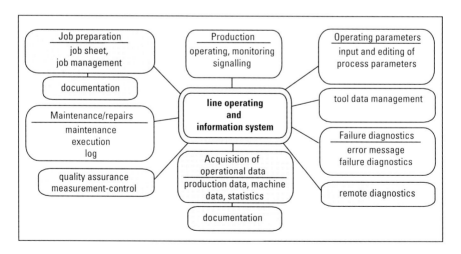

▲ **Fig. 4.6.34** Demands made on modern operating and visualization systems

The widespread popularity of MS Windows as a graphic user interface has provided a platform for many software suppliers to offer efficient software solutions – also in the field of process visualization, which provides users with an ergonomically configured representation of production sequences. Through integration in MS windows, communication is possible with other Windows applications via the "dynamic data exchange (DDE)" interface. Drivers for the relevant PLC systems are required for communication between the process visualization system and the "programmable logic controller (PLC)".

Line operating and information system
The line operating and information system (ABI-Plus) is a modular, open-ended and flexible software system designed to address present and future press automation requirements. ABI-Plus helps reduce machine and production line standstill periods, and so contributes towards improved productivity as well as greater flexibility.

Process visualization
For ABI-Plus, a standard commercially available process visualization system was selected as the core of the overall system. The operator is able to monitor and control the process either interactively or automatically through the control system. The man-machine interface provides an optical representation of the machine status with the aid of dynamic graphics and real images *(Fig. 4.6.35)*.

Intervention in the process is possible using the softkeys. The user is guided through the various operating modes such as set-up, automatic continuous operation or die change on a step-by-step basis. The process parameters are entered using an alphanumeric keypad and feasibility tested by the program.

If a machine error occurs, the relevant error message with date and time is superimposed on the screen mask currently being used. The operator is able to access the troubleshooting module in order to localize the fault and remedy it if necessary. The need of a machine maintenance interval is also indicated to the operator by means of a flashing maintenance symbol.

Tool data management
Another module of the ABI-Plus is a comprehensive tool data management system which allows all die-specific data and machine settings to

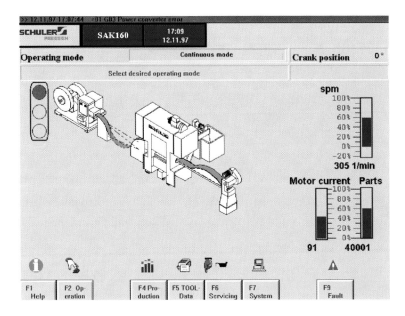

▲ **Fig. 4.6.35 Machine status display**

be entered, edited, stored and transferred to the PLC *(Fig. 4.6.36)*. For all of these activities, the operator is offered support in the form of symbols and graphic images, as well as being able to delete, rename and copy tool data records at will. The number of tool data records which can be stored depends on the storage capacity of the industrial PC.

Failure diagnostics module

If a machine failure is signalled, the operator is able to access the failure diagnostics module by pressing a softkey and have the possible error causes and the relevant repair process with the necessary work stages displayed on screen *(Fig. 4.6.37)*. For each work stage, an instruction text with a real image and, where necessary, also circuit diagrams, hydraulic plans or technical drawings are displayed. The relevant section of the illustration is highlighted by an icon.

For each error message, the operator is able to enter the date, the name of the processing staff member and a commentary for the records.

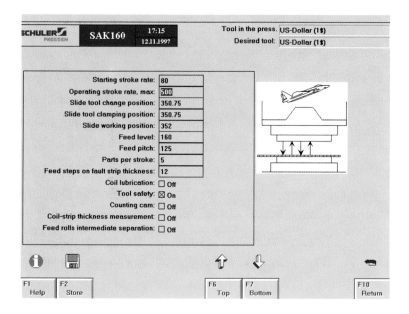

	Tool in the press: US-Dollar (1$)
SAK160 17:15 12.11.1997	Desired tool: US-Dollar (1$)

Starting stroke rate:	80
Operating stroke rate, max:	500
Slide tool change position:	350.75
Slide tool clamping position:	350.75
Slide working position:	352
Feed level:	160
Feed pitch:	125
Parts per stroke:	5
Feed steps on fault strip thickness:	12
Coil lubrication:	☐ Off
Tool safety:	☒ On
Counting cam:	☐ Off
Coil-strip thickness measurement:	☐ Off
Feed rolls intermediate separation:	☐ Off

F1	F2		F6	F7		F10
Help	Store		Top	Bottom		Return

▲ **Fig. 4.6.36** Tool data display

Maintenance module

The maintenance signal is provided in the form of a flashing maintenance symbol to indicate to the operator that maintenance is due. A table with color-coded bars, which can also be accessed during production, indicates all the maintenance intervals *(Fig. 4.6.38)*. The operator accesses the maintenance module using a softkey. The individual work steps of the maintenance activities are indicated by an instruction text with a real image assigned to it *(Fig. 4.6.39)*. In addition, every work stage can be further illustrated by accessing a circuit diagram, hydraulic plan or technical drawing. Here too, it is possible to store production log information such as date, name of the operator and commentary.

Electrical equipment

As a rule, the electrical equipment of the press is broken down as follows:

– operating and information system with swivel-mounted pendant control unit
– supply, distribution and electrical control in the switch cabinet

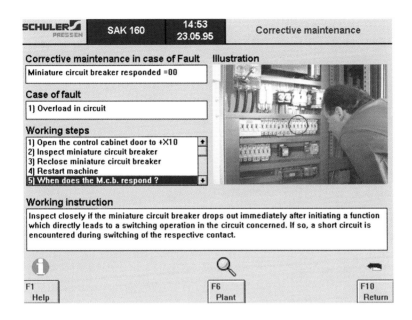

▲ **Fig. 4.6.37** Failure diagnostics module

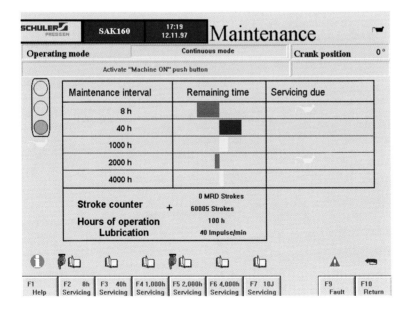

▲ **Fig. 4.6.38** Overview of the maintenance module

- decentral I/O module at the press, at the feed unit, at the hydraulic and lubrication systems and in the pendant control unit
- sensors and actuators at the press, at the feed unit and at the hydraulic and lubrication systems
- main press drive system

Depending on the configuration of the press line, a range of peripheral devices and units such as coil lines with straightening machines (cf. Sect. 4.3) and recoilers or coil lubrication devices are additionally used. These are equipped with their own control systems, sensors and actuators.

Structure of the operating unit

The operating unit comprises an industrial PC for line operation and monitoring, tool and machine data management, as well as information system functions for the elimination of failures as well as backup during maintenance and repairs *(Fig. 4.6.40)*. Basic and safety functions are supported by keys, switches and lamps.

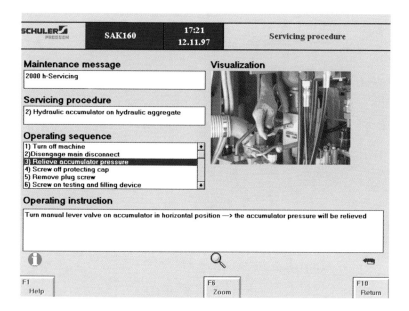

▲ **Fig. 4.6.39** Maintenance instructions

Structure of the electrical control system

Some of the special functions not described in the Section "Press control systems" will be illustrated here using the example of a high-speed blanking line control system (cf. Sect. 3.5). *Figure 4.6.41* provides an overview of the system structure.

The sensors and actuators integrated in the press line are connected via the I/O system of the central PLC unit. The wiring of the power supply to the switch cabinet is added to this.

Depending on the scope of the press line in question, the transition from central to decentral I/O modules allows the wiring input to be substantially reduced: Retrofitting, modifications, installation and troubleshooting are considerably simplified. In addition, there is a marked improvement in electromagnetic compatibility (EMC).

Communication between the PLC and the I/O units at the press line runs via the serial field bus with transmission speeds of up to 12 Megabits/s. The transmission medium is a shielded twisted-pair copper cable. For applications in environments subject to marked electromagnetic

◀ **Fig. 4.6.40**
Control unit

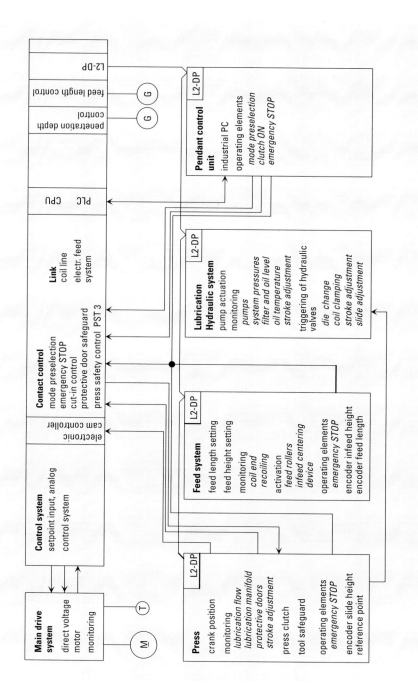

▲ **Fig. 4.6.41** Structural breakdown of the control system in a high-speed blanking line

interference, or where both increased transmission ranges and speeds are involved at the same time, fiber optic cables can be used.

Data communication to peripheral equipment such as coil lines, coil stock thickness gauges or coil lubrication systems is performed via the I/O interface of the PLC. Between the industrial PC and the PLC, data communication is implemented via an active serial interface with 20 mA. For data communication with a higher-level computer system, an additional serial interface is available in the industrial PC.

■ 4.7 Fine blanking

In practice, fine blanking is also known by the terms precision blanking or fine stamping. The production process involves shear blanking for the manufacture of parts with a smooth cut surface. Like standard shear blanking, according to DIN 8580 fine blanking falls under the main category of material separation or parting processes, and is defined according to DIN 8588 under the sub-category dividing *(cf. Fig. 2.1.2)*. Compared to standard shear blanking, also known as normal blanking *(cf. Fig. 4.5.1)*, however, different process parameters apply, resulting in a higher workpiece quality.

The basic concept of fine blanking is that, in contrast to standard blanking, the material is separated after being clamped on all sides only as a result of material flow, i.e. without a fractured surface *(cf. Fig. 2.1.30)*.

■ 4.7.1 Fine blanking process

Advantages of fine blanking
Apart from the quality-related advantages of fine blanking, cost-effectiveness is also a decisive factor in favor of the application of this process. Specific blank-related advantages compared to normal shear blanking include:

– smooth blanked surfaces free from fracture and tearing which are able to fulfil functions without the need for reworking,
– extended field of application for shear blanking,

- smaller dimensional tolerances,
- consistent dimensions throughout a complete series run, due to breakthroughs in the cylindrical blanking plate.

Process-related advantages are:

- no blanking shock and the consequent reduction of noise levels and vibrations,
- fewer production steps in the processing of blanked and combined blanked/formed parts.

As an example, *Fig. 4.7.1* illustrates the production of a chain wheel for a motor cycle using the conventional and fine blanking methods. The nine work operations required during conventional production, including blanking on individual presses and machining through deburring and grinding, can be reduced by using the fine blanking technique in a complete blanking die to three operations. Fine blanking is also more advantageous here if we consider the need for time-consuming and cost-intensive machine changeover using conventional methods.

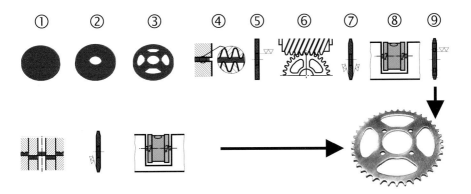

▲ **Fig. 4.7.1** Conventional chain wheel production for a motor cycle (top row):
1. blanking the outside contour; 2. piercing the inside hole; 3. piercing four windows;
4. levelling; 5. turning the inside hole and chamfering on both sides;
6. milling the teeth in the stack; 7. bevel turning of tooth tips on both sides;
8. boring holes and deburring; 9. grinding
Production using fine blanking (bottom row):
1. fine blanking; 2. bevel turning of tooth tips on one side, the second side already
has the die-roll by the fine blanking process; 3. deburring

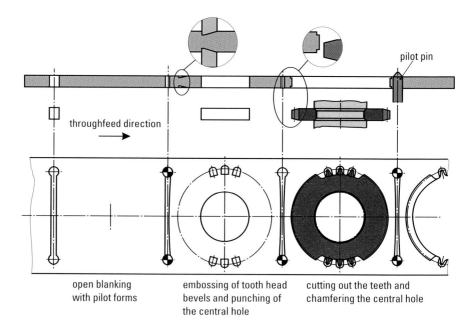

pilot pin

throughfeed direction

| open blanking with pilot forms | embossing of tooth head bevels and punching of the central hole | cutting out the teeth and chamfering the central hole |

Fig. 4.7.2 Strip layout for a three-station compound progressive die when manufacturing a chain wheel

The fine blanking process using compound progressive dies (cf. Sect. 4.1.1) comprises only three stages *(Fig. 4.7.2):*

- free blanking with pilot pins,
- embossing the addendum flanks and piercing the center hole,
- cutting out the toothed area and chamfering the center hole.

This method can lead to production cost savings of around 80% compared to conventional methods.

Die principle
The advantages of fine blanking are based upon differing process parameters. During fine blanking, three forces act via the die on the punched material *(Fig. 4.7.3 left)*. Before the start of blanking, a vee-ring is pressed over the pressure plate outside the blanking line with the vee-ring force F_R. Inside the blanking line, the material is pressed by the counterforce F_G via the ejector onto the blanking punch. In the clamped status, the blanking process is carried out by the blanking force F_S.

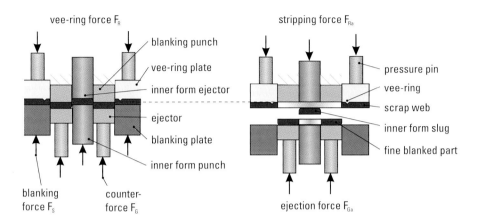

▲ **Fig. 4.7.3** *left:* Tooling principle of a fine blanking die during blanking
right: Tooling principle of a fine blanking tool after ejection of the blank, stripping of
the inner form slug and the scrap web

Once the blanking process is completed, the vee-ring force and counterforce are both deactivated, the die opens and after a certain opening stroke the scrap web and the inner form slugs are stripped off the punch by the stripping force F_{Ra} *(Fig. 4.7.3 right)*. The fine blanked workpiece is ejected from the blanking plate by the ejection force F_{Ga}.

Supplementary forces and functions
In fine blanking presses, the slide always works from the bottom upwards. The slide stroke is divided into a rapid closing, touching, blanking and a rapid return travel *(Fig. 4.7.4)*.

During the blanking process, the vee-ring piston and counterforce piston are displaced until the completion of the blanking process. During the return movement of the slide, the part is ejected from out of the blanking plate by the ejector punch, the inner form slug is stripped by the ejector pin and both are then removed from the die mounting area by means of an air jet or removal arm. The coil feed cycle also takes place during the return movement of the slide. Just before the slide reaches the bottom dead center, the cropping shear separates the scrap web.

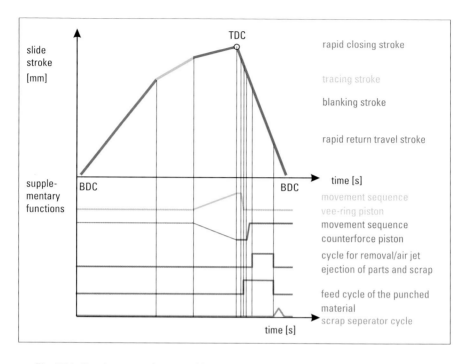

slide
stroke
[mm]

TDC

rapid closing stroke

tracing stroke

blanking stroke

rapid return travel stroke

supple-
mentary
functions

BDC

BDC

time [s]

movement sequence
vee-ring piston

movement sequence
counterforce piston

cycle for removal/air jet
ejection of parts and scrap

feed cycle of the punched
material

scrap seperator cycle

time [s]

▲ **Fig. 4.7.4** Supplementary forces and functions during the slide stroke

Blanking force-stroke diagram
Figure 4.7.5 compares the variation of the blanking force over the slide
stroke or the time required for normal blanking to that required for fine
blanking. During normal blanking, the sheet metal is initially elastical-
ly deformed on impact of the blanking punch (I) and then cut (II). In
phase III, the part tears through, and in phase IV the system vibrations
in the machine and die decay in all three directions *(cf. Fig. 4.5.1)*. This
is the phase in which the greatest die wear takes place. During fine
blanking, the tearing and the vibration phases are eliminated. The special process parameters involved in fine blanking prevent the sudden
breakthrough of the part, ensuring a smooth cut over the entire sheet
metal thickness.

Blanking clearance
The blanking clearance is the dimensional difference or gap between
the blanking punch and the blanking die measured on a single side.

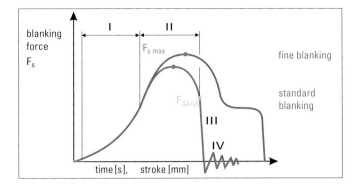

▲ **Fig. 4.7.5** Press force-stroke curves for standard and fine blanking

The smooth section of the cut surface and the blanking clearance are closely related: the greater the blanking clearance, the lower the smooth cut proportion. In addition to its size, the uniformity of the gap along the part contour is also important. The gap should remain constant when subjected to load during the blanking operation.

The size of the blanking clearance is primarily determined by the sheet metal thickness to be cut, but also by the nature and strength of the workpiece material. In contrast to the blanking clearance occurring during standard blanking of around 5 to 10% of the sheet metal thickness, the corresponding value for fine blanking lies at around 0.5% *(Fig. 4.7.6)*. In certain cases, the blanking clearance can also vary over the part contour. Parts with toothing, for example, are manufactured with a blanking clearance which is greater in the area of the tooth root than it is at the tooth tip.

Vee-ring
The vee-ring is a characteristic feature of fine blanking dies. It runs in the form of a raised serrated jag at a defined distance from the blanking line and is pressed into the scrap web. Its function is to hold the punched material outside the blanking line and so prevent lateral flow of the material during the blanking process. In addition, it serves to apply compressive stress to the sheet metal, so improving the flow process.

To simplify maintenance, where sheet metal thicknesses of up to 4.5 mm are being processed, the vee-ring is mounted on the pressure

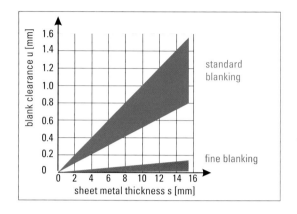

◀ **Fig. 4.7.6**
Guideline values for the blanking clearances for standard and fine blanking in function of sheet metal thickness s

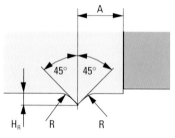

blanking plate

sheet metal thickness [mm]	A [mm]	H_R [mm]	R [mm]
2.8 - 3.2	2.5	0.6	0.6
3.3 - 3.7	2.5	0.7	0.7
3.8 - 4.5	2.8	0.8	0.8

vee-ring plate

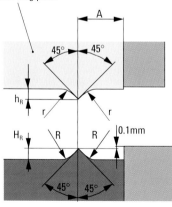

sheet metal thickness [mm]	A [mm]	H_R [mm]	R [mm]	h_R [mm]	r [mm]
4.5 - 5.5	2.5	0.8	0.8	0.5	0.2
5.6 - 7.0	3.0	1.0	1.0	0.7	0.2
7.1 - 9.0	3.5	1.2	1.2	0.8	0.2
9.1 - 11	4.5	1.5	1.5	1.0	0.5
11.1 - 13	5.5	1.8	2.0	1.2	0.5
13.1 - 15	7.0	2.2	3.0	1.6	0.5

▲ **Fig. 4.7.7** Guideline values for vee-ring dimensions

plate of the die, while for thicker materials a second vee-ring is required on the blanking plate. Where the inner forms are particularly large, the ejector also features a vee-ring, which is pressed into the inner form slugs inside the blanking line. Depending on the thickness and shape of the part, the vee-ring is being increasingly placed on the die plate. In individual cases, fine blanking also takes place without the use of a vee-ring.

The dimensions of the vee-ring and the distance from the blanking edge depend on the thickness of the sheet metal *(Fig. 4.7.7)*. In some cases, vee-rings are mounted only partly in the die area where a smooth cut is required for functional reasons.

Work result
Generally, normal blanked parts feature a smooth and a fractured cut surface *(Fig. 4.7.8)*. The proportion of the smooth portion relative to the fractured surface depends on a wide number of different influencing variables, with for example as little as $1/4$ smooth to $3/4$ fractured surface. The smooth area features the die-roll, while on the opposite side is the blanking burr (cf. Sect. 4.5). The flatness and dimensional tolerances of the standard blanked part are inferior. The fine blanked surface, in contrast, is tear and fracture-free over the entire thickness of the sheet metal and

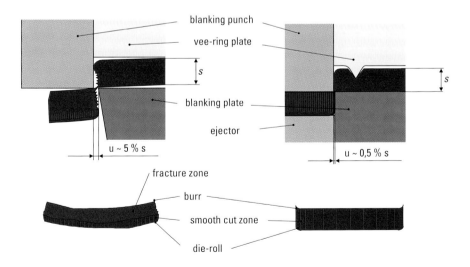

▲ **Fig. 4.7.8** Work result for standard blanking *(left)* and fine blanking *(right)*

approximately vertical relative to the surface of the sheet metal. The fine blanked part is also flatter than a standard blanked part.

■ 4.7.2 Fine blanking materials, forces, quality characteristics and part variety

Material selection

Steel is the most commonly used material in the field of fine blanking technology. Non-ferrous metals such as copper, aluminium and their alloys account for no more than 10%, although the tendency to use these is on the increase. Due to the high proportion of steel materials used, these are described in greater detail here.

The steel types used are broken down into a number of groups: soft unalloyed steels, general structural steels, fine-grained structural steels,

Table 4.7.1: Selection of fine blanking materials

Steel group	Group as per DIN	Abbreviation	Material no.		as per EURO standard
soft, unalloyed steels					
hot strip	1614	StW22	1.0332	FeP11	EN111
metal sheet	1623	St14	1.0338	FeP04	EN10130
cold strip	1624	St4	1.0338		
general structural steels	17 100	St37-3N	1.0116	Fe360D1	EN10025
		St44-3N	1.0144	Fe430D1	EN10025
fine-grain structural steels	17 102	StE420	1.8902	S420N	EN10113-2
		StE460	1.8905	P460N	EN10028-3
case hardening steels	17 210	Ck15	1.1141		
		16MnCr5	1.7131		
heat-treatable steels	17 221	Ck45	1.1191	2C45	EN10083-1
	17 222	41Cr4	1.7035	41Cr4	EN10083-1
	17 200	42CrMo4	1.7223	42CrMo4	EN10083-1
tool steels	17 350	100Cr6	1.2067		
		C85W	1.1830		
stainless steels	17 440				
ferritic		X6Cr13	1.4000		
		X30Cr13	1.4028		
austenitic		X5CrNi18.10	1.4301		

Aluminium and aluminium alloys			Copper and copper alloys		
Abbreviation	Material no.	as per DIN	Abbreviation	Material no.	as per DIN
AlMg3	3.3535	1745	CuZn37	2.0321	17670
AlMgSi1	3.2315	1745	CuSn8	2.1030	17670
			CuNi25	2.0830	17670

case hardening nitrited steel, heat-treatable steel, tool steel, rust-proof ferritic steels, rust-proof austenitic steels and special steels (cf. Sect.4.2.2). These steels can be processed in the form of hot or cold strip or flat bar steel. Flat products must be free of scale (Table 4.7.1).

Material stress and properties

There is an underlying difference in the material stress which occurs during fine blanking and standard blanking. This corresponds to a flow shearing process which is only possible if certain specific material properties are fulfilled. The grains of the metal microstructure are subjected to a marked cold forming process by the blanking punch prior to the point of separation. This grain deformation calls for a formable material with a high formability, i.e. with a high degree of ultimate elongation and fracture necking.

The flow processes taking place in the shearing zone and the work hardening are illustrated in *Fig. 4.7.9.* The cold forming of the grains increases from the die-roll to the burr side of the part, and decreases from the cut surface towards the center of the workpiece. At a certain distance from the surface, the grains which make up the microstructure are no longer affected by the fine blanking process.

The fine blanking capability of steels is determined on the basis of their chemical composition, their degree of purity, their microstructure, their treatment and the prevailing mechanical and technological con-

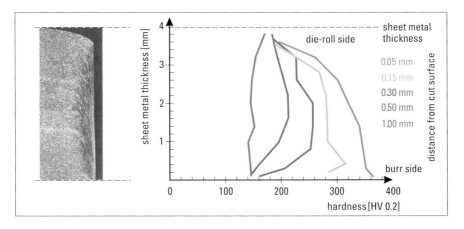

▲ **Fig. 4.7.9** Material flow *(left)* and work hardening *(right)* in the shearing zone

ditions. Starting with soft, unalloyed steels with a low carbon content, fine blanking capability increases on principle with an increasing proportion of carbon and a higher alloy content. There is no precise limit based on the content of carbon and alloy materials from which fine blanking capability can be said not to exist.

The microstructure of a material has a particularly marked influence on the properties of the cut surface, the dimensional stability of the part and the service life of dies. The illustration on the top right of *Fig. 4.7.10* is a schematic representation of the microstructure of a C45 steel which has

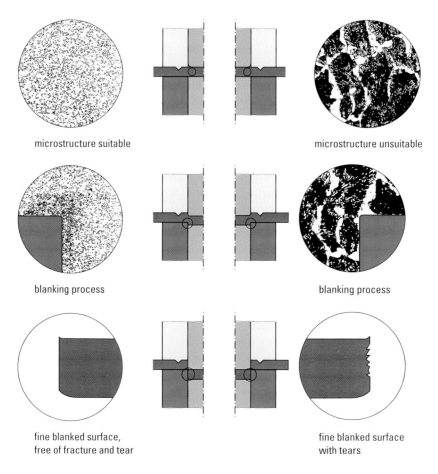

microstructure suitable   microstructure unsuitable

blanking process   blanking process

fine blanked surface,   fine blanked surface
free of fracture and tear   with tears

▲ **Fig. 4.7.10** Schematic representation of the fine blanking process with a suitable *(left)* and unsuitable *(right)* material microstructure

not been soft annealed: The microstructure comprises ferrite and perlite. The hard cementite plates have to be broken through when the blanking punch penetrates the material. The result is tearing of the cut surface.

The illustration at the top left of *Fig. 4.7.10* shows the C45 in a soft annealed state, spheroidized: The microstructure comprises a ferrite matrix with spheroidal cementite embedded in it. Here, the spheroidal cementite grains are not divided during the blanking process, but pressed into the soft ferrite matrix: The blanking process takes place without tear formation.

With an increasing carbon and alloy content, the tensile strength of the material in the non-annealed and soft-annealed state *(Fig. 4.7.11)* increases. C45 which is not soft-annealed has a hot forming microstructure with a strength of around 700 N/mm². Where C45 has an optimum soft annealing microstructure, i. e. extra soft spheroidized, its tensile strength lies at around 480 N/mm², while the corresponding value for the standard spheroidized quality more commonly used for fine blanking is 540 N/mm².

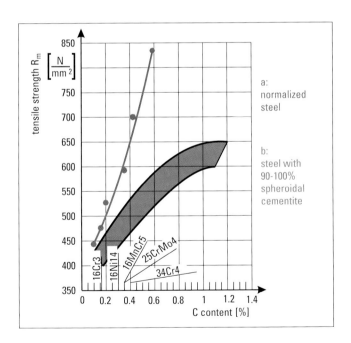

▲ **Fig. 4.7.11** Dependency of tensile strength upon the proportion of carbon/alloys in unannealed **a** and annealed **b** steels

Determining the degree of difficulty

The slide force F_{St} [N] during fine blanking is

$$F_{St} = F_S + F_G \ [N],$$

with the blanking force F_S [N]:

$$F_S = l_S \cdot s \cdot R_m \cdot f_1 \ [N]$$

To calculate the blanking force, the sheet metal thickness s [mm], the tensile strength of the material R_m [N/mm²] and the length of the cut contour l_S [mm] is required (cf. Sect. 4.5). For circular hole punching, for example the following results:

$$l_S = d \cdot \pi \ [mm]$$

Depending on the prevailing conditions, factor f_1 [–] can fluctuate between 0.6 and 1.2. In order to ensure a sufficient blanking force, in practice 0.9 is taken for f_1. This takes into consideration the influences of blanking edge properties (blunting and surface roughness of the blanking elements, sheet metal thickness tolerance and alteration of the blanking clearance as a result of abrasive wear).

The counterforce F_G [N] is calculated from the surface area A_G [mm²] under pressure by the ejector and the counterpressure q_G [N/mm²]:

$$F_G = A_G \cdot q_G \ [N]$$

The value q_G lies between 20 N/mm² for thin parts with a small surface area and 70 N/mm² for larger parts. The counterforce F_G must be selected in such a way that the required cut surface quality and optimum evenness of the part are achieved. As the counterforce must be overcome directly by the blanking force, an excessively high counterforce exercises the same effect as if the sheet were of a higher strength level or thickness. In this way, the high counterforce also influences the service life of the die. The punch stress and the force exerted by the punch both increase. Depending on the part geometry, the counterforce amounts to between 10 and 25 % of the blanking force.

With the help of this information and the surface area A_{St} [mm²] of the hole punch with the diameter d [mm], which is:

$$A_{St} = \frac{d^2 \cdot \pi}{4} \ [mm^2],$$

it is also possible to determine the mean pressure p_m [N/mm²] applied on the punch:

$$p_m = \frac{F_S + F_G}{A_{St}} \left[\frac{N}{mm^2} \right]$$

or when punching a hole, whereby the counterforce is generally taken as 10% of the punch force:

$$p_m = \frac{4.4 \cdot s \cdot R_m \cdot f_1}{d} \left[\frac{N}{mm^2} \right]$$

The mean pressure p_m must not exceed the 0.2% compression limit $R_{p0.2}$ [N/mm²] of the perforating punch ($p_m \le R_{p0.2}$). Accordingly, on the basis of the previous equation for the maximum ratio of sheet metal thickness s to perforating punch diameter d, the following equation results:

$$\frac{s}{d} \le \frac{R_{p0.2}}{4.4 \cdot R_m \cdot f_1} \ [-]$$

If we assume that the perforating punch is made of high-speed steel S6-5-2 with $R_{p0.2} = 3,000$ N/mm² and HRC 63-64 and that the tensile strength of the fine blanking material is 500 N/mm², the following s/d ratio results when blanking with counterpressure:

$$\frac{s}{d} \le \frac{3,000}{4.4 \cdot 500 \cdot 0.9} = \frac{3,000}{1,980} = 1.50$$

and when blanking without counterpressure:

$$\frac{s}{d} \le \frac{3,000}{4 \cdot 500 \cdot 0.9} = \frac{3,000}{1,800} = 1.67$$

This maximum s/d ratio is generally assumed to be 1 in normal shearing practice.

Part configuration: flat and formed
The geometric shape of a part, the thickness of the sheet metal and the characteristics of the material determine the production possibilities available by fine blanking. In order to ascertain whether a part can be manufactured using fine blanking, its degree of difficulty is determined: S1 (easy), S2 (medium), and S3 (difficult). Here, the various formed elements such as slot widths, section widths, hole diameters, tooth forms, corner angles and radius must be evaluated with the aid of *Fig. 4.7.12.*

The highest single degree of difficulty determines the overall difficulty level of the part. Under the limiting line S3, fine blanking does not offer the necessary process reliability using classical tooling technology.

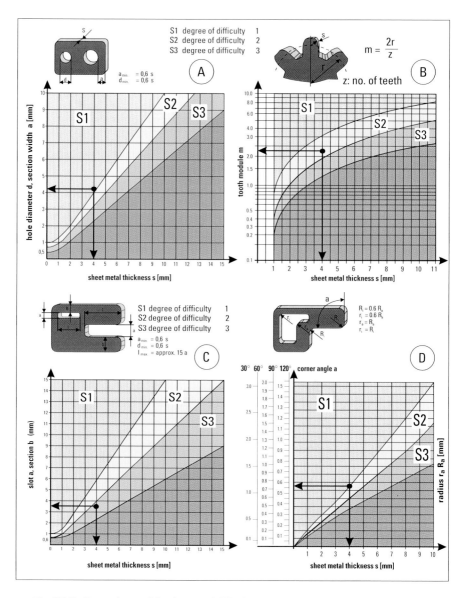

▲ **Fig. 4.7.12** Dependency of the degree of difficulty of a fine blanking part upon the thickness/ geometrical shape: hole diameter/section width (A); tooth module (B); slot/section (C); radii (D)

Example:
We wish to produce an index cam with a thickness of 4 mm in C15 (spheroidized) with a tensile strength of 420 N/mm² *(Fig. 4.7.13)*. The following forming elements are included and assigned to their respective degrees of difficulty in accordance with *Fig. 4.7.12:*

– hole diameter d (mm):	4.1	S1
– section width b (mm):	3.5	S3
– module m:	2.25	S2
– radius R_a (mm) with an angle of 80°:	0.75	S1/S2

The greatest degree of difficulty is presented by the section width (S3). This sets the total degree of difficulty of the part at S3, which means that the part can be produced.

By constructing *compound progressive dies* and *transfer dies with part transfer,* it is possible to combine forming processes such as deep drawing, countersinking, semi-piercing, offsetting, bending and embossing on workpieces with fine blanking *(Fig. 4.7.14* and *4.7.15,* cf. Sect. 4.1.1).

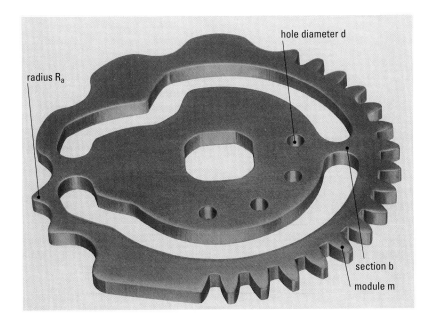

▲ **Fig. 4.7.13** Indexing cam: fine blanked part with differing degrees of difficulty

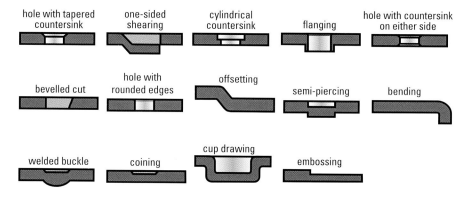

hole with tapered countersink	one-sided shearing	cylindrical countersink	flanging	hole with countersink on either side
hole with rounded edges	offsetting		semi-piercing	bending
bevelled cut				
welded buckle	coining	cup drawing	embossing	

▲ **Fig. 4.7.14**   Examples of different forming techniques which can be combined with fine blanking

Properties of the cut surface

The cut surface of fine blanked parts can be blanked smooth over the entire workpiece thickness (100 % of s). However, at times tearing and fracture may occur. While tearing depends mainly on the microstructure of the material *(Fig. 4.7.10),* fracture behavior is influenced by the magnitude of the blanking clearance *(Fig. 4.7.16).*

For configuration of the dies and on-line in-process quality control, the cut surfaces of a part must be described and defined in accordance with the functional requirements.

◀ **Fig. 4.7.15**
Fine blanked part featuring the forming processes semi-piercing, offsetting, cup drawing, coining, bending, taper sinking and flanging

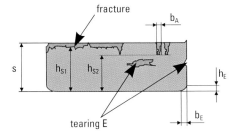

achievable smooth cut ratio					
h_{S1} /s in %	100	100	90	75	50
h_{S2} /s in %		90	75		

▲ **Fig. 4.7.16** Terms defining the fine cut surface:
h_{S1} min. smooth cut section in % of the sheet metal thickness s in case of fracture; h_{S2} min. smooth cut section in % of the sheet metal thickness s in case of shell-shaped fracture; b_A shell-shaped fracture width (sum of all b_A can be defined as required by the user); h_E die-roll height; b_E die-roll width; E admissible tear in accordance with VDI Guidline 3345/2906 part 5 relative to size no. 1, 2, 3 or 4

In the part drawing, the designer is required to define the necessary smooth cut ratios, surface roughness characteristics, admissible tearing, die-roll heights and widths and also blanking burr formation. The specifications are made either on the basis of the cut surface standard according to VDI 3345

$$R_a \sqrt{\dfrac{h_{S1}}{h_{S2}} E}$$

or as per VDI 2906 page 5: $\dfrac{\frac{h_{S1}}{s} / \frac{h_{S2}}{s} E}{\sqrt{R_{ZDIN}}}$ whereby E represents the admissible

degree of tearing according to VDI 3345 in accordance with no. 1, 2, 3 or 4. In order to avoid the need for repeated complex specifications, the blanked surfaces are marked with simple symbols $\left(\sqrt[z]{\ },\sqrt[y]{\ },...\right)$, starting with the last letter of the alphabet (z). The length of the cut surface is indicated by a dotted line *(Fig. 4.7.17)*.

Dimensional and form tolerances
The achievable tolerance levels depend on the material, the workpiece thickness and geometrical shape of the part. The blanking press, the die and the lubricant used are also significant in determining the achievable part quality. The guideline values for achievable tolerances are indicated in Table 4.7.2.

Table 4.7.2: Achieveable tolerances in fine blanking

Sheet metal thickness [mm]	Tensile strength up to 500 N/mm²			Tensile strength over 500 N/mm²		
	Inside contours ISO quality	Outside contours ISO quality	Hole distance tolerances [mm]	Inside contours ISO quality	Outside contours ISO quality	Hole distance tolerances [mm]
0.5 to 1	6 to 7	7	± 0.01	7	8	± 0.01
1 to 2	7	7	± 0.015	7 to 8	8	± 0.015
2 to 3	7	7	± 0.02	8	8	± 0.02
3 to 4	7	8	± 0.02	8	9	± 0.03
4 to 5	7 to 8	8	± 0.03	8	9	± 0.03
5 to 6	8	9	± 0.03	8 to 9	9	± 0.03
> 6	8 to 9	9	± 0.03	9	9	± 0.03

The *die-roll* with its width b_E and height h_E depends on a variety of blank and material-related factors (*Fig. 4.7.8, 4.7.16* and *4.7.18*). The angle and radius of inward and outward pointing corners, the material and the microstructure, strength and sheet metal thickness, for instance, exert a considerable influence. Also important in determining the die-roll amount is the edge preparation of the blanking plate and inner form

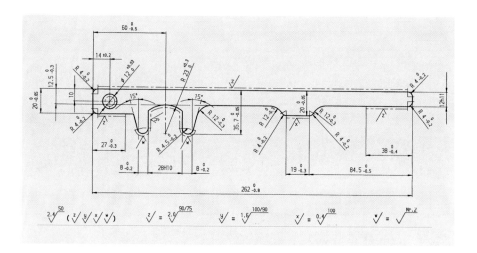

▲ **Fig. 4.7.17** Example dimensions for a fine blanked part in accordance with VDI 3345

punches, as well as the effect of the vee-ring. The die-roll width b_E depends on the die-roll height h_E. In most cases, the following applies:

$$b_E \approx 5 \cdot h_E \; [\mathrm{mm}]$$

Burr-free fine blanking is not possible. The *blanking burr* is located opposite to the die-roll, and occurs as soon as the first part is blanked due to edge preparation at the blanking plate and the blanking punch. As a result of wear of the active elements, the height and width of the burr increase with the increasing number of blanking operations. The blanking burr is generally removed by belt or flat grinding.

The *cut surfaces* are not at absolute right angles to the plane of the sheet metal. The outside contours of a blank on the burr side are greater than at the die-roll side – inner contours are smaller on the burr side than on the die-roll side. As a guideline, the difference amounts to 0.0026 mm per 1 mm of blank thickness, and depends on a number of influencing variables, such as dimensioning of the blanking plate, con-figuration of the blanking plate with or without shrink ring, prepara-tion and coating of active elements.

Application examples
The following pictures provide examples of fine blanked parts. In *Fig. 4.7.19*, an automatic car transmission is illustrated to indicate the

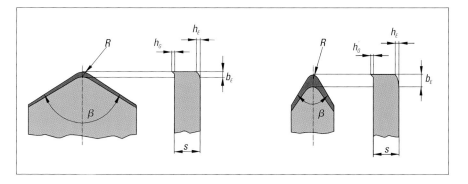

▲ **Fig. 4.7.18** Geometry and die-roll on a fine blanked part:
R corner radius; b_E die-roll width; β corner angle;
s blank thickness; h_G burr height; h_E die-roll height

▲ **Fig. 4.7.19** Fine blanked components in the automatic transmission of a passenger car

variety of fine blanked parts. High demands concerning cut surface quality, dimensional tolerances and flatness are required from the laminations, drive boxes and intermediate plates. Extremely small holes in the intermediate plate are also fine blanked, although the s/d ratio lies at around 2.

Several fine blanked parts are also required for subassemblies used in the brake and drive system *(Fig. 4.7.20),* for instance brake disks with their many holes which have a diameter smaller than half the sheet metal thickness. These small perforations could not be executed using standard blanking methods. In addition, a tear and fracture-free cut surface is essential in order to avoid notching effects, as the disks are subsequently quenched and tempered. Gear teeth and various chain wheels can also be highly economically fine blanked.

■ 4.7.3 Fine blanking tools

Tool types

Internally and externally contoured parts are produced in a single stroke of the slide using *complete blanking dies*. The blanking burr is located on the same side for both the inside and outside contour. Parts manufactured in this way exhibit a high degree of flatness. The dimensional tolerances of the parts depend largely on the manufacturing quality of the die. Tolerances in the coil feed do not influence the dimensional tolerances of the blank.

Due to the complex part geometries involved, in the case of *progressive blanking* dies, blanking of part contours takes place in a number of stages (cf. Sect. 4.1.1). For internal transfer from one station to the next within the die, the part remains in the strip or scrap web. These are referred to as pre-piercing dies whereby the pre-piercing process can take place in one or more stages and is performed prior to blanking of the outside contour. Here, the differences in feed exert an influence on

▲ **Fig. 4.7.20** Fine blanked parts used in the automobile industry

the position of the inside contours relative to the outside contour. The blanking burr on the inside contours is located on the opposite side to that of the outside contour. The compound progressive die also permits forming operations such as bending, offsetting, drawing, etc. in the consecutive stations *(Fig 4.7.21* and *4.7.22)*. Compared to complete blanking, parts produced using this method exhibit generally larger dimensioned tolerances and lower flatness.

In the case of transfer dies, the part is not held in the scrap web. The blanks are transported from one workpiece station to the next by means of a gripper rail, which is frequently coupled to the press *(cf. Fig. 4.4.24* or as illustrated in *4.7.22,* by means of a transverse feeder).

Die systems
In fine blanking technology, the dies belonging to the various die types are subdivided into "moving punch" and "fixed punch" systems.

The system making use of a moving punch is used mainly for complete blanking dies in the production of small to medium-sized parts with few inner forms *(Fig. 4.7.23)*. The blanking plate 2 is mounted in the upper die-set 16 on the die block. For the purpose of positioning, it is positioned and fastened and additionally supported by the base plate 14, piercing punch retaining plate 10, back-up plate 8 and holding ring 9.

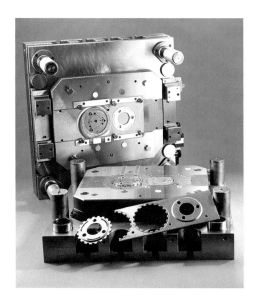

◀ **Fig. 4.7.21**
Two-station compound progressive fine blanking die:
1st *station:* semi-piercing from below and hole punching,
2nd *station:* embossing the cylindrical countersinks at the inside hole, blanking the slots and outside contour.
Tooling system "fixed punch", four-column steel frame with ball gibs, active elements are constructed and coated in segments

▲ **Fig. 4.7.22** Two-station compound progressive fine blanking die with part transport by means of transverse shear pusher:
1st station: processing sequence in vertical direction – cup drawing by blanking and drawing
2nd station: processing sequence in vertical direction – setting cup radii, piercing inner form, trimming outside toothing. Active elements are structured in ring formation and coated

The ejector 3 is guided in the blanking plate. The ejector in turn guides the inner form punch 5. Via pressure pins 7 and pressure pad 11, counterpressure, i.e. ejection force, is applied to the ejector 3.

In lower die-set 17, the guide plate or vee-ring plate 4 is mounted directly on the frame, positioned and fastened. The moving punch 1, which is connected to the press via the punch base 12, runs in guide plate 4. The inner form slugs are ejected by the press hydraulic system via pressure pins 7 of ejector bridge 13, and the inner form ejector pin 6 from punch 1. The upper die-set 16 and lower die-set 17 are precisely

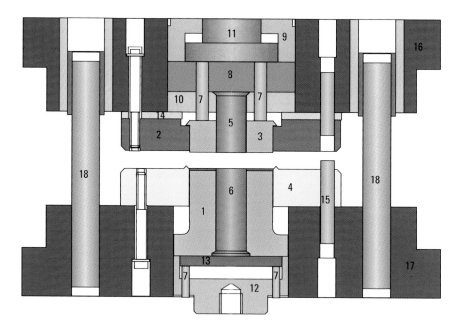

▲ **Fig. 4.7.23** Moving punch system, complete blanking die:
*1 blanking punch; 2 blanking plate; 3 ejector; 4 guide/vee-ring plate;
5 inner form punch; 6 inner form ejector pin; 7 pressure pins; 8 back-up plate;
9 holding ring; 10 piercing punch retaining plate; 11 pressure pad;
12 punch base; 13 ejector bridge; 14 base plate; 15 latch bolt;
16 upper frame; 17 lower frame; 18 gib unit*

positioned by means of the press frame's column gib 18. The blanking
and guide plate are fixed using the latch bolt 15.

The *fixed punch system* is suitable for all die types – in particular for the
manufacture of thick, large parts *(Fig. 4.7.24)*. The blanking punch 1 is
positioned on a hardened pressure plate 9 on the upper die-set 16, and
is permanently positioned and fastened to the upper die-set 16. The precisely fitted guide plate 15 and the vee-ring plate 8, which bear the latch
bolts 14, guarantee the fitting precision of the blanking punch 1 relative
to the blanking plate 2 by means of the column gib unit 18 of the frame.

The inner form slugs are stripped from the blanking punch 1 by the
inner form ejectors 6 and 7. Pressure is applied on the ejectors 6 and 7
and the vee-ring plate 8 by the vee-ring piston of the press via pressure
pins 13. Pressure pins 13 transmit the force of the vee-ring during the
blanking process and the stripping force during the stripping cycle.

Blanking plate 2 is located, positioned and fastened on the lower die-set 17. In special cases, for example where a split blanking plate is used or extremely thick materials are being processed, the blanking plate 2 accommodates a shrink ring 19. The ejector 3, which is guided in the blanking plate 2, is also responsible for guiding the inner form punches 4, and transmits the counterpressure and ejection force from the counterpressure piston of the press via the pressure pins 13. The inner form punches 4 are mounted on the piercing punch retaining plate 10 and supported by the back-up plate 12.

The underlying principle of the fixed-punch die system is also used for progressive blanking dies, compound progressive dies and transfer dies *(Fig. 4.7.25)*. In the case of progressive and compound dies, an additional coil guidance system 1, an initial blanking stop 3, as well as pilot

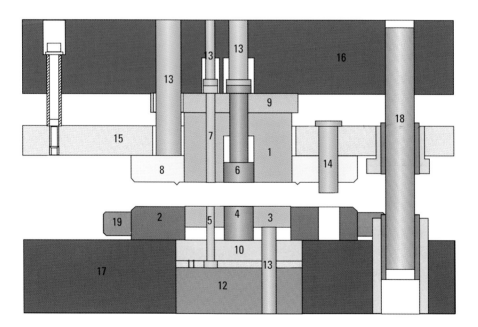

▲ **Fig. 4.7.24** Fixed punch system, complete blanking die:
1 blanking punch; 2 blanking plate; 3 ejector; 4 inner form punch;
5 piercing punch; 6 inner form ejector; 7 ejector pins; 8 vee-ring plate;
9 pressure plate; 10 piercing punch retaining plate; 11 intermediate plate;
12 back-up plate; 13 pressure pins; 14 latch bolt; 15 guide plate; 16 upper frame;
17 lower frame; 18 gib unit; 19 shrink ring

pins 7/8 and positive press-off pins 4 are required. The coil guide not only ensures coil guidance but also stripping of the scrap web from the pilot pins 7/8. The initial blanking stop 3 ensures precise initial blanking of the sheet metal at the start of a coil or strip. The pilot pins 7/8 are responsible for precisely positioning the sheet metal in each die station, i.e. ensuring that the prescribed feed step is adhered to precisely. The task of the positive press-off pins 4 is to compensate for the counter-pressure so that it does not have to be absorbed fully by the individual piercing punches 6 during initial blanking in the first die station.

Die design

The factors which determine the design layout of a fine blanking die are the geometrical shape and size of the part and the type and thickness of the sheet metal being processed. These parameters determine both the type of die used and also the die system in general.

Further definition of the type of die design must be carried out according to the following conditions, which are basically the same as those also valid in standard blanking. In order to achieve optimum blanking conditions, the line of application of the slide force must coincide as far as possible with that of the blanking or forming force (cf. Sect. 4.5). The more precisely this condition is complied with, the better the blanking result and the lower the degree of die wear.

The part drawing must contain all the dimension and tolerance specifications, and the cut surface properties of the part must be indicated (Fig. 4.7.17). Depending on the type of die used, the strip layout must indicate the position of the part in the sheet metal or the processing sequence (*Fig. 4.7.2* and cf. Sect. 4.1.1). The most important aspect of the strip layout on the sheet metal is always the achievement of optimum material utilization (*cf. Fig. 4.5.2 to 4.5.7*). The strip layout is also used to define the strip width and feed step.

Calculation of press forces

The following forces are exerted during fine blanking: the blanking force, vee-ring force, counterforce, stripping force and ejection force (*Fig. 4.7.3*). The first three forces determine the configuration of the press, adding up to create the total machine force:

$$F_{Ges} = F_S + F_R + F_G \ [N]$$

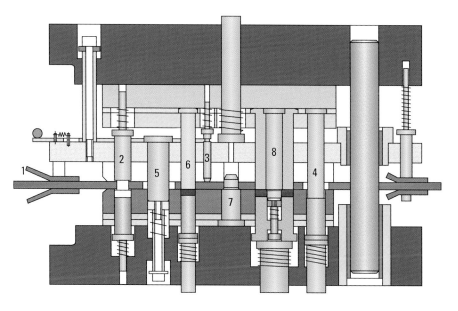

▲ **Fig. 4.7.25** Compound progressive die
  1 coil guidance unit; 2 coil guidance bolts; 3 initial blanking stop;
  4 positive press-off pins; 5 latch bolts; 6 piercing punch;
  7 feed step pilot/embossing punch; 8 feed step pilot/embossing punch

The blanking force F_S [N] and the counterforce F_G [N] were already calculated when establishing the degree of difficulty. The vee-ring force F_R [N] depends on the tensile strength of the material R_m [N/mm²], the length l_R [mm] and the height h_R resp. H_R [mm] of the vee-ring (Fig. 4.7.7) as well as an offset factor f_2 [–]:

$$F_R = f_2 \cdot l_R \cdot h_R \cdot R_m \quad [N]$$

For f_2, experiments have produced the value 4. This value only applies in cases where the geometrical conditions of the vee-ring described in sect. 4.7.1, are adhered to.

The stripping force F_{Ra} [N] is calculated from the blanking force and the offset factor f_3 [–], which lies between 0.1 and 0.2 :

$$F_{Ra} = F_S \cdot f_3 \quad [N]$$

The ejection force F_{Ga} [N] is approximately the same as the stripping force:

$$F_{Ga} = F_S \cdot f_3 \; [N]$$

With a cut contour of the same length, a toothed shape requires a greater ejector force than a round contour. Both forces are additionally influenced by the lubrication, the surface roughness of the blanking elements and the elastic recovery of the sheet metal.

Die lubrication
Fine blanking is impossible to perform without lubrication. The process would lead to cold welding of the active elements to the blank and rapid wear and blunting of the die. Accordingly, the necessary precautions must be taken to ensure that the active elements are supplied with sufficient lubricant at every point of the cut contour and forming areas in the die.

Fine blanking oils can be applied to the sheet material using *rollers or spray* jets. Both the top and the bottom surface of the sheet metal must be evenly wetted by a film of lubricant. For this purpose wetting substances are added to the fine blanking oils.

Where a spray system is used, oil mist is created which must be removed using a suitable extraction device. If high-viscosity oils are used for thick, higher strength materials, it is also possible to apply lubricant through the rollers.

To ensure that the fine blanking oil applied to the sheet metal surface also reaches the friction partners during the blanking process, special design measures are called for at the die. The guide plate, the ejector and the inner form ejection pins must be chamfered or prepared along the cut contours in order to create lubrication pockets *(Fig. 4.7.26)*. Recesses are provided to expose the vee-ring and blanking plate, so ensuring that oil-wetted sheet material reaches the active part of the die with each feed step. In the case of closed dies, the blanking oil on the sheet metal is pressed into the lubrication pockets and serves as a supply for lubrication of the punch and blanking plate.

No universal fine blanking oil exists which satisfies equally all requirements. The fine blanking oil used in each case must be specially adjusted according to the material, the thickness and strength of the part. The viscosity of the oil must be coordinated to the specific types of stress occurring during fine blanking.

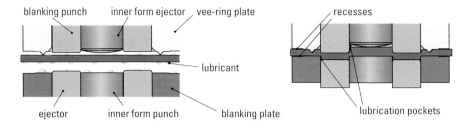

blanking punch   inner form ejector   vee-ring plate   recesses

lubricant

ejector   inner form punch   blanking plate   lubrication pockets

▲ **Fig. 4.7.26** Formation of the lubricant film in the die

Depending on the sheet metal thickness and material strength, the oil is subjected to different pressure and temperature levels. The oil must be capable of withstanding these levels of stress to the extent that the lubricant film remains intact during the blanking process. A low viscosity oil, for example, is suitable for low material strengths of 400 N/mm² up to a thickness of 3 mm, while for materials with a strength of 600 N/mm² the same oil would only be suitable up to a sheet thickness of 1 mm. A high-viscosity oil can be used for a blank thickness of 14 mm and a strength of 400 N/mm², or for a thickness of 10 mm with a strength of 600 N/mm².

In the past, organic chloride compounds were used as extreme pressure (EP) additives in fine blanking oils. However, due to the toxicological effects of the chlorine additives, chlorine-free oils are now increasingly used. The present developments in this field only permit this requirement to be fulfilled up to a sheet metal thickness of around 6 mm and a strength of around 450 N/mm² without detriment to the service life of dies. Where high temperatures and pressure levels occur, the use of chlorine-free oils involves a significant compromise in terms of tool life, although coating the active die elements can help to counteract this effect.

■ 4.7.4 Fine blanking presses and lines

Requirements
Fine blanking is only possible through the effect of three forces: the blanking force, the vee-ring force and the counterforce *(Fig. 4.7.3)*. Accordingly, special triple-action presses are used for fine blanking

operations. The machines require a controlled movement sequence with a precise top dead center *(Fig. 4.7.4)*. The narrow blanking clearance of the dies must not change even under high levels of stress. Fine blanking presses are, therefore, required to comply with stringent precision requirements for example regarding slide gibs, high frame rigidity and the parallelism of die clamping surfaces. Both mechanical and hydraulic systems are used for the main slide drive.

Machine layout and drive system

In standard configuration, mechanical presses *(Fig. 4.7.27)* are equipped with a "combination bed" for the tool systems of moving and fixed punch *(Fig. 4.7.23* and *4.7.24)*. Making use of the straight-side principle, the monobloc press frame as a welded construction offers good dimensional rigidity and freedom from vibrations *(cf. Fig. 3.1.1)*. A clearance-free, pretensioned slide gib is used *(cf. Fig. 3.1.5)*. A central support in the upper and lower die clamping plate ensures optimum die support and introduction of forces to the die. A controlled, infinitely variable DC motor drives the press via the flywheel, a disk clutch and a worm gear pair on two synchronously running crankshafts with different eccentricity *(Fig. 4.7.27)*. The crankshafts drive a double knuckle-joint system which generates the movement sequence of the slide required for fine blanking.

This drive system is particularly suited for material thicknesses between approx. 1 and 8 mm, and total press forces of up to 2,500 kN. The vee-ring force and counterforce are applied by hydraulic systems. The mechanical drive system is characterized by the following factors:

– a fixed slide movement sequence with constant stroke and precise position of the top and bottom dead center,
– low energy consumption,
– high output depending on the size of the press with stroking rates of up to 140/min,
– minimal setting and maintenance input.

The press body of *hydraulic presses* comprises a robust monobloc welded straight-side construction with high rigidity *(Fig. 4.7.28* and *cf. Fig. 3.1.1)*. This is achieved by implementing suitable design measures such as selection of large upright cross-sections and large ribs. The slide and main piston together form a unit which is integrated in the press bed. The slide is guided in the press body and the lower part of the slide piston.

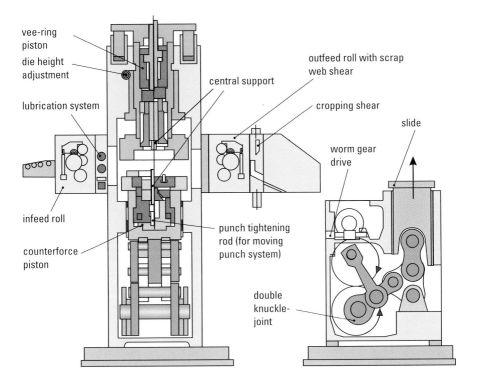

▲ **Fig. 4.7.27** Layout of a mechanical fine blanking press

The gib geometry comprises a high-precision eight-track sliding gib. This construction is able to largely eliminate slide tilting even under major eccentric loads.

Efficient hydraulic fine blanking presses with a total force ranging between 2,500 and 14,000 kN are equipped with a pressure accumulator drive. The counterpressure piston is integrated in the slide, and the vee-ring piston is mounted in the press crown. Both pistons are equipped with support bolts, in order to provide support over the entire surface of the dies. Hydraulic fine blanking presses are characterized by:

– application of force independent of slide position,
– processing of material thicknesses up to 12 mm,
– integration of forming operations into compound progressive and transfer dies,
– a high degree of operational reliability in the work process.

Due to the design principle of the fine blanking dies, not only the blank itself but also all trimmings and inner form slugs are ejected or stripped of the die into the die area. The workpieces and scrap are then removed from the die mounting area by means of air jets or removal arms. If there are parts still remaining in the die area before the next blanking cycle, the slide movement must be halted in time to avoid damage to the die. For this purpose, fine blanking presses are equipped with *tool breakage safety systems*. Systems used include die stroke-dependent sensors, die-independent sensors in mechanical presses and die-independent pressure sensing in hydraulic presses.

Die-independent pressure sensing offers the greatest degree of safety, but can only be used in hydraulic presses *(Fig. 4.7.29)*. The system works

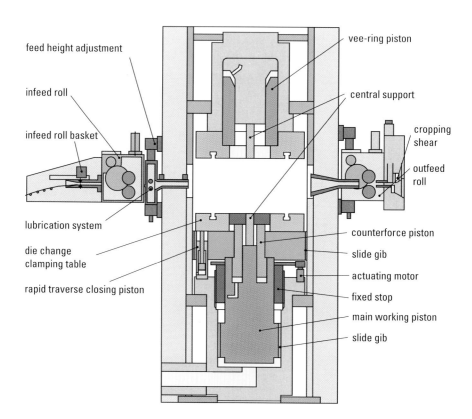

▲ **Fig. 4.7.28** Layout of a hydraulic fine blanking press

measuring the pressure in the rapid traverse closing cylinders. Pressure is applied to the rapid traverse closing cylinders from the low pressure system in such a way that the weight of the slide and the bottom die can just be raised at tracing speed *(Fig. 4.7.4)*. If the operation is uninterrupted, the slide switch activates the high-pressure accumulator, and the main piston initiates the blanking movement.

If there are foreign objects located in the die, the contact between the foreign object and sheet metal or upper die initiates an increase of pressure in the rapid traverse closing cylinder before the slide switch is reached. The pressure increase – a maximum of 2 bar – is detected by a pressure switch, causing the press control system to interrupt the closing movement immediately. This safety system offers the following advantages over other systems:

– structurally cost-effective solution,
– monitoring of the entire die space,
– no moving parts in the die space,
– tracing speed and pressure can be set depending on part sensitivity.

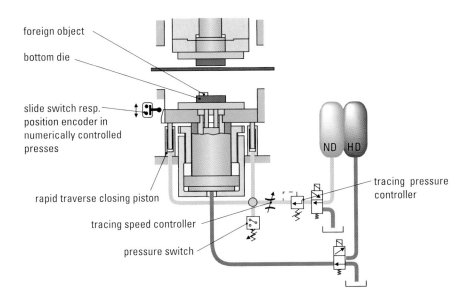

▲ **Fig. 4.7.29** Die-independent pressure tracing

Examples of production lines

Figure 4.7.30 shows a CNC-controlled hydraulic fine blanking press with a total press force of 4,000 kN. Its maximum part-dependent stroking rate is 50/min.

Figure 4.7.31 illustrates an automatic production line for the manufacture of synchronous tapered plates in three stages. The nerve center of the line, which produces some 4,500 parts per shift, is a CNC-controlled hydraulic fine blanking press with a total force of 14,000 kN. The complete line can be operated by a single person.

The blanks produced on a blanking press are stacked on pallets and then deposited on the height-adjustable pallet platform. From here, the

▲ **Fig. 4.7.30**   Fine blanking press with a nominal press force of 4,000 kN

pallets reach the stacking feeder. An accumulating roller conveyor transports the pre-perforated blanks to the indexed press transfer unit which transports them through three die stations. On the outfeed side of the press, the finished synchronous tapered plates are fed via a buffer conveyor towards the stacking unit and the scrap is disposed of separately. The loaded pallets then travel via an accumulating roller conveyor towards the height-adjustable removal point.

Outside the press, the dies are prepared on a die change cart *(cf. Fig. 3.4.2)*. The three individual die modules are mounted on a common die change plate. This system permits automatic die change to be completed in a matter of minutes.

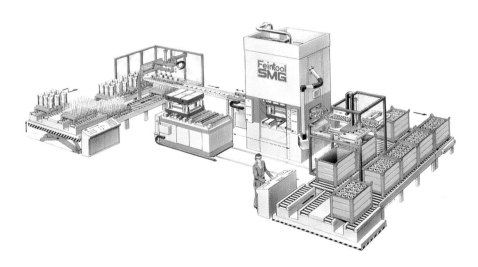

▲ **Fig. 4.7.31** Overall line for the manufacture of synchronous tapered cups with a 14,000 kN CNC controlled fine blanking press

■ 4.8 Bending

■ 4.8.1 Bending process

Classification of bending processes
Bending by forming as a production process is subdivided according to DIN 8586 into two groups: bending by forming using linear and rotating die motions *(cf. Fig. 2.1.3)*. Sections 4.8.2 and 4.8.3 deal with bending using rotary die motion, i.e. roll forming and roll straightening. These fall under the category of roll bending.

Bending radius and bending angle
Bending dies should be designed so as to avoid sharp bent edges. The inside bending radius r_i [mm] depends on the sheet metal thickness s [mm] and should be selected to be as large as possible, because sharp bent edges may lead to material failure. On principle, the bending radius should assume the values recommended by DIN 6935, i.e. they should be selected from the following series (preferably using the values in bold type):

1 1,2 **1,6** 2 **2,5** 3 **4** 5 6 8 **10** 12 **16** 20 **25** 28 **32** 36 **40** 45 **50** **63** **80** 100 etc.

When bending sheet metal, particular attention should also be paid to the rolling direction. The most suitable direction for bending is transverse to the direction of rolling. If the bending axis is positioned parallel to the rolling direction of the sheet metal, $r_{i\,min}$ must be selected higher than when bending at right angles to the direction of rolling.

Table 4.8.1 gives the smallest admissible bending radii for bending angle α up to a maximum of 120°. For bending angles α > 120°, the next highest value applies: When bending Q_{St} 42-2 sheet steel materials with a thickness s = 6 mm at right angles to the rolling direction, for example, the smallest admissible bending radius r = 10 mm for α ≤ 120° and r = 12 mm for α > 120°.

Springback

When designing a bending die, it is necessary to consider springback that occurs after unloading. Springback characteristics differ depending on the material type. Springback occurs with all types of forming by bending, when bending in presses, folding, roll forming and roll bending.

As a result of springback, the bending die angle α does not correspond precisely to the angle desired at the workpiece $α_2$ *(Fig. 4.8.1)*. The angle ratio is the so-called springback factor k_R, which depends on the material characteristics and the ratio between the bending radius and sheet metal thickness (r/s):

$$k_R = \frac{\alpha_2}{\alpha_1} = \frac{r_{i1} + 0.5 \cdot s}{r_{i2} + 0.5 \cdot s} \ [-],$$

With $α_1$: angle at the die (required bending angle) [°],
$α_2$: desired angle at the workpiece (after springback) [°],
s: sheet metal thickness [mm],
r_{i1} : inside radius at the die [mm],
r_{i2} : inside radius at the workpiece [mm].

The springback factor k_R for various materials is given in Table 4.8.2.

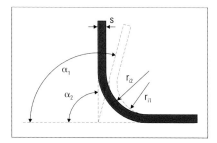

◀ **Fig. 4.8.1**
Elastic recovery after bending:
s sheet metal thickness, $α_1$ required
bending angle, $α_2$ desired angle,
r_{i1} inside radius of die, r_{i2} inside
radius of workpiece

Table 4.8.1: Minimum bending radius $r_{i\,min}$ for bending angles α under 120°

Steel sorts with a minimum tensile strength [N/mm²]	Bending direction compared to roller direction of sheet	Smallest admissible bending radius $r_{i\,min}$ for sheet metal thickness s [mm]							
		1	> 1 to 1,5	> 1,5 to 2,5	> 2,5 to 3	> 3 to 4	> 4 to 5	> 5 to 6	> 6 to 7
to 390	transverse	1	1.6	2.5	3	5	6	8	10
	longitudinal	1	1.6	2.5	3	6	8	10	12
over 390 to 490	transverse	1.2	2	3	4	5	8	10	12
	longitudinal	1.2	2	3	4	6	10	12	16
over 400 to 640	transverse	1.6	2.5	4	5	6	8	10	12
	longitudinal	1.6	2.5	4	5	8	10	12	16

Steel sorts with a minimum tensile strength [N/mm²]	Bending direction compared to roller direction of sheet	Smallest admissible bending radius $r_{i\,min}$ for sheet metal thickness s [mm]						
		> 7 to 8	> 8 to 10	> 10 to 12	> 12 to 14	> 14 to 16	> 16 to 18	>18 to 20
to 390	transverse	12	16	20	25	28	36	40
	longitudinal	16	20	25	28	32	40	45
over 390 to 490	transverse	16	20	25	28	32	40	45
	longitudinal	20	25	32	36	40	45	50
over 400 to 640	transverse	16	20	25	32	36	45	50
	longitudinal	20	25	32	36	40	50	63

Table 4.8.2: Springback factor k_R

Material	Springback factor k_R	
	$r_{i2}/s = 1$	$r_{i2}/s = 10$
St 0-24, St 1-24	0.99	0.97
St 2-24, St 12	0.99	0.97
St 3-24, St 13	0.985	0.97
St 4-24, St 14	0.985	0.96
stainless austenitic steels	0.96	0.92
high temperature ferritic steels	0.99	0.97
high temperature austenitic steels	0.982	0.955
nickel w	0.99	0.96
Al 99 5 F 7	0.99	0.98
Al Mg 1 F 13	0.98	0.90
Al Mg Mn F 18	0.985	0.935
Al Cu Mg 2 F 43	0.91	0.65
Al Zn Mg Cu 1.5 F 49	0.935	0.85

Accordingly, the necessary angle on the die is

$$\alpha_1 = \frac{\alpha_2}{k_R} \ [°]$$

The required inside radius at the die can thus be calculated as

$$r_{i1} = \frac{r_{i2}}{1 + \dfrac{r_{i2} \cdot R_m}{s \cdot E}} \ [mm]$$

with tensile strength R_m [N/mm²] and elasticity module E [N/mm²].

Bending causes residual stresses in the workpiece. The smaller the bending radius relative to the sheet metal thickness, the greater these stresses are. When working with materials which are sensitive to stress-corrosion, workpiece failure is therefore possible within a relatively short period of time after forming. Coining following the bending process helps to reduce residual stresses. When a subsequent heat treatment is used to reduce residual stresses in the workpiece, it is important to remember that heat treatment alters the workpiece radii and the angles.

Unwanted deformation during bending operations
If a sheet metal strip has a rectangular cross section, the sides of the rectangle are determined by the coil width and the material thickness. In the case of thick coil stock and sharp bends, i.e. small inside curvature radius, this rectangle assumes a trapezoidal shape *(Fig. 4.8.2)*.

If several operations are executed on a single workpiece, i.e. several folding processes are carried out simultaneously, steps must be taken to ensure that sufficient material flow is available to replace the material displaced during the forming process. Otherwise, under certain circumstances, it is possible that significant weakening or fracturing takes place at the workpiece corners. In addition, the force required to achieve the final shape increases.

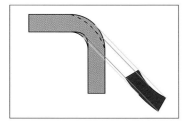

Fig. 4.8.2
Deformation of the cross section during bending

Determining the blank length for bent workpieces

The blank length of the part to be bent is not equal to the fiber length located at the center of the cross section after bending. The extended length of bent components 1 [mm] is calculated as

$$l = a + b + v \quad [mm] \text{ for opening angles } 0° \text{ to } 165° \text{ and}$$

$$l = a + b \quad [mm] \text{ for opening angles } > 165° \text{ to } 180°$$

whereby a [mm] and b [mm] stand for the lengths of the two bent legs, and v [mm] is a compensation factor which can be either positive or negative *(Fig. 4.8.3)*.

It should be noted that α is the bending and β the opening angle of the bent part. Values for v are contained in DIN 6935 or can be calculated for every required angle:

– for β = 0° to 90°

$$v = \pi \cdot \left(\frac{180° - \beta}{180°} \right) \cdot \left(r + \frac{s}{2} \cdot k_R \right) - 2 \cdot (r + s) \quad [mm]$$

– for β > 90° to 165°

$$v = \pi \cdot \left(\frac{180° - \beta}{180°} \right) \cdot \left(r + \frac{s}{2} \cdot k_R \right) - 2 \cdot (r + s) \cdot \tan \frac{180° - \beta}{2} \quad [mm]$$

with correction factor k [–]:

– k = 1 for r/s > 5 and

– $k = 0.65 + \frac{1}{2} \cdot \log \frac{r}{s}$ for r/s to 5

The calculated extended length dimensions should be rounded up to the next higher full millimeter.

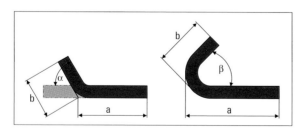

◀ **Fig. 4.8.3**
Geometry of bent legs

Example:

You wish to bend a piece of 3 mm thick sheet metal made of St 37-2, material number 1.0037, in accordance with *Fig. 4.8.4*. The minimum tensile strength for this material is $R_m = 360$ N/mm². Using Table 4.8.1 it is accordingly possible to determine the smallest permissible bending radius at $r_{i\ min} = 3$ mm. To calculate the blank length of the sheet metal part, we require the sum of the leg lengths and all the relevant compensation values v, using the appropriate correction factor k for each case.

Sum of the leg lengths:

$$a + b + c + d = 30 + 60 + 75 + 20 = 185 \text{ mm}$$

for $\beta = 90°$, $r = 5$ mm, $s = 3$ mm:

$$k_1 = 0.65 + \frac{1}{2} \cdot \log \frac{5}{3} = 0.761$$

$$v_1 = \pi \cdot \frac{90}{180} \cdot \left(5 + \frac{3}{2} \cdot 0.761\right) - 2 \cdot (5 + 3) = -6.35 \text{ mm}$$

for $\beta = 45°$, $r = 3$ mm, $s = 3$ mm:

$$k_2 = 0.65 + \frac{1}{2} \cdot \log \frac{3}{3} = 0.65 + \frac{1}{2} \log 1 = 0.65 + 0 = 0.65$$

$$v_2 = \pi \cdot \frac{135}{180} \cdot \left(3 + \frac{3}{2} \cdot 0.65\right) - 2 \cdot (3 + 3) = -2.63 \text{ mm}$$

for $\beta = 135°$, $r = 10$ mm, $s = 3$ mm:

$$k_3 = 0.65 + \frac{1}{2} \cdot \log \frac{10}{3} = 0.911$$

$$v_3 = \pi \cdot \frac{45}{180} \cdot \left(10 + \frac{3}{2} \cdot 0.911\right) - 2 \cdot (10 + 3) \cdot \tan \frac{45°}{2} =$$

$$= \pi \cdot 0.25 \cdot 11.366 - 26 \cdot \tan 22.5° = -1.84 \text{ mm}$$

The sum of the compensation values amounts to:

$$v_{ges} = v_1 + v_2 + v_3$$

$$v_{ges} = -6.35 - 2.63 - 1.84 = -10.82 \text{ mm}$$

The total extended length adds up to:

$$l = a + b + c + d + v_{ges} = 185 - 10.82 = 174.18 \text{ mm}$$

The result is rounded up to the next higher millimeter, resulting in a final blank dimension of 175 mm.

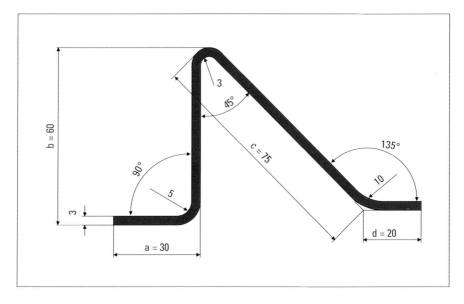

Fig. 4.8.4 Bent workpiece (cf. example)

Bending force and bending work during V-die bending
V-die bending is a production technique that is widely used. Here, bending takes place between the bending punch and die *(Fig. 4.8.5)*. This is generally followed by the application of pressure on the die, which serves to relieve residual stresses (reducing springback).

When bending angles in the V-die, the bending force F_b [N] required for forming depends on the die width w [mm], as this determines the bending moment. In contrast to this, the magnitude of the bending radius plays only a minor role, provided that the corresponding die width w has been correctly selected. Generally speaking, the dimension l [mm] of the die is selected in accordance with the sheet metal thickness s [mm]:

$$l = 6 \cdot s \quad [mm]$$

When using a conventional die in which the sheet metal is positioned at a distance w on both sides while the punch presses centrally onto the sheet metal, with a coil stock width of b_S [mm] the following bending force is obtained:

$$F_b = \frac{b_s \cdot s^2 \cdot R_m}{w} \quad [N] \quad \text{for } w/s \geq 10$$

or

$$F_b = \left(1 + \frac{4 \cdot s}{w}\right) \cdot \frac{b_s \cdot s^2 \cdot R_m}{w} \quad [N] \quad \text{for } w/s < 10$$

These are simplified formulae which, however, are sufficient in practice for approximate calculations. The bending work W_b [Nm] amounts to:

$$W_b = x \cdot F_b \cdot h \quad [Nm],$$

if the path of the punch from impact with the sheet metal until the end of the bending process is designated as h [m]. The constant value x takes into account the uneven force progression along the path of the punch and generally lies between x = 0.3 and 0.6, depending on bending requirements and machine setting.

Example:

If a 2 mm thick aluminium sheet AlMg5Mn measuring 50 mm wide with a minimum tensile strength of R_m =280 N/mm2 is bent in a V-die with a width of w = 15 mm using a punch displacement of h = 7 mm, the resulting bending force is:

$$F_b = \frac{50 \text{ mm} \cdot 4 \text{ mm}^2 \cdot 280 \text{ N/mm}^2}{15 \text{ mm}} = 3{,}733 \text{ N}$$

When bending using a V-die, the process factor x = 1/3. The bending work thus amounts to:

$$W_b = \frac{1}{3} \cdot 3{,}733 \text{ N} \cdot 0.007 \text{ m} = 8.7 \text{ Nm} = 8.7 \text{ J}$$

■ 4.8.2 Roll forming and variety of sections

Process

Roll forming is categorized according to DIN 8586 as a method of forming by bending using a rotary die motion *(cf. Fig. 2.1.3)*. During roll forming, the sheet coil or blank is formed in a stepless and continuous

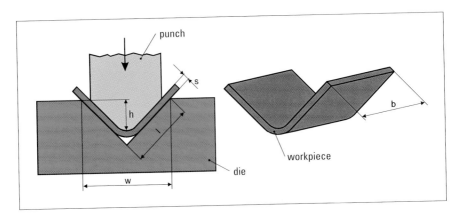

▲ **Fig. 4.8.5** Bending in the V-die

process to create profiles with the aid of rollers *(Fig. 4.8.6)*. One pair of rollers is required per forming stage *(Fig. 4.8.7)*. The advantage of the shaping process, compared to other bending methods, is that sheet strips of any length can be formed at a relatively high speed.

Basic section geometries
Sections are manufactured in steps using a combination of simple basic roll forms:

– *Rolling:* The edge of a profile is shaped by more than 180°. Three different configurations are most commonly used *(Fig. 4.8.8)*.

◀ **Fig. 4.8.6**
Pair of forming rolls:
last forming stage

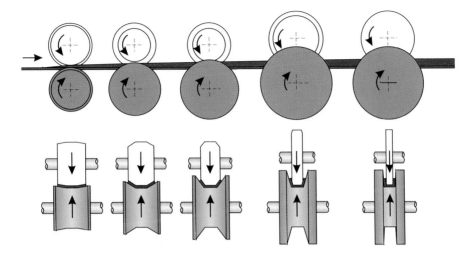

▲ **Fig. 4.8.7** Roll forming: roll bending in several stages

- *Folding:* step-by-step bending of straight legs around a common point (Fig. 4.8.9).
- *Beading:* recesses in the profile, e.g. for reinforcement *(Fig. 4.8.10).*
- *Doubling:* The material is doubled back on itself, for example at the edges *(Fig. 4.8.11).*

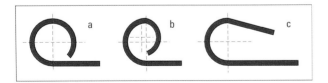

▲ **Fig. 4.8.8** Rolling: **a** *eye:* circular shape with only one radius; **b** *oval:* circular shape, made up of several radii; **c** *loop:* material not entirely bent over

◀ **Fig. 4.8.9** Folding

▲ **Fig. 4.8.10** Beading

Application examples
By combining the basic shapes, it is possible to manufacture any optional section *(Fig. 4.8.12 to 4.8.17). Figure 4.8.18* illustrates the step-by-step
manufacture of a special section geometry.

Roll bending radius
During the roll bending process, the actual bending radius must be
between the maximum and the minimum admissible bending radius:

$$r_{i\,max} = \frac{s \cdot E}{2 \cdot R_{eL}} \quad [mm]$$

$$r_{i\,min} = s \cdot c$$

with the modulus of elasticity E [N/mm²], the yield strength R_{eL} or $R_{p0,2}$
[N/mm²] (cf. Sect. 4.2.2) and the material coefficient c [–] in accordance
with Table 4.8.3.

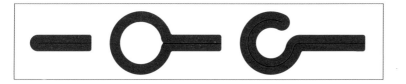

▲ **Fig. 4.8.11** Doubling

▲ **Fig. 4.8.12** U/C and rail sections

◀ **Fig. 4.8.13** Hat sections

◀ **Fig. 4.8.14** Angular, L and Z sections

◀ **Fig. 4.8.15**
T and omega sections

◀ **Fig. 4.8.16** Closed sections

◀ **Fig. 4.8.17**
Wide strip sections

Example:

When roll forming with 2 mm thick strain hardened sheet steel in St 37-2 (material number 1.0037) with the modulus of elasticity $E = 2.1 \cdot 10^5$ N/mm² for steel and yield strength $R_{eL} = 220...240$ N/mm² the useful radius range is calculated as follows:

$$r_{i\,max} = \frac{2 \text{ mm} \cdot 2.1 \cdot 10^5 \text{ N/mm}^2}{2 \cdot 240 \text{ N/mm}^2} = \frac{210{,}000}{240} \text{ mm} = 875 \text{ mm}$$

$$r_{i\,min} = 2 \text{ mm} \cdot 0.5 = 1 \text{ mm}$$

The actual radius when bending at right angles to the rolling direction should thus be at least 1 mm in size.

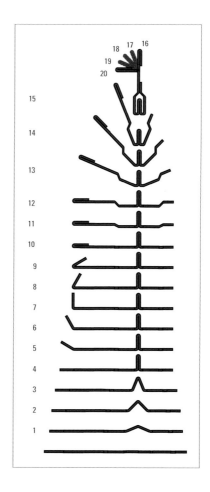

Fig. 4.8.18 ▶
Step-by-step manufacture of
a section using 20 rolls
(flower diagram)

Table 4.8.3: Material coefficient c for the calculation of $r_{i\,min}$

Material	Bending direction compared to rolling direction of sheet	Material coefficient c	
		Material soft annealed	Material strain hardened
Al	transverse	0.01	0.30
	longitudinal	0.30	0.80
Cu	transverse	0.01	1.00
	longitudinal	0.30	2.00
Ms 67, Ms 72, CuZn 37	transverse	0.01	0.40
	longitudinal	0.30	0.80
St 13	transverse	0.01	0.40
	longitudinal	0.40	0.80
C 15 – C 25 St 37 – St 42	transverse	0.10	0.50
	longitudinal	0.50	1.00
C 35 – C 45 St 50 – St 70	transverse	0.30	0.80
	longitudinal	0.80	1.50

Roll forming lines

Roll forming lines are classified according to two different operating modes. *Start-stop* lines consist in their simplest form of a decoiler, a coil loop, a roll forming machine and a parting die. The product length is determined by the roll forming machine *(Fig. 4.8.19)*. Due to the use of the start-stop method, blanking and embossing operations are also possible using the roll forming machine's positioning system.

Continuous running roll forming lines operate at a constant speed. This type of line is used to process sheet metal blanks or strips to create long sections. While on blank processing lines cut-to-length blank metal

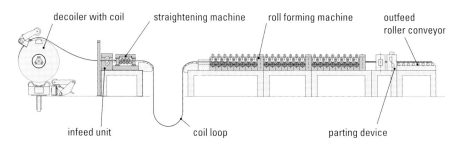

decoiler with coil   straightening machine   roll forming machine   outfeed roller conveyor

infeed unit   coil loop   parting device

▲ **Fig. 4.8.19**   Start-stop roll forming line

sheets are processed, when using a sheet strip processing line the produced profile is parted after roll forming using automatic cut-to-length devices.

The advantage of a blank processing line is that the complex profile parting die can be substituted by a simple cross-cutting shear. This eliminates the need for costly die inserts, a particularly important consideration where different profile shapes or widths are being processed. However, a further point of consideration is that this method requires a certain minimum sheet length, as the feed force of the rollers must always be greater than the torque necessary for the forming process which has to be overcome. Furthermore, adequate provision must be made to ensure efficient blank guidance.

The benefit of a continuously operating line producing off the coil is its higher output when processing greater profile lengths. This production method is essential if continuously running processes such as longitudinal seam welding, gluing, foaming, etc. have to be integrated during profiling, or where with respect to the machine size excessive masses would have to be accelerated.

In practice, the basic forms of this processing technique are frequently upgraded, either due to insufficient output or because additional punched patterns have to be taken into consideration. In this case, double loop lines with a separate punching device are used.

To keep set-up times to a minimum, roll forming lines are frequently equipped with dual swivel-mounted decoilers capable of accommodating two coils and coil end welding machines. The coil end welding machines join the end of the previous coil to the start of the next one, saving not only the time required for feeding in a new coil but also reducing the amount of scrap generated at the start and end of each coil.

Die changeover techniques with roll stand sets and rolling dies
To allow different section shapes to be manufactured on a single production line, various roll changeover techniques are available. The rolling dies are mounted on rigid stands or on stands with telescopic roll set or directly mounted on the gear blocks *(Fig. 4.8.20)*. The stands are connected to the gear units of the roll forming machine by means of universal joint driven cardan shafts.

Width adjustable roll sets permit the manufacture of products with different web widths and leg heights using a common roll set *(Fig. 4.8.21)*.

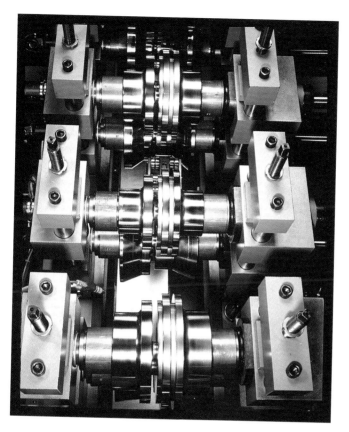

◀ **Fig. 4.8.20**
Roll stands

The roll stand for *profile roll changeover* is used for the manufacture of compact sections with fixed dimensions. However, if spacers are manually inserted into each stand, it is also possible to manufacture sections with different widths *(Fig. 4.8.22)*. Both stand types – rigid and with telescopic roll set – permit the rolling dies to be exchanged by detaching the front stand upright. However, this involves long set-up times.

Drive shaft changeover offers a compromise between exchanging individual rollers and exchange of a complete stand. At a special work station equipped with a suitable crane harness, the upper and lower drive shafts are exchanged together *(Fig. 4.8.23)*. This helps to avoid set-up errors and saves the need for a second stand set. The time required for this changeover process lies somewhere between that required for exchanging rollers and that required when using pallets for the stands.

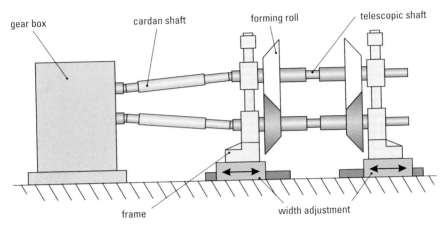

gear box • cardan shaft • forming roll • telescopic shaft

frame • width adjustment

▲ **Fig. 4.8.21** Width adjustable telescopic roll set

It is also possible to use a separate set of stands for each different section. This permits not only fast resetting but also a high degree of process reliability and repeat accuracy *(Fig. 4.8.24)*. For *stand changeover* there are two different changeover techniques possible:

– Exchange of individual pallets using a crane: Pallet changeover is advisable where short set-up times are required. The roll forming

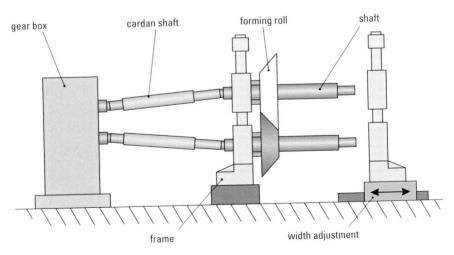

gear box • cardan shaft • forming roll • shaft

frame • width adjustment

▲ **Fig. 4.8.22** Roll stand for exchange of section rolls

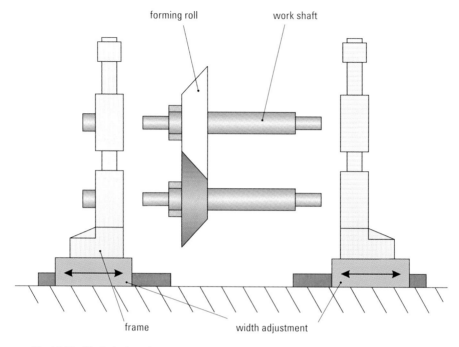

forming roll   work shaft

frame   width adjustment

▲ Fig. 4.8.23 Work shaft exchange

stands, for example three or four, are mounted in a fixed setting on pallets together with the rolls. The pallets are raised using a crane. For fastening and centering, mechanical or hydraulic systems are used depending on the level of automation.

– Exchange of individual pallets or complete sets of rolls and stand bearings: For fully automatic resetting, the tool pallets are exchanged using a rack handling system. This method is more economical where frequent resettings are required.

■ 4.8.3 Roller straightening

Straightening is required wherever plastically deformed sheet metals, bars, wires or pipes have to be straightened. A typical application is the straightening of sheet metal in a straightening machine for further processing in a forming press after being taken off the coil (cf. Sect. 4.3).

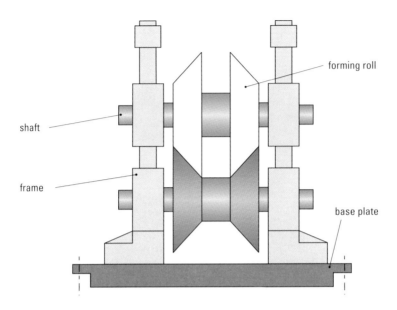

shaft

frame

forming roll

base plate

▲ **Fig. 4.8.24** Stand for a complete stand exchange

In accordance with DIN 8586, roller straightening is classified as a bending process using rotating tool motion *(cf. Fig. 2.1.3)*. Bending back unwanted curvature by means of a precisely defined bending process which must be exactly coordinated with the starting radius has proven to be impossible in practice. Instead, the technique used straightens the metal by imparting alternating bending operations using gradually increased bending radii. This technique is known as roller straightening *(Fig. 4.8.25)*.

Bending and counter bending must each take place within the plastic range of the material in order to ensure that the bending direction is retained following elastic recovery, i.e. the yield strength of the material must be exceeded. On the other hand, care must be taken to ensure that the straightened material does not sustain damage. If bending is too pronounced, brittle materials can develop slight cracks at the surface. The last roller must be set in such a way that it causes the bending process which generates a flat sheet following elastic recovery *(Fig. 4.8.26)*.

As a result of alternate bending, the residual stress in the sheet material is reduced as the number of bending processes and the bending

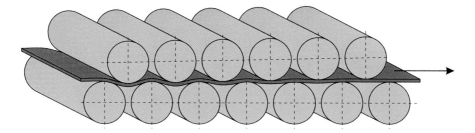

▲ **Fig. 4.8.25** Roller straightening with 13 rollers

radii increase. The reduction of residual stress is highly beneficial for further processing *(Fig. 4.8.27).* The larger the number of rollers used, the lower is the residual stress in the sheet metal after straightening.

In all roller straightening machines, there are two rows of straightening rollers arranged in offset formation relative to each other. These are called the upper and lower rollers. The distance between roller centers is known as the roller pitch t_w and varies in different machines, as does the roller radius R_w and the number of straightening rollers. Straightening machines with a minimum of five rollers are used. Five straightening rollers are sufficient to straighten thick plates or coils, or where no special demands are made on straightening accuracy. Where thinner materials or more stringent straightening requirements are involved, for example

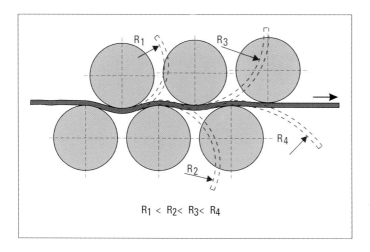

$R_1 < R_2 < R_3 < R_4$

Fig. 4.8.26
Alternate
bending

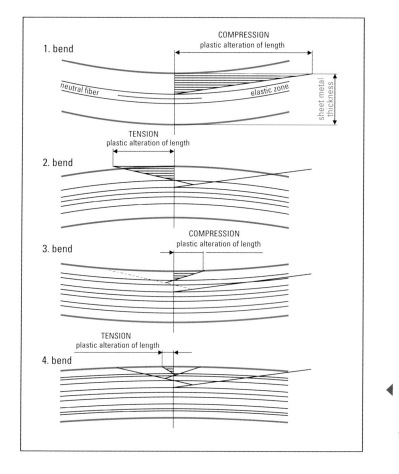

1. bend

COMPRESSION
plastic alteration of length

neutral fiber

elastic zone

sheet metal thickness

TENSION
plastic alteration of length

2. bend

COMPRESSION
plastic alteration of length

3. bend

TENSION
plastic alteration of length

4. bend

Fig. 4.8.27
Residual
stresses in
the sheet
metal

for fine blanking, machines with up to 21 rollers are used. The feed value z_w for the rollers is limited by the geometry of the straightening machine *(Fig. 4.8.28):*

$$z_{w\,max} = 2 \cdot R_w + s - \sqrt{\left(2 \cdot R_w + s\right)^2 - t_w^{\,2}} \quad [mm]$$

with the roller radius R_w [mm], sheet metal thickness s [mm] and roller pitch t_w [mm].

In addition to the limits imposed by the roller radius and pitch, the material properties also have to be taken into account; when processing brittle aluminium, for instance, in order to avoid cracking the material it is important not to select the smallest bending radius.

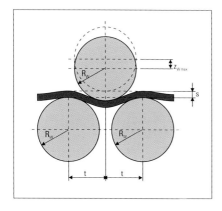

◀ **Fig. 4.8.28** Straightening roller geometry

Example:

Sheet steel St 37 (material number 1.0037) with a thickness of 2 mm has to be straightened. The straightening roller pitch of the straightening machine is $t_w =$ 22 mm, the roller diameter $R_w = 20$ mm. The maximum feed value is calculated as

$$z_{w\,max} = 2 \cdot 20 + 2 - \sqrt{(2 \cdot 20 + 2)^2 - 22^2} = 6.2 \text{ mm}$$

Roller straightening machines

Generally speaking, roller straightening machines for straightening sheet metal comprise an upper and a lower row of straightening rollers with the same diameter. The two rows are generally arranged horizontally. The rollers in the top row are generally arranged precisely above the gaps between the rollers of the bottom row.

To allow modification of the rolling gap, the upper rollers must be adjustable in height. The type of adjustment characterizes the straightening machine type. While in one case each upper roller can be separately height adjustable, in the other type of machine all the upper rollers are grouped together in a block to form an assembly, and are fed towards the lower rollers by a tilting motion of the block.

The thickness range, the yield strength and the modulus of elasticity of the coil stock to be processed determine the diameter and pitch of the straightening rollers. In the case of materials with particularly sensitive surfaces, intermediate rollers embedded between the straightening and support rollers are used in order to avoid damaging the sheet surface. With the help of straightening rollers with crowning capability, waviness at the edges and center of the material can be eliminated.

The first pair of rollers must cause plastic deformation of the material, but a certain roller diameter must not be exceeded. The thickness of a straightenable sheet metal is accordingly calculated from the maximum straightening force and the torque, which in turn is determined by the size of the cardan shafts between the gear and rollers and the roller diameter. As neither the gear nor the cardan shafts can be enlarged at will for space reasons, the sheet thickness range and accordingly also the flexibility of straightening machines with rollers of equal size are restricted. This range can be extended by approximately one and half times if individual rollers can be disengaged. Using the further developed roller system depicted in *Fig. 4.8.29* with different roller diameters, it is possible to straighten sheet metals from 0.5 to 20 mm in thickness.

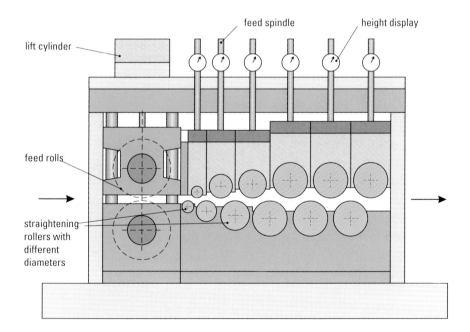

▲ **Fig. 4.8.29** Roller straightening machine for larger sheet metal thickness range

■ 4.9 Organization of stamping plants

■ 4.9.1 Design

Sheet metal stamping plants are among the most capital-intensive of all manufacturing installations in industry. The high degree of investment necessary and long-term commitment to selected manufacturing methods mean that careful investment and production planning are essential. In the production of sheet metal parts using deep drawing, stretch drawing and blanking processes, major criteria include part accuracy, die life, line output and its availability. The implementation of measures for continuous in-process quality assurance using suitable sensor technology allows objective assessment of part quality (cf. Sect. 4.9.3).

Stamping plants generally comprise different sheet metal processing equipment, each of which manufactures a certain range of parts *(Fig. 4.9.1)*. The aim of efficient stamping plant layout is to ensure the most economical operation possible. The selection of the individual equipment depends primarily on the size, shape and metal forming requirements imposed on the components, as well as the planned batch sizes.

The sheet metal is supplied from the rolling mill in the form of coil. When producing medium-sized to large parts such as bathtubs, sinks, car doors or roofs, a separate blanking line cuts the blanks from the coil stock and stacks them *(cf. Fig. 4.6.5)*. The blanks are destacked by the blankloader *(cf. Fig. 4.4.15 and 4.4.16)* and fed individually and automatically into the press where they are formed. Due to its high capacity, a blanking line is able to service several press lines *(cf. Fig. 4.4.19)* or large-panel transfer presses *(cf. Fig. 4.4.27 and 4.4.38)*.

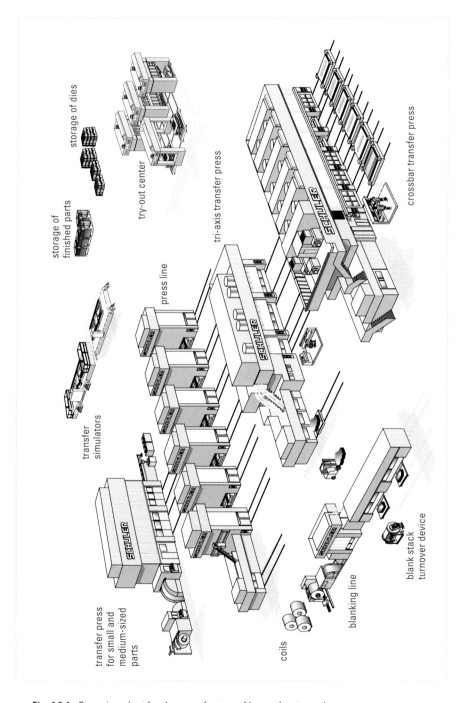

▲ **Fig. 4.9.1** Stamping plant for the manufacture of large sheet metal parts

When producing small sheet metal components such as drawer slides, disc carriers for automatic transmissions, wishbones, bicycle brake levers etc., the coil stock can be fed directly to the press by a coil line *(cf. Fig. 4.3.1)*. The sheet metal forming process takes place in a compact piece of equipment, generally a high-speed press *(cf. Fig. 4.6.6)*, a universal *(cf. Fig. 4.4.1)* or transfer press *(cf. Fig. 4.4.23)*.

The finished parts are palletized either manually or by using a stacking device *(cf. Fig. 4.4.41)* and made ready for downstream processing, for example welding, painting or hemming. At the same time, scrap disposal takes place.

For the introduction of new dies for large parts and transfer presses, try-out presses *(cf. Fig. 4.1.19* and *4.1.20)* and simulators *(cf. Fig. 4.1.24)* are employed. These are used by the die shop to prepare tooling as well as try-out of the dies before the final try-out on the actual production press.

This shortens the die try-out time on the production presses substantially for obtaining the first OK parts. Consequently, this procedure shortens the time necessary to produce the first OK parts and increases the machine utilization rate.

■ 4.9.2 Layout

The following provides a detailed description of a typical example stamping plant layout. To simplify matters somewhat, our explanation is based on the assumption of ideal conditions and new installation of the complete equipment (Greenfield plant). This is a typical task to establish manufacturing facilities for the production of medium-sized and large car body panels such as doors, roofs, hoods, floor panels, side panels, fenders etc. for a particular car type. There are three press systems available: tri-axis transfer presses *(cf. Fig. 4.4.27)*, crossbar transfer presses *(cf. Fig. 4.4.38)* and press lines *(cf. Fig. 4.4.19)* of various bed sizes.

Before preparing the plant layout, it is first necessary to determine which parts are to be outsourced and which are produced in-house. The following parts are generally produced in-house:

– outer panels whose surface could be damaged during transport,
– joining parts with narrow gap tolerances for precise matching,
– parts of large volume resp. size which are difficult to stack or transport.

These so-called A-parts – there are generally between 30 and 40 of them per vehicle type – must be grouped together to families of parts. These are then allocated to the different press systems on which they can be economically manufactured.

Figure 4.9.2 and Table 4.9.1 illustrate the correlation between part size and press system. When processing medium-sized panels, a press line with a medium bed size or a tri-axis transfer press will be employed. Press lines with large bed sizes or crossbar transfer presses offer the ideal conditions for processing large panels with low inherent stiffness or double panels. *Figure 4.9.3* offers a more precise allocation of parts to both available transfer press systems. Some parts can be handled by crossbars with suction cups as well as by gripper rail systems. They represent the overlap between two families of parts. Some of these parts, such as doors, are produced as single parts using gripper rail transfer systems *(cf. Fig. 4.4.29)* or as double parts using a crossbar transfer system *(cf. Fig. 4.4.34)*.

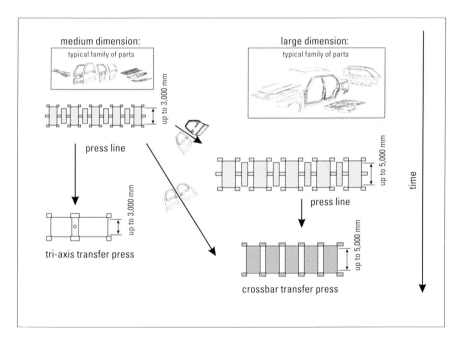

▲ **Fig. 4.9.2** Press systems for the manufacture of car body parts

Table 4.9.1: Criteria for press selection

	large-panel transfer presses		press lines with differing bed sizes
	crossbar transfer	tri-axis transfer	
part spectrum	large, unstable parts and double parts	medium-sized parts	all parts
complete part production	++	+	0
low investment	0	+	++
low space requirement	+	++	0
die change time [min.]	10	10	30

The selection and loading of presses depend primarily on the number of vehicles to be produced. The number of vehicles determines the production requirement of all parts per day. If the A-parts of a vehicle type are manufactured on large-panel transfer presses, the stamping plant must have a capacity of between 3 and 4 million parts per year (basis of planning: 1 vehicle type = 100,000 vehicles per year).

The production scope per year, i.e. the number of parts which can be produced on the press system per year, must be co-ordinated with the production requirement. This calls for an output analysis of the individual press system. The basis for calculation is the production stroking rate. In transfer presses, this is equivalent to the stroking rate set at the press. In the case of press lines, the production stroking rate is reduced to some 50% to 60% of the set press stroking rate due to standstill periods caused by the single stroke operation. The number of parts which can be produced during press operation also depends on the forming requirements of the individual parts. A complex part requires a longer processing time. In addition, the time necessary for transportation of

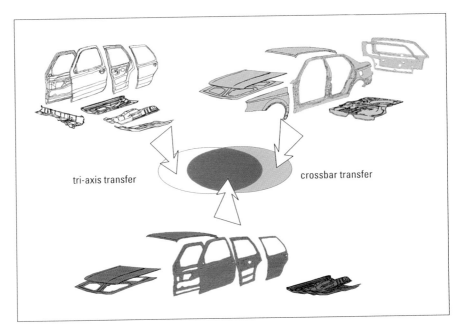

▲ **Fig. 4.9.3** Families of parts for large-panel transfer presses with different transfer systems

large, unstable parts increases. Because of its range of parts and longer feed pitch, a crossbar transfer press is not able to run at the same high stroking rates as a tri-axis transfer press. In general, it is true to say that the output of a press system must always be considered in the light of the part family planned for production on the press.

In case of double part production in a crossbar press the processing time per part is reduced. When double dies are used, the average output of a crossbar transfer press is larger than the output of a tri-axis transfer press using single dies, despite of the fact that the average production stroking rate is lower. Double dies are preferred for the manufacture of outer and inner doors, front and rear, left and right. Where 10 doors per minute can be produced on a tri-axis transfer press, for instance, the output of a crossbar transfer press can be increased by 50% to 15 parts per minute.

The number of hours worked also plays an important role in calculating the production level, based for example on the number of days worked per week, the length of the seasonal shut downs and shut downs due to public holidays which vary from country to country. Let us assume between 200 and 300 effective days' work per year and daily working periods of between 840 and 1,260 min depending on two or three-shift operation. Then the annual production capacity fluctuates between 100% at the lower end and 225% at the higher end. Major influencing factors which cannot be freely selected by the manufacturer, such as statutory labor legislation, collective agreements etc. may determine the possible framework. This means that production capacity must always be calculated relative to personnel requirements.

Other influencing factors include for example necessary production system maintenance intervals. If the stamping plant operates at full capacity, these periods may have to be deducted from the production capacity. In case a five day week is worked, maintenance can be done on Saturdays and Sundays without loss of production.

If we reduce the working time by the periods in which no production takes place, we arrive at the press availability or uptime. Particular factors which must be planned for during working hours include for example die change, die setting or die cleaning times. In addition to scheduled press stops, unscheduled downtime must also be allowed for. This can be caused, for example, by failure of the blank supply, press or dies. Overall, the uptime of modern large-panel transfer presses is generally in the range of 70% of the total available work time *(Fig. 4.9.4).* The actual average line output is thus calculated from the production stroking rate, the press operation time and the proportion of press running time during which double parts are produced. Extrapolated over a whole year, values of up to 4.3 million parts may result, depending on the press system and working hours. Table 4.9.2 illustrates an example of output for a production system comprising a blanking line and two large-panel transfer presses. Providing that double parts are produced for 50% of the time on the crossbar transfer press, with a total of around 3.9 million production strokes of the transfer presses per year, we arrive at an output of 4.8 million sheet metal parts per year.

When generating a press load schedule, batch sizes and the production sequence are defined. The sequence is repeated within each complete cycle, whereby a part can be produced several times during any one cycle. *Figure 4.9.5* illustrates an example of a press load schedule and the relevant inventory in the intermediate storage of finished parts for the production of four different sheet metal parts. Parts A and B are

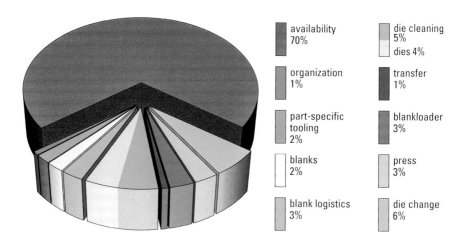

availability 70%

organization 1%

part-specific tooling 2%

blanks 2%

blank logistics 3%

die cleaning 5%

dies 4%

transfer 1%

blankloader 3%

press 3%

die change 6%

▲ **Fig. 4.9.4** Availability and downtimes of large-panel transfer presses

Table 4.9.2: Actual part output of a blanking line and two large-panel transfer presses

	set stroking rate [1/min]	average production stroking rate [1/min]	availability [%]	net output [parts/min]	planned utilization rate [%]	output with 1,260 min working period on 222 days/ year	strokes/year resp. blanks/year
blanking line	20 ... 55	27	65	17.5	80	3,916,000	3,916,000
tri-axis transfer press	12 ... 16	14	70	9.8	80	2,193,000	2,193,000
crossbar transfer press 50 % single dies, 50 % double dies	10 ... 12	11	70	single dies 7.7 double dies 15.4 average 11.55	80	2,584,500	1,723,000

each manufactured twice in one cycle with batch sizes of 5,000 and 3,000, while type C and D parts are only produced once in a cycle with batch sizes of 4,800 and 3,300 respectively. During production, the inventory of the relevant part increases in the intermediate storage. However, at the same time parts are removed (illustrated in *Fig. 4.9.5* for part A). The steeper the production curve, the shorter the processing

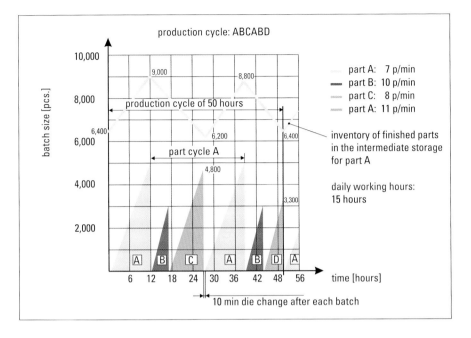

▲ **Fig. 4.9.5** Load schedule and parts inventory for production on a large-panel transfer press

time per part. Die changes and setting of new dies takes place within press stops between production, whereby the die changing time plays an important role (cf. Sect. 3.4).

Production must be resumed once the storage drops to the defined safety inventory level, generally one day's production. If we take type A, for example, this level would be reached at 6,200 pieces, while for type B the corresponding figure is 2,800 pieces. The 5,000 or 3,000 parts produced, less the parts used during manufacture, must last until the next planned production period – around 14 resp. 22 hours later. When compiling the load schedule, a fine balance must be drawn: On the one hand the production period for a batch size has to guarantee sufficient time for set-up of the die set for the next part. At the same time, however, excessively large batch sizes violate the principle of just-in-time production in terms of capital costs for work-in-process and costs for intermediate storage of finished parts.

Example:

A production unit has to be configured in which 44 pressed parts – primarily A parts – have to be produced every year for 100,000 vehicles. The unit comprises a blanking line, a crossbar transfer press and a tri-axis transfer press. A possible configuration of this stamping plant with optimized material flow and peripheral devices is illustrated in *Fig. 4.9.6*. The material is supplied in the form of coil stock and stored in the coil storage, from where it is fed to the blanking line. This supplies through the blank storage both large-panel transfer presses with a total of 4 million blanks a year. Following production on the transfer presses, the parts are transported in part-specific racks to the intermediate storage for finished parts.

Table 4.9.3 lists the part spectrum for both forming presses. The crossbar transfer press is run using eight individual and eight double die sets, whereby the double dies in this example always produce two different parts. Their die change time amounts to 10 min. The tri-axis transfer press manufactures 20 different parts using 20 die sets. The die change time is also 10 min. Overall, the 44 parts are produced using 36 sets of dies – 20 on the tri-axis transfer press, and 24 on the crossbar transfer press.

If 222 working days are available per year, parts for 450 vehicles must be produced every day. In three-shift operation with working periods of 1,260 min per day, the total of approx. 4.8 million parts indicated in Table 4.9.2 (achieved with 3.9 million strokes) can be produced every year on the transfer presses. This is based on a planned capacity utilization of 80 %. Three days are selected as a production cycle. If we take the simplified assumption that every part is only run once per cycle, three times 450 parts must be produced per cycle. The intermediate storage for finished parts has a theoretical maximum inventory of 1,800 pressed parts of each part type: 1,350 parts from running production plus 450 parts as a minimum safety inventory. On average, however, as body-in-white production runs on a parallel basis, only around half of the parts are located in the intermediate storage. At any point in time, the inventory comprises 900 × 44, i.e. approx. 40,000 parts. As a result, the part storage and the capital investment this involves are considerably lower than when producing on conventional press lines.

■ 4.9.3 Quality assurance through quality control

The aim of achieving zero-defect production calls for continuous quality control. This involves the detection of any changes in the production process which could precede a defect, and the segregation of individual defective parts. Conventional methods of quality assurance in stamping plants include visual inspections – in the case of deep drawn

Table 4.9.3: Production data for crossbar and tri-axis transfer presses

Crossbar transfer press

Production cycle: 3 days
Availability: 70%
Die change time (min): 10
Die sets: 16 (24)
Planned utilization rate: 80%

Part name	Parts/ stroke	Part require- ment/day	Strokes/ parts/cycle	Part batch size
body side, outer, right	1	450	1350	1350
body side, outer, left	1	450	1350	1350
door, outer, front, right + left	2	450	1350	2700
trunk lid, inner	1	450	1350	1350
roof	1	450	1350	1350
trunk lid, outer	1	450	1350	1350
door, inner, front, right + left	2	450	1350	2700
body side, inner, rear, right + left	2	450	1350	2700
door, outer, rear, right + left	2	450	1350	2700
door, inner, rear, right + left	2	450	1350	2700
hood, outer	1	450	1350	1350
hood, inner	1	450	1350	1350
fender, outer, front, right + left	2	450	1350	2700
wheel house, inner, front, right + left	2	450	1350	2700
floor, mid	1	450	1350	1350
floor, front	2	450	1350	2700

Tri-axis transfer press

Production cycle: 3 days
Availability: 70%
Die change time (min): 10
Die sets: 20
Planned utilization rate: 80%

Part name	Parts/ Stroke	Part require- ment/day	Strokes/ parts/cycle	Part batch size
floor, rear	1	450	1350	1350
lock reinforcement, front, right	1	450	1350	1350
lock reinforcement, rear, right	1	450	1350	1350
rib, reinforcement, rear, right	1	450	1350	1350
hinge pillar, front, right	1	450	1350	1350
hinge reinforcement, front, right	1	450	1350	1350
rib, reinforcement, front, right	1	450	1350	1350
lock reinforcement, front, left	1	450	1350	1350
lock reinforcement, rear, left	1	450	1350	1350
rib, reinforcement, rear, left	1	450	1350	1350
hinge pillar, front, left	1	450	1350	1350
hinge reinforcement, front, left	1	450	1350	1350
rib, reinforcement, front, left	1	450	1350	1350
dash panel	1	450	1350	1350
shelf, rear	1	450	1350	1350
seat, reinforcement, rear	1	450	1350	1350
wheel house inner, rear, left	1	450	1350	1350
wheel house inner, rear, right	1	450	1350	1350
sunroof	1	450	1350	1350
sunroof aperture panel	1	450	1350	1350

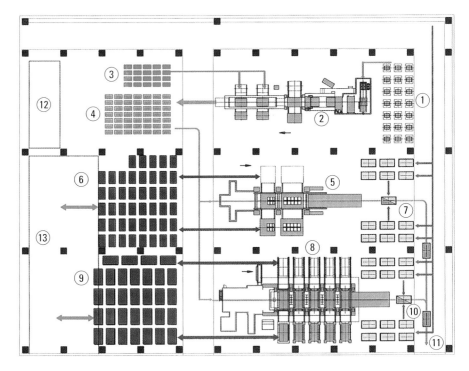

▲ **Fig. 4.9.6** Layout of a stamping plant with material flow
1 coil storage; 2 blanking line; 3 intermediate pallet storage;
4 intermediate blank storage; 5 tri-axis transfer press; 6 die storage for 5;
7 racks for finished parts for 5; 8 crossbar transfer press; 9 die storage for 8;
10 racks for finished parts for 8; 11 to intermediate finished part storage;
12 offices and recreation facilities; 13 die repair and cleaning

parts, for example, for wrinkles and cracks (cf. Table 4.2.2) – and also the testing of workpieces for dimensional accuracy using measuring devices or gauges. However, this type of control process is both time-consuming and labor intensive, and is generally only used in the case of large, complex workpieces such as body panels or for random sampling of smaller parts.

Where small punched parts are involved, the preferred system is one of on-line controls in which quality assurance is integrated in the manufacturing process itself, i.e. in the press. This permits monitoring of each individual part and of process parameters such as die wear.

Visual inspections or workpiece gauging are increasingly being replaced by optoelectronic and image processing systems. Simple optoelectronic systems are used, for instance, for measuring displacement: An LED or laser source emits light beams which are picked up by a detector on the other side of the workpiece. The received beams are evaluated and used by the computer to recognize and indicate any possible die offset. More efficient monitoring systems consist of a video camera with workpiece illumination, a digitalization unit, a computer for image processing and a screen to permit monitoring by the operator *(Fig. 4.9.7)*. The camera is located at the press, the geometry data of the current part is compared during the production process with the specified data for a good part. In the case of deviations beyond a certain tolerance, the defect is displayed and production stopped. The image processing system is used for the detection of wrinkles, scratches, dents, beads, cracks and also for testing holes and determining dimensions.

In case of a gradual deterioration of part quality, for example as a result of die wear, a control device for the process parameters (forming force, blank holder force, etc.) is beneficial. For this purpose, quality monitoring can be coupled with the press control system.

Another quality assurance system is based on monitoring of the press force. A sheet forming process always produces the same force time curve when manufacturing defect free parts. The forming force acting

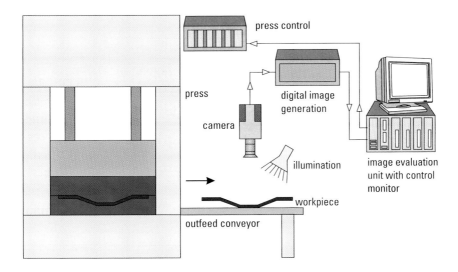

▲ **Fig. 4.9.7** Quality assurance through digital image generation and processing

on the punch is obtained by measuring the elastic deflection of loaded components. Load cells or strain gages can be located directly in the primary force path or indirectly in the secondary force path, for example in the press frame. The positioning possibilities of individual sensors for force, acoustic emission and displacement measurement are indicated in *Fig. 4.9.8.* A defect in the production process is indicated by a deviation in the force curve, for example a rapid drop in the forming force during the deep drawing process as soon as a crack appears in the sheet metal. Furthermore, controlling the blank holder force during press setting also helps to avoid formation of wrinkles in drawn parts.

A method similar to force measurement is acoustic emission monitoring. The dynamic processes taking place at the die during a blanking or forming operation cause high-frequency acoustic emissions, so-called structure-born noise. Measurements of the acoustic emission are carried out by measuring the degree of vibration using accelerometer. In case of troublefree production, always the same vibration curve is produced. Production of a defective part can be detected by the presence of an

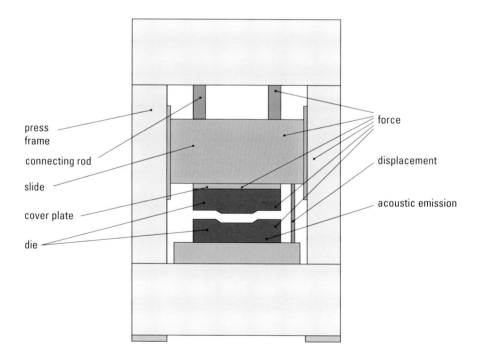

press frame
connecting rod
slide
cover plate
die

force

displacement

acoustic emission

▲ **Fig. 4.9.8** Positioning of sensors at the press for process monitoring

unexpected acoustic emission signal. The amplitude of the vibrations increases in comparison to the usual vibration curve, for instance, when a crack occurs in a drawn part. Monitoring of acoustic emission signals are used mainly for the detection of cracks or punch breakage.

The efficiency of quality assurance systems increases with the number of sensors used simultaneously or by combining different systems. When using combined monitoring systems for operation and surveillance, whenever possible it is recommended to utilize the same computer and display unit. Process data acquisition and documentation are essential to ensure continuous quality assurance. Systems with learning capability for diagnosis and the avoidance of defects have been shown to offer an optimum solution.

Bibliography

Bareis, A., Braun, R., Hoffmann, A.: Hochleistungs-Stanzanlage für die Verarbeitung dickerer Bleche, VDI-Zeitschrift-Special-Rationelle Blechbearbeitung (1990) 4.

Bareis, A., Dommer, H.-M., Hoffman, H.: Flexibilität in der Stanztechnik, VDI Journal 131 (1989) 9.

Beisel, O.: Tuschieren und Probieren ohne Produktionsausfall, Werkstatt und Betrieb (1993) 10.

Birzer, F.: Fine blanking does even better, Swiss quality production (1993) 4.

Bräutigam, H.: Richten mit Walzenrichtmaschinen, Eigenverlag, (1987), 2nd edition

Breuer, W.: Multimediafähiges Informationssystem ergänzt herkömmliche Pressensteuerung, Maschinenmarkt (1995) 16.

DIN 1623: Steel flat products; Cold rolled sheet and strip; technical delivery conditions – part 1 (02.83): Cold reduced sheet and strip, part 2 (02.86): General purpose structural steels, Beuth Verlag, Berlin.

DIN 1745: Wrought aluminium and aluminium alloy plate, sheet and strip greater than 0.35 mm in thickness - part 1: Properties, Beuth Verlag, Berlin (02.83).

DIN 1788: Wrought aluminium and aluminium alloy sheet and strip with thicknesses between 0.021 and 0.350 mm, properties, Beuth Verlag, Berlin (02.83).

DIN 5512: Werkstoffe für Schienenfahrzeuge; Stähle - Teil 1 (02.84): Allgemeine Baustähle - part 2 (09.89): Flacherzeugnisse aus unlegierten Stählen bis 3 mm Dicke – part 3 (01.91): Flacherzeugnisse aus nichtrostenden Stählen, Beuth Verlag, Berlin.

DIN 6935: Cold bending of flat rolled steel products, Beuth Verlag, Berlin (1975)

DIN 8586: Fertigungsverfahren Biegeumformen, Beuth Verlag, Berlin (1971).

DIN 17 440: Stainless steels; technical delivery conditions for plate and sheet, hot rolled strip, wire rod, drawn wire, steel bars, forgings and semi-finished products, Beuth Verlag, Berlin (07.85).

DIN 17 441: Stainless steels; technical delivery conditions for cold rolled strip and slit strip and for plate and sheet cut therefrom, Beuth Verlag, Berlin (07.85).

DIN 17 670: Wrought copper and copper alloy plate, sheet and strip; - part 1: Properties, Beuth Verlag, Berlin (12.83).

DIN 17 860: Sheet and strip of titanium and wrought titanium, Beuth Verlag, Berlin (11.73).

Doege, E., Besdo, D., Haferkamp, H., Tönshoff, H. K., Wiendahl, H.-P.: Fortschritte in der Werkzeugtechnik, Meisenbach Bamberg Verlag, Hannover, (1995).

Fritz, W.: Herstellen einbaufertiger Umformteile aus Stahlblech vom Band, Bänder Bleche Rohre (1994) 7-8.

Frontzek, H.: Beitrag zur Bestimmung der Reibungsverhältnisse in der Blechumformung, Dissertation, Darmstadt, (1990).

Guericke, W.: Theoretische und Experimentelle Untersuchungen der Kräfte und Drehmomente beim Richten von Walzgut auf Rollenrichtmaschinen, Dissertation, Magdeburg (1965).

Habig, K.-H.: Verschleiß und Härte von Werkstoffen, Hanser-Verlag, Munich (1980).

Hoffmann, H., Bareis, A.: Höhere Teilegenauigkeit – Wirtschaftliches Nuten von Elektroblech, Industrie-Anzeiger (1987) 77.

Hoffmann, H., Bareis, A.: Rotor- und Statorbleche für Elektromotoren flexibel fertigen, Werkstatt u. Betrieb 123 (1990) 7.

Hoffmann, H., Bareis, A., Waller, E.: Optimierte Schnelläuferpressen-Regelung der Stempeleintauchtiefe verlängert die Werkzeugstandzeit, Bänder Bleche Rohre (1988) 3.

Hoffmann, H.: Großteilpressen mit Saugerbrückentransfer, VDI Berichte Nr. 946, VDI-Verlag, Düsseldorf (1982).

Hoffmann, H., Schneider, F.: Wirtschaftlich Fertigen auf Großteil-Stufenpressen, Entwicklung, Bauarten, Produktionsmerkmale, Werkstatt und Betrieb 122 (1989) 5.

Hoffmann, H., Schneider. F.: Wirtschaftlich Fertigen auf Großteil-Stufenpressen, Werkstatt und Betrieb 122 (1989) 6.

Kellenbenz, R.: Anforderungen an schnellaufende Schneidpressen bei der Fertigung von Elektroblechen, Neuere Entwicklung in der Blechumformung, DGM Verlag, Oberursel (1990).

König, W.: Fertigungsverfahren, Volume 5: Blechumformung, VDI-Verlag, Düsseldorf (1990).

Lange, K.: Umformtechnik, Volume 3: Blechbearbeitung, Springer-Verlag, Heidelberg, (1990).

Lepper, M.: Flexible automatisierte Blechverarbeitung, Werkstatt und Betrieb (1992) 6.

Lepper, M.: Robot interlinked press line for car body panels, Sheet Metal Industries (1995) 5.

Mang, T.: Schmierung in der Metallbearbeitung, Vogel-Verlag, Würzburg (1983).

Oehler, G., Kaiser, F.: Schnitt-, Stanz- und Ziehwerkzeuge, 7th edition, Springer-Verlag, Berlin (1993).

Oehler, G.: Biegen, Carl Hanser Verlag, Munich (1963).

Pischel, H.: Neue Gesamtfertigung von Längsträgern für LKW-Chassis, Werkstatt und Betrieb (1995) 5.

Pischel, H.: Pressenstraßen optimieren komplexe Fertigungsabläufe, Werkstatt und Betrieb (1995) 3.

Reiss, A.: Walzrichten, Eigenverlag, 2nd edition (1990).

Rossner, S.: Vom Blechband zum Kompressor-Gehäuse, Werkstatt und Betrieb (1993) 6.

Siegert, K.: Innovative Pressentechnik, DGM Verlag, Oberursel (1994).

Stahl-Eisen-Werkstoffblatt 093: Kaltgewalztes Band und Blech mit höherer Streckgrenze zum Kaltumformen aus mikrolegierten Stählen, Verein Deutscher Eisenhüttenleute, Verlag Stahleisen, Düsseldorf (2nd edition 03.87).

Stahl-Eisen-Werkstoffblatt 094: Kaltgewalztes Band und Blech mit höherer Streckgrenze, Verein Deutscher Eisenhüttenleute, Verlag Stahleisen, Düsseldorf (1st edition 07.87).

Tschätsch, H.: Handbuch Umformtechnik, Hoppenstedt Technik Tabellen, Verlag, Darmstadt (1993).

Tschätsch, H.: Taschenbuch Umformtechnik, Carl Hanser Verlag, Munich (1977).

Weimar, G.: Zum gegenwärtigen Stand des Walzprofilierens, Bänder Bleche Rohre, 8 (1967) 5.

5 Hydroforming

5.1 General

One of the aims of the sheet-metal processing industry is the minimization of costs and the optimization of its products concerning weight, strength characteristics and rigidity. In search for alternative production processes, hydroforming – the manufacture of hollow bodies with complex geometries by means of fluid pressure – has been shown to offer an interesting technical and economic potential to sheet metal manufacturers. The achievement of beneficial component characteristics using this process is only possible where component and process configuration are selected by considering the overall system design.

5.2 Process technology and example applications

5.2.1 Process technology

The hydroforming technique is based on the inflation of, for instance, a tube, coupled with axial or radial compression and by subsequent expansion and sizing against the die wall. This process comprises in general of the following steps: expansion, compression, calibration.

Hydroforming is categorized as a cold forming process. It is used for the manufacture of geometrically highly complex hollow bodies from tubular or preforms. An underlying difference is drawn between free hydroforming and hydroforming using dies. The latter is subdivided again according to processes involving longitudinally and transversely divided dies *(Fig. 5.2.1)*.

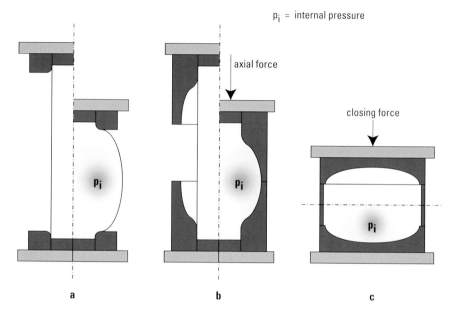

p_i = internal pressure

axial force

closing force

p_i

p_i

p_i

a   b   c

▲ **Fig. 5.2.1** Process variants of hydroforming: **a** free hydroforming; **b** tool-dependent hydroforming, transversely divided die; **c** tool-dependent hydroforming, longitudinally
divided die

The example used here to explain the hydroforming process is the
production of a T-fitting *(Fig. 5.2.2)*. This type of component can be
either a copper fitting or a joint element in a load-bearing structure. A
special hydraulic press is equipped with a two-part multiple purpose
die. Depending on the workpiece, the dies have two seal punches (horizontal cylinders) positioned axially relative to the tube ends and a
counterpressure punch. The tubular preform is in the bottom die and
the die is closed. The ends of the tube are sealed by the axial punches,
and the tube is filled with pressure medium. In the actual forming
process, the punches compress the tube, while the pressure medium is
fed to inflate the part until the part wall rests against the die contour.
The counterpressure punch controls additionally the material flow.
The calibration pressure forms the workpiece in such a way that its
contour corresponds to that of the die accurately and reproducibly. The
die is finally opened and the formed component is ejected.

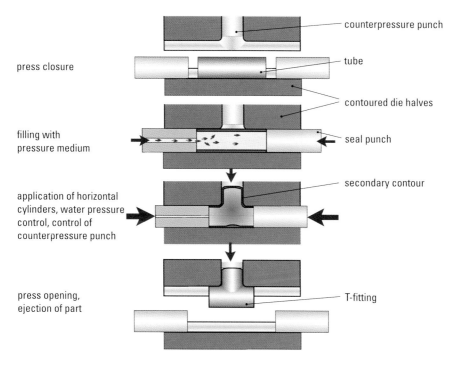

counterpressure punch

press closure

tube

contoured die halves

filling with
pressure medium

seal punch

application of horizontal
cylinders, water pressure
control, control of
counterpressure punch

secondary contour

press opening,
ejection of part

T-fitting

▲ **Fig. 5.2.2** Production of a T-piece

The component and material-specific process parameter curves *(Fig. 5.2.3)* are used to control the forming process. If an existing hydroforming line is intended for the manufacture of a different workpiece, the die is changed and a different program accessed in the machine's control system.

Expansion takes place under axial material flow, which is accomplished by compression in the axial or radial direction *(Fig. 5.2.4)*. Subsequently, the component which has already been largely finish formed is calibrated against the wall of the die.

The active forming forces are applied both by the moving die elements and also by the pressurized medium. In the die-related method, the formed part is pressed against the wall of a contoured die which thus determines the final shape of the part.

After hydroforming, the components possess a high degree of dimensional stability and high strength values. As a result of the three-dimen-

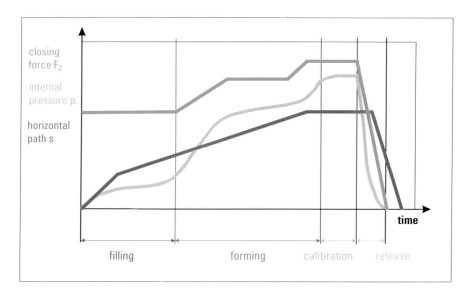

closing
force F$_Z$

internal
pressure p,

horizontal
path s

time

filling   forming   calibration   release

▲ **Fig. 5.2.3** Hydroforming process sequence

sional state of stress – with a relatively high hydrostatic stress component compared to deep-drawing and stretch drawing processes – relatively high degrees of deformation can be achieved (cf. Sects. 4.2.1,
4.2.4 and 4.2.5).

5.2.2 Types of hydroformed components

A basic difference is drawn between hydroformed components which
permit the following process conditions, in particular based on part
geometry and friction:

– continuous expansion and compression or
– only partial expansion and compression or
– only calibration

The differences are explained below:

These *components (Fig. 5.2.5)* are characterized by a high degree of
deformation due to the axial material flow which takes place over the
entire contour of the workpiece. They are expanded and compressed

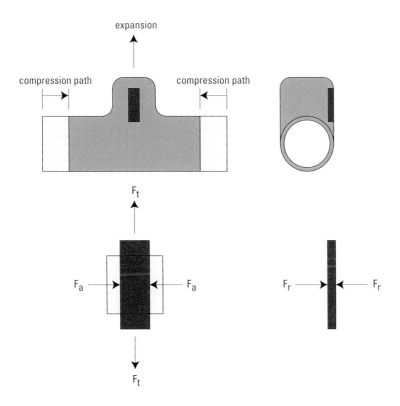

Fig. 5.2.4 Forces exerted on a tube section with a branch (T-fitting)

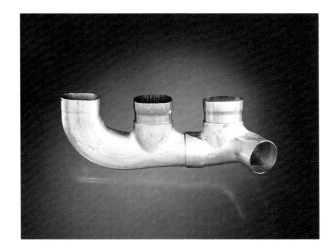

Fig. 5.2.5
Pure expansion/com-
pression operation

during the process. As a result of the axial material flow, higher degrees of deformation are possible than when forming only by changing the wall thickness. In some cases, local material compression results.

These *components* are characterized by a number of locally distributed bulges over the component length with or without intermediate bends *(Fig. 5.2.6)*. Axial material flow as a result of compression has an effect only at the component ends. In the case of shaped elements following bends or a marked bulge, positioned close to the horizontal cylinders, only expansion and calibration take place, but hardly any axial compression.

In the case of such *components,* axial compression is not possible for various reasons *(Fig. 5.2.7)*. A calibration process is applied here, whereby forming only takes place mainly by changing the wall thickness.

■ 5.2.3 Fields of application

The design and structural possibilities offered by hydroforming are used successfully in a number of applications for the manufacture of high-strength components and assemblies with optimized service life and weight characteristics.

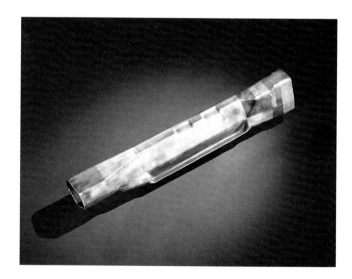

Fig. 5.2.6
Expansion/
compression and
local calibrating

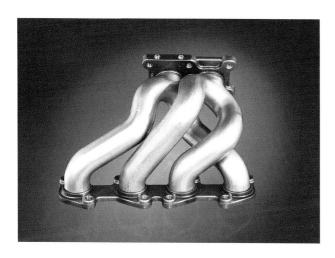

◁ **Fig. 5.2.7**
Pure calibration

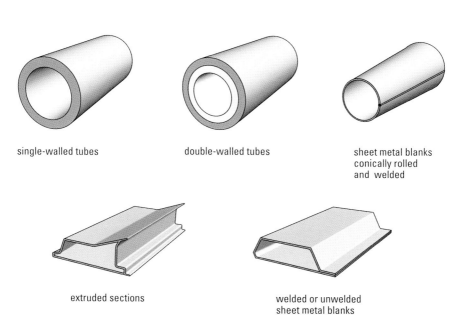

single-walled tubes   double-walled tubes   sheet metal blanks
                                                          conically rolled
                                                          and welded

          extruded sections   welded or unwelded
                                             sheet metal blanks

▲ **Fig. 5.2.8** Blank shapes

On the basis of the various semi-finished product shapes *(Fig. 5.2.8)*, hydroforming can be applied to a wide variety of fields. Three main user groups have emerged as the most significant over recent years (Table 5.2.1):

– the automotive engineering and supply industry,
– sanitary and domestic installation industry,
– pipe and pipe component manufacturers.

Table 5.2.1: Fields of application for the hydroforming technique

Sector	Assembly	Component (examples)
automotive industry vehicles – road – water – air – rail	chassis, exhaust and intake systems, add-on parts, drive system, seats, frames/bodywork, steering	cross members, side members, manifolds, roof rails, spoilers, gear shafts, seat frame components, A/B/C pillars, roof frame profiles, steering column with compensation
chemical, gas, oil industry, power station construction	piping and tank components, pipe fittings	T-fittings, reducers, housings, panelling
domestic appliance industry	fittings, machines	tube bends, T-fittings, reducers, cross pieces, bends
bicycle industry	pipe fittings	pedal bearings, joints, frames
processing of sections	frame construction, semi-finished pipes, rail-bound vehicles, utility vehicles	joints, girders, calibrated tubes, sections, roof arches, frames, structural members
pumps and fittings	housings	intake pipes
heating, ventilation, air conditioning	fittings	tube bends, reducers, T-fittings, manifolds
furniture industry	frames, moulded elements	legs, structural members, joints, shells, shelves
lighting industry	street lighting	light masts, lamp shells
optics	telescopes, torches	housings

5.3 Component development

5.3.1 User-oriented project management

In practical industrial application, it is increasingly important to design a component not only for functionality but also for producibility. "Simultaneous engineering" and "early involvement" have become important watchwords in the field of product design, with customer and process developer cooperating from an early stage in determining component configuration. Only in this way is it possible to optimize functional and process engineering related factors within a short time-frame. A component development activity can be, for example, carried out as follows:

1. preliminary meeting (customer and developer)

2. technical and economic feasibility study (customer and developer)

3. component configuration suitable for hydroforming (customer and developer) and definition of a production sequence

4. prototype development: CAD surface data generation (customer) and CNC milling data generation (developer), manufacture of a forming die and production of prototypes using different materials and wall thicknesses (developer)

5. try-out of prototypes (customer)

6. definition of quality-relevant (QA) data (customer)

7. series production (customer or supplier)

5.3.2 Feasibility studies

When developing hydroformed components, the precise analysis of boundary conditions is essential. Optimum design of components taking into consideration special process-specific factors enhances safety and also the cost-effectiveness of series production.

The feasibility study, the component configuration and definition of a production sequence are closely interlinked. Once these processes have produced a positive result, it is possible to start with prototype development of the components. When working through the individual steps, different boundary conditions must be taken into account. These are briefly outlined below.

Review of feasibility
When reviewing feasibility, a difference is drawn between two groups of components: new developments with a geometry that has been configured specifically with hydroforming in mind, and the production of existing components using hydroform technology.

Significant factors here include the geometry, wall thicknesses, materials and procurement of the semi-finished parts or preforms. Initially, the geometry data is evaluated with the aid of a CAD/CAM system. Circumferences, radii, cross sections and transitions are all determined. A decision must also be taken regarding the characteristics of the preform. In the case of a tube, for instance, these factors would include the material to be used, the diameter, the wall thickness and also any possible need for pre-forming.

The expected circumferential expansion is then determined. Taking into account the material data, the wall thickness, pre-forming of the tubular blank and other boundary conditions, it is possible to assess the manufacturing feasibility. Subsequently, the internal pressure, closing force, forces and movements of the horizontal cylinders have to be calculated. Since the correlations between all the different influencing variables are highly complex, specific expertise is an essential component of successful part configuration.

Part configuration
If it is possible to manufacture the part, the next stage is to design the details, the position in the die and the slide channels. In the case, when

a component does not appear to be producible using hydroforming technology, an attempt is made to modify the existing conditions in such a way that the part can be produced without modifying its functional characteristics.

The part geometry must be optimized taking into account the process limits. Where assemblies are produced, it may under certain circumstances be possible to group several parts into a single one or to split the assemblies up differently. A different material can also be selected. When making this type of modification, the question of reliability in series production must not be neglected, as this is of decisive importance concerning the cost effectiveness and process engineering related issues.

Configuration for series production

This stage includes the definition of a production sequence and configuration of the necessary production lines. Consistent implementation of the prototyping results is of particular importance here. An optimum product is only achieved through effective interaction of theoretical analysis and practice-oriented testing. Another aspect is the definition of quality-relevant QA data.

Forming simulation

The Finite Element Method (FEM) has become an established feature of metal forming technology. The objective of FEM is to replace costly and elaborate experimental testing by fast, low-cost computer simulation.

Any practical implementation of preliminary considerations must take into account the special features of hydroforming:

- consideration of partial strain hardening on pre-formed tubes,
- correlation between process parameters: motions and forces of the horizontal cylinders, variation of internal pressure in time, motion of additional die components such as hole punches and counterpressure punches, closing movement of the die,
- functional coordination of process parameters: internal pressure, different additional die elements such as hole punches and counterpressure punches and axial cylinders,
- springback,
- failure criteria: buckling and bursting,
- material behavior: anisotropy and flow curve and
- description of the contact (friction) conditions.

The following information can be computed using a suitably configured FEM program:

- equivalent plastic strain,
- effective stress,
- thickness distribution,
- distribution of plastic strain rate,
- principal strains and
- distance between outer wall of workpiece and the die wall.

5.3.3 Component design

In addition to the use of pre-bent tubes, the process technologies specific to hydroforming include the generation of cross-sectional changes,
flanges, breakthroughs, simple and multiple branches or the creation of
surfaces for centering welding operations. The key data indicated in
Fig. 5.3.1 describes the current state of the art for using tubular-shaped
blanks. The maximum achievable height of a bulge at a straight leg is
markedly higher than one on a bend, as the material is prevented from
further displacement by the geometry of the bend *(Fig. 5.3.2)*. A larger
branched tube height can be achieved if the branch is located near a
horizontal cylinder. In principle, sharp corners and edges should be

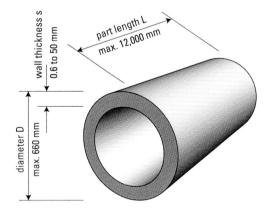

Fig. 5.3.1
Present state of the art in hydroforming using tubes as blanks

▲ **Fig. 5.3.2** Examples of achievable branched tube heights: the achievable height decreases with an increasing degree of difficulty

avoided when designing a component. Radii must be adjusted to the wall thickness *(Fig. 5.3.3)*. It is beneficial to provide gentle round transitions between different cross sections when this is allowable by the functional requirements.

Some of the fundamental guidelines to be observed when designing hydroformed components are outlined below.

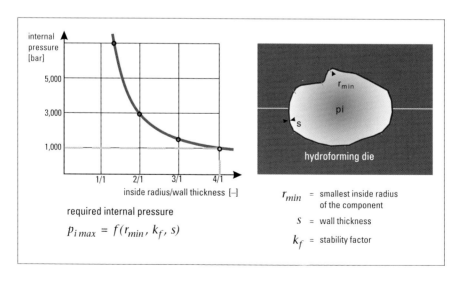

required internal pressure

$$p_{i\,max} = f(r_{min}, k_f, s)$$

r_{min} = smallest inside radius of the component

s = wall thickness

k_f = stability factor

▲ **Fig. 5.3.3** Internal pressure as a function of the inside radius

Straight component axis

The inflation limit is restricted by the risk of bursting and buckling. The likelihood of buckling increases with the length of the expanded section. With a buckling length ≤ 2 · the diameter of the tubular blank (starting diameter), and if a rotationally symmetrical cross section is being processed, the achievable deformation may be assumed to lie well beyond the ultimate elongation of the material.

Axial curvature

In the case of components with a curved axis in the inflation contour, axial displacement of material is possible without problems, provided the curvature radius is R ≥ 5 · the diameter of the initial preform. Partial inflation takes place here at the inner shell of the bend.

The free buckling length should not exceed the starting diameter, extend over no more than half the circumference of the tubular blank and at the same time it should be continuously decreasing.

In free forming without support, the component buckles if axial pressure is applied. The inflation capacity is limited by the ultimate elongation, as forming takes place purely as a result of the internal pressure.

Cross sections

During free expansion, the maximum deformation is reached when the free buckling length amounts to a maximum of 2 · starting diameter, and the cross section is rotationally symmetrical and located in the tube axis.

An asymmetrical cross section with a deep, narrow contour area can only be formed in cases where material is able to flow axially into the contour. Sharp contour transitions lead to partial and localized stretching in the wall and thus to premature bursting. Asymmetrical cross sections outside the tube axis can be formed in cases where at least half the circumference of the tubular blank is supported by the die contour.

Longitudinal sections

Contour transitions are best designed with large radii. Suddenly changing transitions in the direction of material flow lead to the formation of cross-wrinkles at the component (buckling) and act as a brake to material flow in the axial direction. Beyond these underlying guidelines, the following points should be observed in the configuration of hydroformed components:

Pure expansion/compression

The *optimum ratio between the pipe diameter D and pipe wall thickness* s is

$$D/s = 20 \dots 45$$

Here too, the *free buckling length* should be $\leq 2 \cdot$ the starting diameter D.

If *D/s > 45*, there is a danger of bursting or buckling. This can be prevented by reducing the free buckling length.

If *D/s < 20*, the buckling length can be greater with increasing wall thickness. Thinning of the wall thickness is prevented by axial displacement of the material. However, the same displacement can also lead to compression of the wall. Only a minimal danger of buckling exists.

Expansion/compression and local calibration

The maximum inflation capacity is limited by the ultimate elongation of the material. Half the ultimate elongation represents the limit of achievable deformation where no axial material flow is possible. In this case, the buckling length is of no significance.

Pure calibration

The limit of achievable deformation corresponds to half of the ultimate elongation of the material.

5.4 Die engineering

5.4.1 Die layout

Figure 5.4.1 illustrates the layout of a hydroforming die. The die set which comprises an upper and a lower die is used to accommodate the functional elements. The die holder plates permit adjustment of the assembly height prescribed by the press stroke or height of the cylinder holder brackets. The seal punches which are actuated by the displacement or pressure-dependent and controlled horizontal cylinders, they seal and push the tube ends during the compression process. The piston

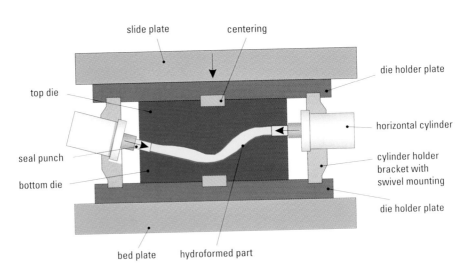

▲ **Fig. 5.4.1** Schematic view of a die layout

diameter and stroke are adjusted to the specific requirements of the component. The cylinder holder brackets absorb the axial forces and align the horizontal cylinders concentrically with the end guides. These can be moved and positioned on the bed or slide plate.

The modular basic structure of the dies permits flexible implementation of the required component geometries. The objective is to minimize production costs, die changing times and the time required to exchange worn parts. In addition, during the development phase it is possible to try out different tube materials and wall thicknesses without changing the dies. Thus, it is possible to test various preform geometries with a minimum of trouble and expense.

Simple changeover from prototype dies to production dies is possible if only cavity or sliding inserts have to be exchanged *(Fig. 5.4.2)*. When designing a prototype die, its service life is only of minor importance. Except for the sliding inserts, it is not necessary to have time-consuming coatings or other surface treatments of the die inserts. Prototyping – and thus step-by-step progress towards a die suitable for series production – takes place in several stages:

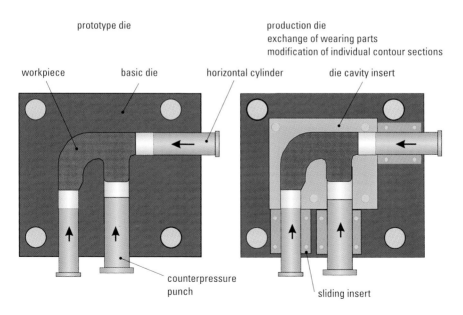

Fig. 5.4.2 Die systems for prototyping and production

- development of the basic component geometry,
- integration of supplementary operations such as hole piercing,
- production of modified inserts.

Here, the following parameters influence the selection of tool materials and the necessary coating or treatment:
- component geometry,
- component material,
- maximum internal pressure,
- sealing system,
- application (prototyping, series operation).

5.4.2 Lubricants

As is the case in other cold forming operations, lubricants are among the most important aids in the hydroforming process, as they minimize friction at locations with relative motion between die and workpiece (cf. Sect. 4.2.3). Many components can only be manufactured in the required shape only through the use of appropriate lubricants.

Lubricants permit to obtain more even wall thickness variation, simplify the forming of tubular branches, reduce partial stretching processes and improve forming in areas that do not allow any material flow by axial feed.

A number of different lubricants can be considered for use in hydroforming. Depending on their composition, these can be assigned to the following groups:

- solid lubricants, generally on a graphite or MoS_2 basis,
- polymer dispersion-based lubricants,
- waxes, oils and
- emulsions.

Lubricant is generally applied by means of spraying or immersion. Except in the case of oils, the applied coating must generally have dried and hardened prior to the forming process. On principle, attention must be paid to ensure an even coating thickness. The optimum thickness of the lubricant layer must be determined by trial and error.

5.5 Materials and preforms for producing hydroformed components

5.5.1 Materials and heat treatment

Basically the same materials which can be used for other cold forming techniques such as deep-drawing (cf. Table 4.2.3) are suitable for hydroforming applications. The material should have a fine-grained microstructure and be soft-annealed.

The best forming results are achieved using materials with a high degree of ultimate elongation, whereby the ultimate elongation at right angles to the longitudinal direction of the tube – i.e. generally at right angles to the rolling direction of the sheet metal – is important for the configuration of hydroform components. A high degree of strain hardening is beneficial.

The tube manufacturing process and pre-forming bring about a strain hardening effect which may have to be eliminated by annealing for certain components prior to the hydroforming process. If the formability of the material is exceeded during a production step in the manufacturing process, forming must be performed using several stages (progressive forming) with intermediate annealing. However, due to the additional costs involved, heat treatment should only be carried out when all other possibilities, for example design or material modifications, have been exhausted.

The preferable method of heat treatment for *austenitic* steels is solutionizing – ideally using inert gas. For small parts, a continuous furnace can be used.

In the case of *ferritic* steel, the type and practicality of heat treatment depend heavily on the respective material and its forming history. On principle, either of two possible processes can be used: recrystallization and normal annealing.

Generally, Mn or Si alloys are used for *aluminium*. The most favorable method of heat treatment must be determined in each individual case, whereby particular attention must be paid to the long-term behavior of the material, and in the case of strain hardening alloys, to rapid further processing.

■ 5.5.2 Preforms and preparation

For hydroforming, parts made of the following semi-finished products are used *(Fig. 5.2.8):*

– drawn or welded tubes,
– double-walled tubes,
– extruded profiles,
– welded or pre-formed tubular blanks.

In general, tubular preforms with circular cross sections are used. Welded and extruded non-round sections are used less frequently.

The production steps required for the preparation of preforms for technically and economically optimum production are described in the following with reference to tubular blanks.

The tube *cutting* process must be extremely precise in order to avoid undesirable effects such as leaks at the beginning of the hydroforming sequence. These would lead to failure of the component as a result of buckling. Fluctuations in the wall thickness along the periphery result in a high scrap level, and in extreme cases even to failure of the component. The following tolerances are generally admissible:

– length: +/– 0.5 mm
– angle of the cross section relative to the longitudinal axis: +/– 0.5°

Concentricity at the tube ends is important in order to avoid unwanted shearing when the die is closed. The tubes must be cleaned of any chips, as these lead to tool damage, forming errors or marks on the surface of the component. Tube ends must be deburred for the same reason, although chamfers must be avoided under any circumstances.

The ideal preform for the hydroforming process is a straight cylindrical tube. However, deviations from this shape are common, as the tube must be laid adequately in the die in order to ensure part quality. This necessitates a *pre-forming* process. On principle, any pre-forming process reduces subsequent formability and leads to differences in the wall thickness and faults in the forming behavior of the parts as a result of local strain hardening. However, suitable configuration of the component and optimum selection of the pre-forming process can counteract these negative effects.

Bending is performed on a mandrel, profile, cam or induction bending machine depending on the bending radius, tube diameter and wall thickness (cf. Sect. 4.8). Because of the notching effects that may occur, surface damage, in particular sharp-edged impressions created by clamping jaws, may lead to premature bursting of the component. The tolerance for the straight ends, as for cutting, is +/– 0.5 mm. Bending errors, such as wrinkles on inside bends, impressions left by clamping jaws or drawing grooves must be avoided, particularly in the case of thin-walled tubular blanks.

Reduction and expansion processes are used where large peripheral differences are required on long elongated components or in case of partial oversize of the used tube in the central section. Here, too, process-specific defects such as grooves, impressions etc. could have a detrimental effect.

Pre-forming in the hydroform die represents an ideal solution if it can eliminate any other pre-forming steps required on the straight tube. Pre-forming is also carried out on parts which have been previously bent, expanded or reduced.

Pre-forming in the die is used in the case of complex geometries, where the above processes fail to produce satisfactory results.

Special processes also exist for tapered, welded tubes, welded tubes with non-round sections, deep-drawn half shells, extruded sections and joined tubes. In the case of complex components, several pre-forming steps can be necessary.

5.6 Presses for hydroforming

Hydroforming presses are specially designed presses that have:

- the electronic system,
- the oil hydraulics,
- the pressure medium hydraulics and control system,
- the bed and slide

adapted to the special requirements of hydroforming. The press types used are either four-upright presses *(Fig. 5.6.1)* or frame presses *(Fig. 5.6.2)*. The four-upright construction guarantees optimum accessibility of dies for loading, process control and die component changeover. The die can be mounted outside the press and fed into the press using a die changing system. Inside the press, the die is locked in position by hydraulic clamps: All that is necessary to create the hydraulic and electrical connections. Die change can be executed depending on requirements at the press location on one of the four sides (cf. Sect. 3.4). The position of the horizontal cylinder is flexible.

Frame presses are used as single-purpose presses for the production of fittings. The horizontal cylinders of this type of press are permanently installed.

The closing force is applied by means of vertical plunger or differential cylinders (cf. Sect. 3.3). User-specific factors are taken as a basis for

- configuration of the plungers and differential cylinders,
- the stroke and diameter of the horizontal and counterpressure cylinders,
- dimensioning the pressure intensifier and
- determining the size of the bed.

Fig. 5.6.1
Four-upright hydroforming press
with a closing force of 25,000 kN

Controllable pumps fill the workpieces up to a defined initial pressure with pressure medium. Pressure levels exceeding this are generated by the pressure intensifier. The line control system is configured for at least four controlled hydraulic axes – one pressure intensifier, two horizontal cylinders and one counterpressure cylinder. All these axes can be controlled to be displacement or pressure-dependent. The closing process for coined bends is provided as the fifth axis. The number of axes can be increased, for example if different components or components with subsequent stroke have to be formed in a single press stroke.

All the necessary process data is entered at the host computer (cf. Sect. 3.5). The sequence of the forming process is executed in the form of a series of steps containing the data on various axes and any additional control commands. The pressure and position of all axes can be monitored during execution of the process on-line on the screen. They can also be recorded. The programmed data from the step chain can be viewed at any time in the form of a diagram. It is possible to draw up a

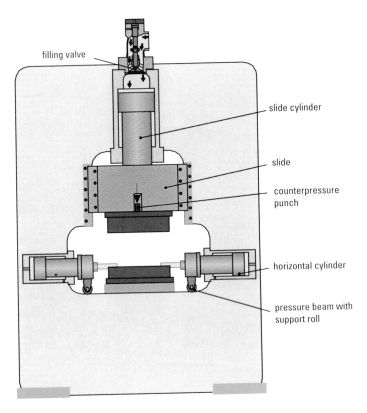

▲ **Fig. 5.6.2** Frame-type hydroforming press

response plan which corrects certain errors automatically using the pre-scribed measures where these occur.

Taking into account the necessary safety measures, the line can be loaded either fully automatically or manually.

5.7 General considerations

5.7.1 Production technology issues

Any comparison between hydroforming and other processes – such as compression moulding, assembly of stamped parts, casting or assembly of bent tubes – must take into consideration all the technical as well as economic aspects. Preceding and subsequent work steps and the properties of the finished component must also be included into any evaluation. Production by means of hydroforming can, for instance, appear uneconomical for small quantities if only the component costs are considered, as the cycle times required are relatively long (Table 5.7.1).

However, if we compare the technical properties of the end product with those produced using different techniques, it is possible to make up for this drawback. Where medium and high quantities of a component suitable for hydroforming are involved, the technical and economic benefits of hydroforming can be brought so effectively to bear that no other manufacturing processes can realistically be considered.

Table 5.7.1: Cycle times for different parts

Component	Automation	Components per stroke	Cycle time
exhaust manifold with dome	no	2	15 ... 25 s
exhaust manifold tube	yes	2 ... 4	15 ... 20 s
side member for pick-up	yes	2	40 s
instrument panel rail	yes	1	35 s
T-fittings	yes	up to 25	13 s

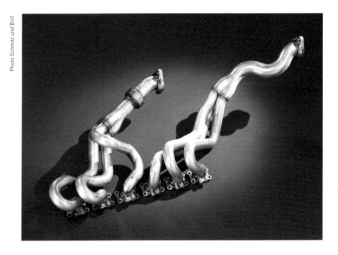

◁ **Fig. 5.7.1**
Hydroforming exhaust
manifold:
series production

Figure 5.7.1 and Table 5.7.2 compare a conventional with a hydroforming construction of an exhaust manifold.

Further processing of parts
Further processing of hydroformed parts includes any production processes which are executed directly on the component following the hydroforming sequence or which are used to join hydroformed parts together. If the geometry of the parts or specific requirements prevent them from being pressed to their final dimension, *finish processing* is necessary. The required final form is achieved using, for example, cutting, stamping or milling processes.

Table 5.7.2: Comparison of conventional and hydroforming production for the exhaust manifold illustrated in Fig. 5.7.1

	Conventional production	**Hydroforming production**
number of individual parts	17	9
service life	700 – 1,000 h	> 1,500 h
manufacturing costs	100 %	85 %
development time	100 %	33 %
flange type	varies	same
weight	100 %	100 %
scrap		< 0.5 %

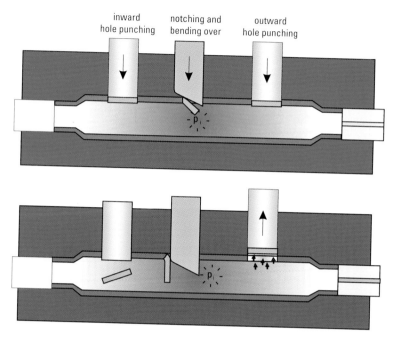

inward
hole punching

notching and
bending over

outward
hole punching

-p-

-p-

▲ **Fig. 5.7.2** Hole punching, notching and bending during hydroforming

In addition, *hole punching, notching, bending or threading* processes can be performed on the part in cases where this has not already been performed as part of the hydroforming sequence itself *(Fig. 5.7.2)*. By calibrating the part, it is possible to produce special contours at the ends of the tubes. A separate process is required, generally by means of welding, in order to *assemble add-on components* or to assemble the part with other components produced using different production processes. In the case of a *production lot-related multiple pressing*, components that are assembled together are formed in a single pressing process and must be separated by cutting.

5.7.2 Technical and economic considerations

The most important technical and economic aspects of hydroforming are that:

– Components with complex geometries can be manufactured which previously could not be produced in a single piece.

- The single-part construction eliminates the need for welding seams.
- The calibration process ensures a high degree of accuracy in the contours and dimensions of hydroformed components.
- Strain hardening generally lends hydroformed components a high degree of torsional rigidity. Their elastic recovery properties are less pronounced than is the case with welded sheet metal shells. For this reason, it is often possible for the wall thickness of hydroformed components to be relatively low compared to cast parts or shell constructions. This effect can be used either to reduce weight or to enhance strength or rigidity.
- Using the same die, it is possible to use and test tubes which, within certain limits, have different wall thicknesses and starting diameters and are made of different materials (provided the ultimate deformation is not exceeded). This allows the weight and strength properties of the part to be optimized.
- Experience has shown that lower flow resistance and larger fatigue strength are achieved in exhaust systems.

By selective component and process configuration, it is possible to save both energy and material and to reduce development and production times. Thus, generally it is possible to reduce manufacturing costs.

Bibliography

Leitloff, F.-U.: Innenhochdruckumformen mit dem Schäfer-ASE-Verfahren, Stahl (1995) 5.

Malle, K.: Aufweiten - Stauchen - Expandieren, VDI Journal 137 (1995) 9.

N. N.: ASE-Verfahren bringt Hohlkörper in Form, Stahlmarkt (1993) 4.

N. N. : Hochdruckumformen - Fertigungsverfahren mit Zukunft, Stahl (1993) 4.

N. N.: Mit Wasser immer in Form, Automobil-Industrie (1993) 4.

N. N.: Schäfer bringt Hohlkörper in Form, Technische Informationsbroschüre zur Schäfer-ASE-Technologie, Wilhelm Schäfer Maschinenbau GmbH & Co.

Schäfer, A. W., Fröb, K., Bieselt, R., Hassanzadah, G.: Kaltgeformte T-Stücke für den Anlagenbau – Anforderungen und technische Realisierbarkeit, 3R International 29 (1990) 9.

6 Solid forming (Forging)

■ 6.1 General

Processes and parts

The terms sheet metal forming and solid forming are based on practical industrial application, and subdivide the field of forming technology into two distinct areas. This distinction is made in addition to the definitions established by DIN 8582 "Forming" *(cf. Fig. 2.1.3)*. Solid forming entails the three-dimensional forming of "compact" slugs (for example sheared billets), while the sheet metal forming method generally processes "flat" sheet metal blanks to create three-dimensional hollow structures with an approximately constant sheet metal thickness. The forces exerted during solid forming are substantially higher than those necessary for sheet metal forming. As a result, it is necessary to use relatively rigid compact-design machines and dies. Solid forming involves not only the processes of extrusion, indentation and drawing which are dealt with in most detail here, but also for example rolling, open die forging *(cf. Fig. 2.1.4)* or closed die forging *(cf. Fig. 2.1.5)*.

The processes *cold extrusion* and *drawing* involve forming a slug or billet placed in a die, most commonly for the *production of discrete parts.* Hot extrusion uses a basically similar forming mechanism to create *semi-finished products such as rods, tubes and sections.* The basic processes involved in cold extrusion *(Fig. 6.1.1)* are classified depending on their forming direction as forward, backward and lateral extrusion and also according to the final form of the workpiece as rod (solid), tube and cup extrusion. When producing solid parts, the classical operations of cold extrusion technology are supplemented by processes like upsetting and ironing *(Fig. 6.1.2)*. The shape variety of parts formed from slugs or bil-

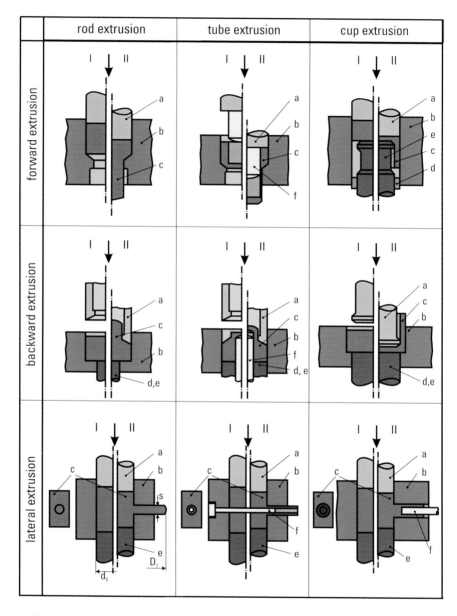

▲ **Fig. 6.1.1** Schematic representation of the cold extrusion (cold forging) processes:
I prior to forging;
II after forging (BDC);
a punch; b container;
c workpiece; d ejector;
e counterpunch; f mandrel

lets that can be produced today is increased substantially by the use of *piercing, flow piercing, trimming and bending processes.*

The range of parts includes not only *fastening elements* (screws, bolts, nuts, rivets, etc.), but parts for the *construction industry* (L and T-fittings for heating systems) and components used in *mechanical engineering* applications (ball bearing races, balls, rollers and cages). Most solid formed parts are used in the *automotive and motor cycle industry*. Engine components (valves, camshafts, cams, connecting rods), gear parts (gear shafts, gear wheels, synchronous rings, differential bevel gears), chassis components (stud axles, ball hubs, stepped housing elements, shock absorbers), steering components (steering column joints, tie rod ends), brake components (brake pistons) and components for electrical motors (starter motor pinions, drivers, pole rotors), can be economically produced by solid forming. In addition, solid formed parts are used in household appliances, office equipment, clocks or medical equipment. *Figure 6.1.3* offers an overview of different shapes which can be produced. The forming techniques described here are currently also used to produce *gears* that can be produced very economically. In many cases, no machining is necessary or at most one final grinding or hardening process may be required at the teeth. Components with straight teeth (spur gears) – such as hub sleeves, bevel wheels and gears –, helical teeth (e.g. planetary gears, small gears) and special teeth forms (e.g. ring gears) can be produced. This type of gears can be formed with any desired tooth that is often involute in shape.

Contours are formed on the surface using the indentation processes (DIN 8583 sheet 5) coining and roll embossing. Here, the complete

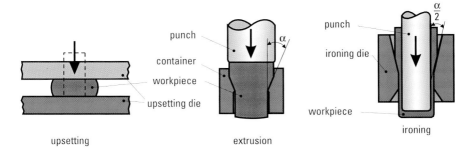

▲ **Fig. 6.1.2** Upsetting, open die extrusion, ironing

▲ **Fig. 6.1.3**　Range of cold, warm and hot forged parts

workpiece or the area to be embossed is enclosed in trapped (closed) dies. During embossing, the part thickness is reduced, in some cases by a considerable amount. Coining is a typical example of an embossing process *(cf. Fig. 2.1.6)*. Bimetal coins are a good example here, as they involve a joining process (DIN 8593) combined with coining and roll embossing. Embossing techniques are also used for example in the production of road signs, and in the cutlery or hand tool industry *(Fig. 6.1.4)*.

Cold coining of hot forged parts (sizing) (DIN 8583 sheet 4) is used to improve dimensional, shape and positioning accuracy and/or to improve surface quality. The objective is to press dimensions and shape to the required tolerances, and thus avoid the need for subsequent metal cutting operations. In contrast to indentation, when using cold coining, the component already has almost its complete geometry, meaning that the height reduction during cold coining is relatively small. Sizing of the bosses in connecting rods (which are required to have a tighter thickness tolerance compared to castings), and sizing of plier halves

▲ **Fig. 6.1.4** Examples of coined parts, such as coins, medals, cutlery and various hand tools, as well as cold-coined hot forged parts

(whose functional surfaces require particular positioning accuracy) are typical examples of this type of application.

Hot, warm and cold forming and process combinations
It is generally true that almost all solid forming processes can theoretically be applied at different temperatures. It is only when different factors come together to create interrelated complexities that either hot, cold or warm forming emerges as the ideal production method in any particular case (cf. Sect. 2.2.1). Above the recrystallization temperature, *hot forming (forging)* takes place. Forging is the oldest known meta forming technique. If the billet or workpiece is not heated prior to pressing, the method used is referred to as *cold forming (cold forging)*. This method did not become feasible for use with steel until the thirties, when a suitable separating layer between the die and the workpiece could be developed. *Warm forming (or forging)*, the last of the three techniques, developed for the industrial practice, was introduced in the eighties. For steels, for example, this method works at a temperature range of between 750 and 950 °C. For many parts and special workpiece materials with a high proportion of carbon, this method combines the technical advantages of hot forging with those of cold forging for optimum economical conditions.

The flow stress *(Fig. 6.1.5)* decreases with increasing temperature. Depending on the material being processed, when using the warm forging method, the flow stress reaches only around one half to a third of the corresponding values for cold forging. When hot forging, even lower flow stress values can be achieved, corresponding to as little as one fifth of the flow stress level during cold forging. Again depending on the material used, in warm forging and hot forging respectively forgeability is two or six times higher than that in cold forging. In the higher temperature range from around 700 °C upwards, the increase in scale formation which has a detrimental effect on dimensional accuracy and surface quality of the parts must be taken into account. The temperature required for warm forging must be carefully selected as in its "lower range" it is limited by the size of the press force and the achievable degree of deformation and in its "upper range" by scale formation.

Table 6.1.1 provides a comparison of the advantages of all three methods on the basis of processing engineering criteria. The weight of the components produced using hot forging ranges between 0.05 and 1,500 kg, in warm and cold forging between 0.001 and 50 kg. Dimensional accuracy and surface quality decrease with increasing

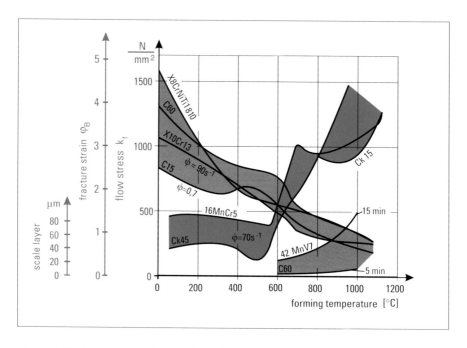

▲ **Fig. 6.1.5** Flow stress, scale formation and achievable degree of deformation (forgeability) in function of temperature

Table 6.1.1: Comparison between hot, warm and cold forging

Forming	Hot	Warm	Cold
workpiece weight	0.05 – 1,500 kg	0.001 – 50 kg	0.001 – 30 kg
precision	IT 13 – 16	IT 11 – 14	IT 8 – 11
surface quality R_z	> 50 – 100 μm	> 30 μm	> 10 μm
flow stress f (T, material)	~ 20 – 30 %	~ 30 – 50 %	100 %
formability f (T, material)	$\varphi \leq 6$	$\varphi \leq 4$	$\varphi \leq 1.6$
"Forming costs" VDW study 1991, Darmstadt	up to 113 %	100 %	up to 147 %
machining required	high	low	very low

temperature. This is due to thermal expansion, which cannot be predicted with accuracy, and also to increased die wear.

As an example, let us consider warm forging as the most economical process for selected parts, subject to technical feasibility. If this process is then assigned a cost factor of 100, the relative costs of hot forging can be up to 13% and of cold forging up to 47% higher. Therefore, in warm and especially hot forging the cost of removing additional material due to larger machining allowances must be considered. In the case of cold forging, in contrast, it is frequently possible to compensate completely for the "highest forming costs" by extremely low or even entirely eliminated finishing costs. In extrusion, therefore, cold and warm forging must be compared as the most economical process alternatives. With an increasing degree of geometrical complexity (higher degree of deformation, for example for steering knuckles), however, hot forging is still an economical process.

In practice, the "ideal processes" of solid forming defined here are also used as process combinations in three different respects. First, almost all complex workpieces are manufactured today by using multiple-stage combinations (process or processing step sequence) of individual operations. To produce a bushing, for example, a combination of the processes upsetting, centering, backward cup extrusion, piercing and ironing are applied in sequence. Second, it is possible for two processes to be carried out simultaneously in a forming operation. Using combined backward and forward cup extrusion, for example, tube-shaped workpieces with a central web *(Fig. 6.1.1)* can be produced. These and similar process combinations reduce the necessary number of processing steps and generally involve lower press forces than the sum of press forces used for individual operations. Such process combinations are more economical, but are only possible using multiple-action presses and dies.

Third, in forming processes where workpieces are initially hot or warm forged and then cold forged, "process combination" has also become an established term. Using this technology, it is possible to benefit simultaneously, for example, from (a) the economic benefits of warm forming to create "near-net shape" components and (b) the technical benefits of cold extrusion or cold forging, in particular its dimensional accuracy and surface quality. In this process, the cold forming stage is not by any means restricted simply to a sizing operation, but can include a multiple-stage forging process.

■ 6.2 Benefits of solid forming

■ 6.2.1 Economic aspects

The benefits offered by solid forming, compared with other production processes, can be summarized as superior quality combined with lower manufacturing costs. The standard of quality is due to the favorable mechanical properties of cold and warm forged parts, such as high strength and toughness, an uninterrupted fiber flow, close tolerances and a good surface quality. The cost benefit achieved using forging techniques can be considerable, but does depend on the specific part considered and on the previously used production method. Individual cost factors include:

– *Low material input:* Almost the complete initial volume of the billet is processed into the finished part. Compared to machining, savings can be as great as 75% *(Fig. 6.2.1).* Where high-alloyed materials are used, the benefit of low material input becomes increasingly significant in relation to overall manufacturing costs.
– *Use of low-cost raw materials:* Strain hardening that occurs during warm and cold forming results on the one hand in higher levels of flow stress, but also in higher ultimate and fatigue strength. Therefore, it is possible to use lower-cost steel grades with lower initial strength characteristics in order to achieve the same mechanical properties obtainable in machined parts. A workpiece with, for example, a hexagonal geometry is produced where machining methods are used from costly extruded material with a hexagonal cross section. By forging instead of machining, this type of part can be formed using substantially less expensive round bar as starting material.

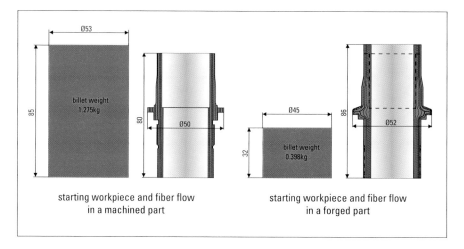

Ø53

85

billet weight
1.275kg

80

Ø50

starting workpiece and fiber flow
in a machined part

Ø45

86

32

billet weight
0.398kg

Ø52

starting workpiece and fiber flow
in a forged part

▲ **Fig. 6.2.1** Comparison of the input weight and achievable geometry for machining and forming processes

– *Reduction/elimination of subsequent machining:* Subsequent metal re-moval processes are frequently only necessary in the case of geometries which present a particular problem for forming processes, for example recesses, undercuts or threads. By using cold forging techniques, substantial savings can be achieved by reducing investment in machine tools for metal-cutting and in metal-operating staff.

– *Minimal process, logistic and transport costs:* This benefit is created as a result of the low cost of automation where transfer systems are used.

– *Generally high productivity:* Small workpieces produced from wire on horizontal forging machines can be manufactured at stroking rates of up to 200/min. For larger parts manufactured by forging from billet in vertical presses from billet, production speeds of up to 50 parts/min are possible.

– *Facility for integration of several functions/geometries in a single component:* Cold extruded parts often provide an opportunity for "re-engineering" to create lower-cost designs. The aluminium oil filter housing illustrated on the left in *Fig. 6.2.2,* for example, is cold extruded in a single piece. This allowed the production of three individual components – flange, drawn cup and connecting piece. Thus, the complex welding operation required previously *(Fig. 6.2.2, right)* could be reduced to a single extrusion operation.

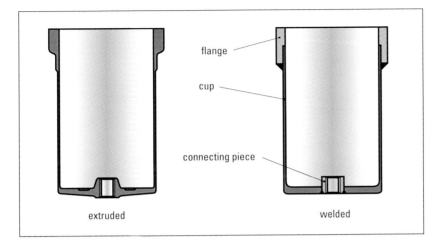

extruded

welded

flange

cup

connecting piece

▲ **Fig. 6.2.2** Oil filter housing – extruded single component compared to three-part welded configuration

■ 6.2.2 **Workpiece properties**

Due to the high degree of geometrical accuracy, surface quality and highly favorable mechanical properties, the workpiece characteristics of cold forged components are generally very good. The mechanical characteristics depend on physical factors and can only be influenced to a limited degree. The factors influencing operating accuracy in the production of precise geometries and surface properties, which are listed in Table 6.2.1, must be taken into consideration in the forge plant. In principle, the primary requirement is for a highly precise and relatively rigid forging press. If such a press is not available, even full compliance with all the other factors such as excellent die design cannot compensate for this deficiency. Next, in order of importance are the volume control of the billet, the process concept, etc. Compliance with and monitoring of all the process parameters thus ensure to achieve the required service characteristics of a component. However, it is necessary to consider that at the same time production costs also increase.

Despite the high degree of cost reduction achieved in automated forging lines, it is only with well trained staff, who are capable of ensuring correct operation of the equipment, that it is possible to achieve a satisfactory return on such a large capital investment. To highlight the importance of the human factor here, this has been included as a separate item in Table 6.2.1. The skill of the press operator is paramount in the successful operation of the production system. He/she takes the influencing factors into account and evaluates them.

Tolerances
Achievable ISO tolerances for different types of cold forging and complementary processes are summarized in Table 6.2.2. The "variation

Table 6.2.1: *Classification of process parameters that affect the geometrical accuracy and surface quality of a forged part*

Line:	Machine	kinematics, stiffness, off-center load capacity, gib precision, heating behavior (warm), natural frequency
	Automation	weight control, careful feed and discharge, positioning accuracy, transfer, press force control
	Heating	type, temperature, temperature control
	Cooling	temperature control
Billet:	Starting material	analysis, strength, microstructure
	Material form	wire, bar, thick sheet; rolled, peeled or drawn
	Manufacturing process	shearing, sawing, blanking
	Pre-treatment	possibly soft-annealing, pickling, phosphating, coating
Process:	Processing sequence	
	No. of steps	single-step, multiple-step, process combination
	Temperature	hot (h), cold (c), warm (w)
	Intermediate treatment	recrystallization, normalization (warm), pickling, phosphating, coating
	Lubrication and cooling	additional lubrication, spraying, flooding (warm), lubricant and coolant (warm)
	Clearance for part transport	
Die:	Precision	die concept, details, production process, adjustment of upper and lower die, mechanical die deflection, thermal die expansion
	Wear	die material, surface, coating, change interval
Human element:	Die designer, die manufacturer, operating staff, training and instruction	specialized skills

range" of a forging process indicated here should be interpreted so that smaller, lighter-weight components fall in general to the left-hand side of the bar (lower tolerance fields) and larger, heavy-duty components fall mainly to the right-hand side (wider tolerance fields). The red sectors are tolerances which can be achieved under normal conditions. The orange areas are ISO tolerances which can be achieved in special cases but with larger effort and cost. This information also indicates that the ironing and reducing processes permit the improvement of accuracy levels by as much as two quality levels. The high degree of dimensional accuracy is not achieved equally for all the dimensions of a component, but should be restricted to dimensions that are critical for functional or material removal considerations.

The achievable dimensional tolerances (normal, narrow) for cold extrusion or the process combination warm extrusion and subsequent cold sizing are indicated for the three most frequent forming processes, solid forward extrusion, cup backward extrusion and ironing (Tables 6.2.3 to 6.2.5). This information is given relative to the main dimen-

Table 6.2.2: ISO tolerances for the cold extrusion and complementary processes (VDI 3138)

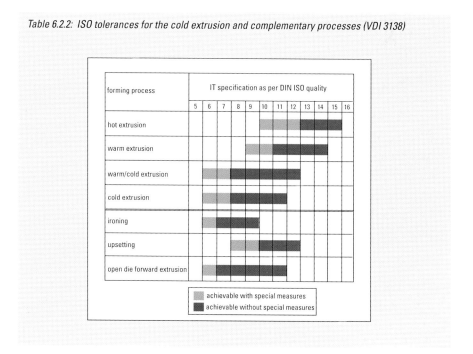

Table 6.2.3: Diameter, length and deflection tolerance of solid forward extruded parts, in mm

D up to mm	diameter D		L up to mm	length		deflection B	
	normal	narrow		normal	narrow	normal	narrow
25	0.10	0.02	150	0.25	0.20	0.15	0.03
50	0.15	0.08	200	0.75	0.50	0.25	0.05
75	0.20	0.15	300	1.50	1.00	0.35	0.07
100	0.25	0.18	450	2.00	1.50	0.50	0.10

Table 6.2.4: Diameter, concentricity, wall and web thickness tolerance of backward extruded cups, in mm

D up to mm	diameter				run-out	
	normal		narrow		normal	narrow
	D	d	D	d		
25	0.20	0.15	0.10	0.03	0.25	0.10
50	0.30	0.20	0.125	0.035	0.30	0.15
75	0.35	0.30	0.15	0.04	0.35	0.20
100	0.40	0.50	0.175	0.06	0.40	0.25

s, h up to mm	wall thickness s		bottom thickness h	
	normal	narrow	normal	narrow
2	0.10	0.075	0.25	0.15
5	0.15	0.10	0.35	0.20
10	0.25	0.15	0.45	0.30
15	0.35	0.25	0.55	0.40

Table 6.2.5: Diameter, concentricity, wall thickness tolerance in mm of tubular components produced by ironing

D up to mm	diameter				run-out		s up to mm	wall thickness s	
	normal		narrow		normal	narrow		normal	narrow
	D	d	D	d					
25	0.15	0.10	0.075	0.025	0.25	0.10	1	0.075	0.035
50	0.25	0.20	0.10	0.03	0.30	0.15	2	0.075	0.04
75	0.30	0.25	0.125	0.35	0.035	0.20	4	0.10	0.05
100	0.35	0.30	0.15	0.45	0.045	0.25	6	0.10	0.07

sions from *Fig. 6.2.3*. As already indicated, closer tolerances can be achieved as a result of special measures such as centering of upper and lower dies.

The length tolerances refer to the shoulder lengths within the upper or lower die, not to the overall length of the workpiece (including excess material).

Surface finish

On net shape and near net shape components which do not need rema-chining, the surface roughness is an important quality characteristic. The contact surfaces of the core of formed claw poles, for example, must not exceed a surface roughness of $R_z = 10\,\mu m$, as otherwise the required magnetic flux characteristics cannot be achieved. By the same token, a similar requirement applies to forged gears. Table 6.2.6 pro-vides a summary of the achievable surface roughness properties R_z and R_a for different forming processes. The orange areas are values which are measured before, the red areas are values which are measured after removal of the lubricant layer. Guideline values on the phosphate layer are $R_a = 0.3 - 0.8\,\mu m$ and after removal approx. $R_a = 0.6 - 2.5\,\mu m$.

The surface quality is positively influenced by a high specific pres-sure, high relative velocities and surface quality prior to forging. Forming processes with high specific pressures and relative velocities

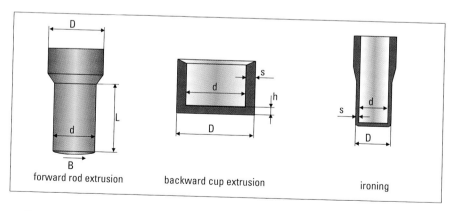

forward rod extrusion   backward cup extrusion   ironing

▲ **Fig. 6.2.3** Main dimensions in solid forward extrusion, backward cup extrusion and ironing process

Table 6.2.6: Surface qualities R_z / R_a achieved using different forging methods

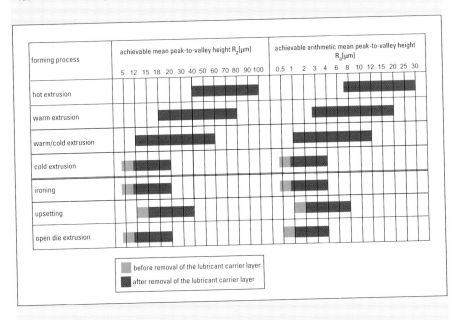

are open die extrusion, ironing and solid forward extrusion. The surface properties are influenced here by the following factors:

starting material: hot rolled, sand blasted, peeled, cold drawn
interface: sheared, sawn
intermediate annealing: scale free: under inert gas
with scale: sand blasted, pickled

Mechanical properties
The hardness, tensile strength and yield strength are increased by plastic flow. However, at the same time the values for notch impact strength, elongation and necking are decreased. Depending on the degree of true strain, in the case of steels with low and medium carbon content, tensile strength can increase by up to 120 %, yield strength by between 100 and 300 %, and hardness by between 60 and 150 %.

Where no heat treatment is carried out, this effect can be utilized in order to replace high-alloy steels by lower-alloy grades. During the development of the process plan, it is possible within certain limits to achieve specified strength values in certain sections of a component by suitable choice of the billet diameter and the process sequence.

6 Solid forming (Forging)

■ 6.3 Materials, billet production and surface treatment

Materials used in solid forming are mainly unalloyed, low-alloy and high-alloy carbon steels, non-ferrous light and heavy alloys such as aluminium, magnesium, titanium and copper and their alloys (cf. Sect. 4.2.2). Significant factors in the selection of materials are formability (flow stress), the path of the flow curve (cf. Fig. 6.1.5), the permissible fluctuations in material composition and issues related to billet preparation including preliminary (heat treatment, coating) and intermediate treatment.

■ 6.3.1 Materials

Steels
Low-carbon and low-alloy steels are particularly suited for *cold forming*. With a carbon content of up to 0.2%, the conditions for forming are very favorable, up to 0.3% favorable and up to 0.45% difficult, which results in low formability, true strain and higher press forces. Phosphorus and sulphur content should remain below 0.035%, as these components tend to reduce formability. The content of nitrogen is restricted to less than 0.01% for reasons of susceptibility to ageing. Pre-heat-treated steels can be used for parts which are sensitive to warping (valve tappet rods) and whose high strength (up to 1,100 N/mm²) permit only minimal deformation. In some cases, micro-alloyed steels containing boron, which permit penetration hardening, are replacing alloyed steels with the prescribed full heat treatment.

All steels which can be cold formed, can also be warm formed. However, it should be emphasized that even steels containing a higher carbon content, for example from 0.45 to 1.0%, can be easily formed, i.e. without costly intermediate heat treatment. Inductive hardening steels such as Cf 53, Cf 60 and the ball bearing steels 100 Cr 6 are classical examples of application for warm forming. For practice-oriented calcu- lation of force and energy requirements, a good approximation for the flow stress curve for hot forging can be obtained by multiplying the cold forging flow curve values with a factor of 0.3 to 0.5. *Figure 6.3.1* indicates the temperature-dependent variation of the flow stress curves for Ck 15 and Cf 53.

Table 6.3.1 provides a summary of the most important steel material groups. It lists not only the DIN and Eurostandard designations but also the abbreviations for similar materials in accordance with the American AISI and A.S.T.M standards as well as with the Japanese JIS standard. The last column provides information on whether the steel is suitable for hot forming (H), warm forming (W) or cold forming (C). For more information on workpiece materials, please refer to DIN worksheets 17006, 17100, 17200, 17210, 17240, 17440. Flow stress curves for cold forming are given in the VDI Guidelines 3134, 3200 sheet 2 (ferrous metals) and sheet 3 (non-ferrous metals).

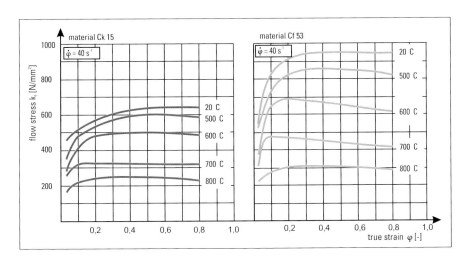

▲ **Fig. 6.3.1** Temperature-dependent flow curves Ck 15 and Cf 53

Table 6.3.1: Selection of steels for solid forming

Material group	No.	DIN 1654	Eurostandard	American standard AISI A.S.T.M.	Japanese standard JIS	Suitable for cold (C), warm (W) and hot (H) forming
general	1.0303	QSt 32-3	C4C	Armco		C
structural steel	1.0213	QSt 34-3	C7C		SS34	C
	1.0204	UQSt 36	C11G1C			C
	1.0160	UPSt37-2				H
	1.0224	UQSt-38	C14G1C		SS41	C
	1.0538	PSt50-2				H
case-hardening	1.0301	C10	C10	C1008	S10C	W, H
steels	1.1121	Ck10	C10E	C1010	S 9CK	C, W, H
	1.1122	Cq10	C10C	C1010	S 9CK	C, H
	1.0401	C15	C15		S15C	W, H
	1.1141	Ck15	C15E	C1015	S15CK	C, W, H
	1.1142	Cq15	C15C	C1015	S15Ck	C
	1.5919	15CrNi6				C
	1.7016	17Cr3				C
	1.7131	16MnCr5			SCr22	C, W, H
	1.7321	20MoCr4				C
heat-treatable	1.6523	21NiCrMo2				C
steels	1.1151	Ck22	C22E	C1020	S20C	C
	1.1152	Cq22	C22C	C1020		C
	1.0501	C35			S30C	W, H
	1.1181	Ck35	C35E	C1035	S35K	C, W
	1.1172	Cq35	C35C		S35K	C, W, H
	1.0503	C45			S40C	W, H
	1.1191	Ck45	C45E	C1045	S45C	C, W, H
	1.1192	Cq45	C45C			C
	1.1193	Cf45	C45G			W
	1.1213	Cf53	C53G			W
	–	Cf60	C60G			W
alloyed	1.5508	22B2				C
heat-treatable	1.5510	28B2				C
steels	1.5511	35B2				C
	1.5523	19MnB4				C
	1.7076	32CrB4				C
	1.7033	34Cr4		5130/-40	SCr1	C, W, H
	1.7035	41Cr4		5130/-40	SCr4	C, W, H
	1.7218	25CrMo4		4120/32		C, W, H
	1.7220	34CrMo4		4135	SCM3	C, W, H
	1.7225	42CrMo4		4140/-42	SCM4	C, W, H
	1.6582	34CrNiMo6				C, H
stainless steels	1.4016	X6Cr17				C, ferritic
	1.4006	X10Cr13	X10Cr13			C, martensitic
	1.4024	X15Cr13	X15Cr13			C, H, austenitic
	1.4303	X5CrNi18-12	X4CrNi18-12			C, austenitic
	1.4567	X3CrNiCu18-9	X3CrNiCu18-9-4			C, austenitic
roller bearing steel	1.3505	100Cr6		52100	SUJ1/2	C, W

Aluminium alloys

Wrought aluminium materials which have been manufactured by forming, not casting, are used for forming processes. Aluminium with a fine-grained microstructure offers a number of advantages over coarser-grained aluminium: Improved formability, superior surface quality of components and greater strength. Ultra-pure and pure aluminium (Table 6.3.2) permit an extrusion ratio (cross-sectional reduction) of up to 95 % (tube production). With an increasing magnesium content, it is possible to set higher strength characteristics. Age-hardenable, corrosion-resistant aluminium alloys such as A1MgSi1 exhibit high strength characteristics as a result of heat ageing which exceed the values achieved when in a strain-hardened state. Flow curves for aluminium materials are also given in the VDI directives 3134 and 3200.

Table 6.3.2: Selection of aluminium materials for cold and hot extrusion

Material group	No.	DIN 1712/25/45	ISO	AA	JIS
pure/ultra-pure aluminium	3.0285	Al 99,8	Al 99,8	1080A	A1080
	3.0275	Al 99,7	Al 99,7	1070A	A1070
	3.0255	Al 99,5	Al 99,5	1050A	A1050
non age-hardenable alloys	3.0515	Al Mn1	AlMn1	3103	A3003
	3.3315	Al Mg1	AlMg1	5005A	
	3.3535	Al Mg3	AlMg2,5	5754	
	3.3555	Al Mg5	AlMg5	5056A	
hardenable alloys	3.3206	Al Mg Si 0,5		6060 (6063)	A6063
	3.2315	AlMgSi 1		6082 (6061)	A6082
	3.1325	AlCuMg 1		2017A	A2017
	3.4365	AlZnMgCu 1,5		7075	A7075

Copper alloys

Like aluminium, copper and its alloys offer outstanding extrusion capability. The yield stress increases with increasing alloying elements, whereby forming should take place while the material is in a soft-annealed state. In the case of technically pure copper, mainly E-Cu and SE-Cu are used, for example, in the production of terminals. Among the variety of different bronzes used, the tin bronze with 1 to 2 % Sn and silicon bronze is most widely formed. If alloy elements such as chrome, zirconium, berylium or silicon (CuNi2Si) are present, many copper materi-

Table 6.3.3: Selection of copper materials for cold extrusion

Material group	Abbreviation	No.	DIN	US designation / UNS	JIS
pure copper	E-Cu F20	2.0060	1787	ETP 99,95 CO-0,040 / C11000	OFCu
	SF-Cu	2.0090	1787	DHP Copper / C12200	DCu
copper alloys	CuZn10	2.0230.10	17660	Commercial copper / C22000	BS( ), RBS ( )
	CuZn28	2.0261.10	17660	Cartridge brass 70 % / C26000	RBS ( )
	CuZn40	2.0360	17660	Muntz Metal / C28000	RBS ( )
	CuNi2Si	2.0855	17666	Silicon bronze / C64700	CN ( ) 1
	CuNi12Zn24	2.0730.10	17663	65Ni-12Ag / C75700	
	CuNi20Fe	2.0878.10	17664	Cu-20Ni / C71000	
	CuAl11Ni	2.0978	17665	Aluminium bronze / C63200	
	CuSn2	2.1010	17662	Phosphor bronze / C50500	

als are age-hardenable through heat treatment. The most important copper alloys are the brass family. The formability of the material improves as the copper content increases and the Zn content decreases. With a Zn content above 36%, the materials become brittle and their cold formability decreases rapidly. As a result, hot forming methods should preferably be used. Flow stress curves for copper alloys are given in the VDI Guidelines 3134 and 3200.

Other materials such as lead, zinc, tin, titanium or zirconium and their alloys can also be successfully formed. However, in terms of quantity and economic significance, these play only a minor role.

■ 6.3.2 Billet or slug preparation

The workpiece materials are specified by the user: Any modifications to the material, to exploit the benefits offered by cold forming (hardening) or to simplify the forming process, must be agreed upon with the die maker and the production engineer. In addition to the workpiece material itself, the nature of the semi-finished product, its separation to prepare individual billets, and the heat and surface treatment are all important points for consideration.

Semi-finished products
Wires, bars and tubes are used as semi-finished products for the production of billets or slug. Wire is delivered in coil form and generally

ranges from 5 to 40 mm in diameter. Wire diameters of up to 50 mm are processed in individual cases, although the handling capability of the wire bundle becomes increasingly difficult as the coils increase in weight.

Low-cost rolled wire in accordance with DIN 59 110 and 59 115 has a wide diameter tolerance and a rough surface finish. It therefore requires preliminary drawing on a wire breakdown machine. The reduction in cross section should be between 5 and 8%. The wire must be annealed, de-scaled and phosphated.

Drawn wire after DIN 668 K (K = cold drawn) is used to produce parts with a lower true strain (small reduction in area), as the material has already undergone a certain degree of preliminary strain hardening during the drawing process. This material also has to be descaled, phosphated and possibly also lubricated (stearate drawn). When a higher degree of true strain is involved (e.g. cup backward extrusion), after preliminary drawing, annealing (G) or spheroidized annealing (GKZ) must be performed in order to restore the original material formability. The form of delivery is specified with the designations K+G or K+GKZ. Light oiling helps prevent rust formation on the wire. In the case of high-alloy steel qualities, a further finish drawing process takes place after annealing with cross section reductions of less than 6%. The form of delivery is then specified as K+G+K or K+GKZ+K.

In practice, bar stock can also be used economically in the larger diameter ranges. With diameters of up to 70 mm, bars of between 6 and 12 m in length are produced. Hot-rolled round steel to DIN 1013 and DIN 59130 is the most favorable selection in terms of cost.

The allowable diameter tolerances cannot be reduced by drawing, leading to volume variations. These must be fully compensated for by excess material during the forming process. For a component weighing around 1.3 kg with a diameter of 46 mm, a weight tolerance of 88 g (average deviation) or 45 g (precision deviation) must be integrated into the die if the admissible tolerances are fully utilized. The transformation clearance in the first stage has to take account both, the diameter tolerance and also the ovality of the billet caused by the shearing process. For a material with a diameter of 46 mm, the first transformation clearance will be approx. 2.5 mm.

Where round steel is being cold forged, the descaling, phosphatizing and lubricating operations take place prior to forging. In the case of warm forging, the billets are fed directly to the heating line.

Bright round steel after DIN 970 and 971 is drawn or peeled following hot rolling, to tolerances h9/h11. The weight tolerance for the above given example is reduced to 5 and 9 g respectively, which, including an admissible length tolerance of 0.2 mm, corresponds to the customary weight tolerance of between 0.5 and 1%. By weighing the billets and adjusting the cut-off length, it is possible to achieve weight tolerances of < + 1%. Peeling of hot-rolled bar allows the elimination of surface inclusions and seams as well as any surface decarburization. Table 6.3.4 gives the diameter tolerances for different round materials.

Alternatively to bar material, tubular or ring-shaped formed parts can also be produced using tube sections as billets or preforms. Criteria to be considered here include wall thickness tolerance, concentricity and economy in comparison to producing these preforms from sheared billets through upsetting, backward cup extrusion, piercing, ironing and by using various intermediate heat treatments.

Where height-to-diameter ratios are h/d ≤ 0.2, production using sheared billets or blanks cut out of hot-rolled coil stock (DIN 1016), flat steel (DIN 1017), thicker sheet metal (DIN 1542/1543) is beneficial. This is the case particularly when irregular starting shapes (preforms) are required.

When processing wire, it is useful to have the preliminary treatment performed by the semi-finished product supplier, as a large investment is required for this type of equipment and the spectrum of required treatment types varies greatly.

Table 6.3.4: Diameter tolerances for round materials with selected diameters in mm

Diameter approx. in mm	Tolerances DIN 668 DIN 970/971 h9/h11	DIN 59110/59115 DIN 59130 DIN 1013 Usual deviation	DIN 1013 Precision deviation
15	– 0.04 / – 0.10	+/– 0.5	+/– 0.20
20	– 0.05 / – 0.13	+/– 0.5	+/– 0.20
25	– 0.05 / – 0.13	+/– 0.6	+/– 0.25
30	– 0.05 / – 0.13	+/– 0.6	+/– 0.25
35	– 0.06 / – 0.16	+/– 0.8	+/– 0.30
40	– 0.06 / – 0.16	+/– 0.8	+/– 0.35
45	– 0.06 / – 0.16	+/– 0.8	+/– 0.40
50	– 0.075 / – 0.19	+/– 1.0	(+/– 0.50)
80	– 0.09 / – 0.19	+/– 1.3	(+/– 0.60)

When using bar and flat material, the billets are generally sheared by the forger and subjected to heat and surface treatment. Strain-hardening effects which occur partially at the flat surfaces of the billet due to the shearing process are thus eliminated prior to forging.

Billet separation

The most commonly used billet separation process is shearing. Slicing on the trimming lathe is used only rarely, for example for test purposes. Blanking and fine blanking are explained in Sects. 4.5 and 4.7.

Shearing is characterized by practically loss-free separation at an extremely high level of output, in terms of quantity. The shear blade plastically deforms the material until its deformation limit in the shearing zone has been exhausted, shearing cracks appear and fracture occurs. With all four shearing principles – without bar and cut-off holder, with bar holder, with bar and cut-off holder and with axial pressure application – plastic flow lateral to the shearing direction is increasingly prevented, while compressive stress increases during the shearing operation. Both tendencies exercise a positive influence over the geometry (ovality, tolerance) of the sheared surface. The most accurate billets are produced using the shearing principle with bar and cut-off holder.

For cold forging, the sheared billets should have the greatest possible rectangularity, volume control and little plastic deformation. The sheared surfaces should be free of shearing defects and exhibit only a moderate amount of strain-hardening. The appearance of the sheared surfaces is the result of interactions between workpiece characteristics, tool, machine and friction. The shearing clearance exercises a major influence here. The greater the strength of the steel, the smaller is the shearing clearance *(cf. Fig. 4.5.12)*. However, aluminium and lead always require a small blanking clearance. The following values may be taken as a guideline for the shearing clearance of steel:

soft steel types	5 – 10 %
(of the starting material diameter in mm)	
hard steel types	3 – 5 %
brittle steel types	1 – 3 %

Rough fractured surfaces, tears and seams indicate an excessively wide shearing tool clearance. Cross fractured surfaces and material tongues

indicate an insufficient tool clearance. With increasing shearing velocity, the deformation zone reduces, the hardness distribution becomes
more uniform and the hardness increase in the sheared surface
becomes less pronounced, i. e. the material characteristics become more
"brittle" (exception: austenitic chrome-nickel and chrome-manganese
steels).

Open shears and closed shearing guides are used as tools *(Fig. 6.3.2)*.
Shearing guides produce very high quality blanks, but involve a number of drawbacks: Sheared billets are more difficult to eject, stops have
to be used and demands on the diameter tolerance of the starting material are more stringent. For round materials, the relative shearing clearance for each billet thickness to be sheared can be kept constant by
grinding an elliptical relief cut on the guides. The parts sheared in this
way demonstrate improved angularity and reduced burr formation at
the edge of the fractured surface. Moreover, the sheared billet quality is
less dependent on variations in material properties, which in turn
reduces variations in volume.

The shearing force F_S and shearing work W_S can be calculated
approximately when seperating round material with the diameter d
using the following formula:

$$F_S = A_S \cdot k_S \qquad\qquad W_S = x \cdot F_S \cdot s,$$

whereby A_S is the sectional surface to be sheared, k_S the shearing resistance of the billet material and s approx. 20% (hard, brittle materials)
to 40% (soft, tenacious materials) of the shearing stroke, i.e. the diameter d. The correction factor x indicates the extent to which the increase
in force deviates from a rectangular force-stroke curve. In general, x is
taken to be between 0.4 and 0.7. The shearing resistance k_S amounts to
approximately 0.7 to $0.8 \cdot R_m$.

Heat treatment of steel
Soft-annealing (spheroidized annealing) plays a particularly important
role in the heat treatment of steel materials used for billet production.
Recrystallization annealing, normalization and under certain circumstances also recovery annealing are typical heat treatments used for
intermediate annealing of already formed workpieces whose formability has been exhausted.

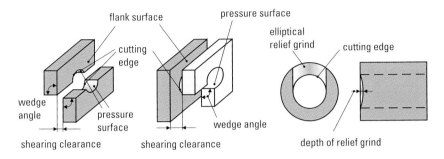

▲ **Fig. 6.3.2** Open and closed shearing knives and a closed shearing knife ground with elliptical relief

The *soft annealing* process spheroidizes any existing harder flaky elements of the microstructure. While the hardness is retained, the flow stress (cf. Sect. 2.2.3) of the material is reduced.

Recrystallization annealing is advantageous for austenitic steels with a low carbon content, as with this heat treatment the lowest flow stress levels are obtained. However, if coarse grain occurs as a result of locally low levels of true strain or deformation (cf. Sect. 2.2.2) during recrystallization annealing, and if this type of coarse grain is not permissible, a normalization process should be carried out.

As a result of two-fold recrystallization (perlite + ferrite <=> austenite) an even, fine-grained and fine-lamellar perlite microstructure is achieved through *normalization*. This can be re-annealed if required to create grained perlite (soft-annealing). A marked difference in microstructure (coarse and fine grains) can also result during warm forming as a result of locally differing cooling conditions and degrees of true strain. These differences can cause cracks during subsequent cold sizing operations (reducing, ironing). This problem can be avoided by normal annealing.

Recovery annealing offers a low-cost alternative to recrystallization or soft annealing, provided the deformation in the subsequent forming operation is minimal, i.e. the level of true strain is not too large.

■ 6.3.3 Surface treatment

A difference is drawn between abrading and depositing surface treatments. Abrading processes include chemical and mechanical cleaning methods, degreasing and descaling. Depositing processes include phosphating, soaping, molycoting and the use of oils. The phosphate layer serves as a carrier layer which facilitates the adherence of lubricants to the surface. Soaping is used for simpler forming operations, molycoting for extremely high degrees of deformation (cf. Sect. 2.2.2) and for parts such as gears. In the case of horizontal multi-station presses operating with wire, oil is used for lubrication and cooling. In vertical multi-station presses, oil is used as a supplementary lubricant in addition to soap/molycote (spray or flood application) (cf. Sect. 6.8).

Cleaning, degreasing and descaling
Chemical cleaning and degreasing processes are performed either through immersion or spraying of solvents, or through the condensation of solvent vapor on the workpieces. Solvent types used include organic solvents (hydrocarbons, petroleum ether, petroleum), chlorinated hydrocarbons (trichloroethylene and perchloroethylene), and water-soluble cleaning agents such as acids, acid saline solutions and alkaline solutions. The mechanical methods used include sand blasting or tumbling using lime, sand, steel shot and other materials. The choice of cleaning methods and agents depends on the type and extent of the contamination, the required degree of purity and the type, shape and quantity of the items to be cleaned. Environmentally friendly water-soluble cleaning agents are frequently used.

Chemical descaling (pickling) is performed using pickling baths containing sulphuric acid or hydrochloric acid solutions. The pickling agents serve to release hydrogen on the substrate metal, causing the oxide layers to split away. As the absorption of hydrogen in the substrate metal leads to embrittlement, as short a pickling time as possible should be used. Crack formation during forging can, for example, be the result of excessive pickling treatment. Sand blasting and tumbling are used for the mechanical removal of thick scale layers. Here, the brittle scale layer is parted off from the tough substrate material through partial application of force.

Phosphating
Phosphating is taken to mean the generation of cohesive crystalline phosphate layers which are firmly adhered to the substrate material. For cold forming processes, mainly zinc phosphate and in some cases

iron phosphate layers are used. Phosphoric acid, zinc phosphate and oxidizing agents are applied in an aqueous solution. On immersion in the solution, the metal surface is pickled by the phosphoric acid, during which process insoluble zinc sulphate is simultaneously precipitated. The chemical reaction does not come to a standstill until the complete surface is coated with zinc phosphate crystals. The oxidizing agent transforms the dissolved metal to a phosphate with low solubility which is precipitated in the form of sludge and has to be removed periodically.

In practice, phosphate layers with a mean thickness of around 8 to 15 mm are applied, which leave behind a shiny coating after forming. The phosphate layer can resist temperatures up to 200 °C. Over 200 °C, a partial transformation takes place. At over 450 °C, complete decomposition of the phosphate layer takes place. Depending on the lubricant used, however, working temperatures of up to 300 °C can be realized during forging.

Soaping and molycoting

Soaping takes place by immersion of the parts in an alkaline soap solution. The main component of these solutions is sodium stearate. A part of the zinc phosphate layer reacts with the sodium soap to form a water-insoluble zinc soap layer. The zinc soap is additionally coated with the sodium soap. This combination of layers has a very low shear strength and thus reduces the effective coefficients of friction (cf. Sect. 4.2.3). Soaping should not take longer than 5 min, as otherwise the entire phosphate layer will be used up and no longer provide an adhesive substrate.

To make sure that the soap coating offers good lubrication properties it is necessary to dry the coated billets completely. A waiting period of one day previous to processing of the treated blanks is usually recommended. The soaps are temperature-resistant to approx. 250 °C, above which chemical transformation processes take place. Excessive soaping results in soap deposits building up on the dies.

Molybdenum disulphide is used under difficult forming conditions. Compared to soap lubricants, molybdenum disulphide causes greater friction losses which mean that larger press and ejector forces have to be taken into consideration. However, it is capable of withstanding far greater levels of stress, i.e. the lubricant film only breaks away under extremely high degrees of surface pressure, and permits considerable surface expansion of the workpiece. Therefore, molybdenum disulphide is used when large deformations, i.e. large true strains, are present. In addition, molybdenum disulphide creates fewer problems as regards the accumulation of surplus lubricant in the dies. The upper application temperature limit is around 350 °C. At higher temperatures, molybdenum trisulphide is created, which is ineffective as a lubricant.

Molybdenum disulphide is also applied in an immersion bath, and in some cases sprayed or applied by tumbling. The coated billets should be stored for one day prior to processing. A storage period of one week should not be exceeded.

As a rough guideline, the overall sequence of surface treatment with the respective functions of different baths, bath composition, times and temperatures is summarized below:

1. Degreasing:	Chemical degreasing in a 5% sodium hydroxide solution at 80 to 95 °C for around 5 to 10 min.
2. Cold rinsing:	Removal of sodium solution residues, cold rinsing in running water for approx. 2 min (water surface must always be clean).
3. Pickling:	Pickling in 12% sulphuric acid or 18% hydrochloric acid at approx. 50 to 65 °C, duration approx. 10 to 18 min (acid content is correct if the pickle effervesces).
4. Cold rinsing:	Removal of acid residues, cold rising in running water for approx. 2 min.
5. Hot rinsing:	Hot rinsing to pre-heat the parts – influences the phosphate layer thickness, 70 – 90 °C, for around 2 to 3 min (at lower temperatures, a lower layer thickness results).
6. Phosphating:	Mainly using zinc phosphate, for rustproof steels with iron oxolate, 70 to 80 °C for around 8 to 10 min (phosphating comprises an initial pickling reaction and a subsequent layer forming reaction, layer thickness approx. 8 to 15 mm).

7. Cold rinsing:	Cold rinsing in running water for approx. 2 min.
8. Hot rinsing:	Hot rinsing to pre-heat the parts at around 60 °C, duration approx. 3 min.
9. Soaping:	Soaping at 60 to 85 °C for approx. 5 min (lower temperatures result in a thicker, lower temperatures in a thinner layer).
10. Drying:	Drying with hot air at approx. 120 °C for approx. 10 min.

In the case of non-ferrous metals, it is not necessary to use lubricant carrying layers. Cleaning and lubrication is all that is required. In the case of copper, where low degrees of deformation are required, lubrication can be eliminated altogether.

Oil as a lubricant
While soaping and molycoting processes are mainly used with billets, in multi-station presses, when forging from wire, mineral-based oils with additives (EP additives) are generally used to improve lubrication properties. The oil also adheres better to the rough phosphate layer. With this type of production, the lubrication of the flat face surfaces presents a problem because these are not coated with a phosphate layer due to the shearing process. The oil is sprayed or flooded into the dies or onto the workpieces.

■ 6.4 Formed part and process plan

Before producing a part by forming, processes (cold, warm, hot), process combinations, processing steps, preliminary treatment, process plan and intermediate treatment stages must all be defined. In addition, the force and energy requirements, as well as relevant die stress levels have to be computed. Only on the basis of these calculations it is possible to define the press configuration concerning functional characteristics such as press force, number of stations, distance between stations, stroking rate, length of slide stroke and automation related questions.

In solid forming applications, processing frequently takes place at the limits of physical feasibility for economical reasons. This applies for example regarding the limits of material formability (crack formation, strain hardening), the load capability of dies (internal pressures of up to 3,000 N/mm^2) and lubrication/cooling (cold welding). For this reason, the economical application of solid forming depends to a large degree on the experience of the designer and the die maker. With an efficient combination of suitable finite element programs for process simulation and existing factory floor experience, in metal forming plants a balanced relationship can be said to exist between theory and practice.

■ 6.4.1 The formed part

Lot size and materials
Solid forming represents a competitive substitute for machining processes and in some cases also for casting processes (cf. Sects. 2.1.1 and 6.2.1). The high development costs and the degree of investment

required for presses and dies mean that certain minimum production quantities are necessary. On the other hand, because of increasing raw material and energy costs, the high degree of material utilization possible using solid forming is expanding the limits of economic viability continuously towards smaller batch sizes. Even for complex parts, this tendency means that batches as small as 10,000 parts per year can still be economically produced.

A precise feasibility study is only possible by considering the individual part concerned and by making cost comparisons with machining processes. Nevertheless, the annual production lot sizes listed in Table 6.4.1 can be considered as the minimum economically viable batch sizes with respect to part weight.

Design rules
When designing a formed part, there are two areas of consideration: a) geometrical aspects of the finished part and b) the limitations of the individual processing operations utilized in the total process sequence. On the basis of the finished (assembly ready) part geometry, first of all a cold or warm forged part is developed, taking into account the following geometrical aspects:

- machining allowances;
- rotationally symmetrical geometries are very well, axially symmetrical geometries well suited for cold and warm forming;
- wall thicknesses can be pressed no lower than 1 mm for steel, no lower than 0.1 mm for aluminium;
- draft angles, used in hot forging, are generally not required for cold and warm forged parts (inside, outside);
- small holes with length to diameter ratios $l/d > 1.5$ cannot be economically produced;

Table 6.4.1: Minimum economical batch sizes for solid forming when compared with machining

Weight	Production quantity per year	
	Cold forming	Warm forming
0.10 – 0.25 kg	200,000	300,000
0.25 – 0.75 kg	150,000	200,000
0.75 – 2.50 kg	25,000	100,000

– small variations in diameter on the inside and outside surface of the formed components should be avoided;
– material overflow must be provided for, due to fluctuations in billet weight (e. g. during forward rod extrusion in the rod length, during backward cup extrusion in the cup height);
– transition radii must be configured as large as possible; sharp edges can only be pressed by using split dies;
– undercuts can be produced, but increase tooling costs;
– it is generally not possible to press plunge cuts, parallel key grooves, transverse bores etc.;
– slip-on gearing can be produced by pressing, high-precision involute gearing (running gears) can be produced with increased die and process development effort.

The process engineering factors depend on the extrusion technique, the selected processing steps, the workpiece material, the permissible die loading and the respective tri-axial stress state. Process engineering considerations may restrict the range of part geometries that can be formed from billet (cf. Sect. 6.5).

Volume calculation
Following transformation of the end product (assembly ready part geometry) into a formed part geometry, volume calculations are made for the relevant geometry. In practice, the volume is calculated by dividing the overall part geometry into simple volumetric elements, such as cylinders, truncated cones or rings, whose calculated individual volumes are added to obtain the total part volume. The calculation must take into account punching and shearing scrap which allows for material overflow during the forming process. Experience has shown that the volume calculation of transition radii can be neglected, as these volumes partially cancel each other out. A far more significant influence is due to volumetric variations that result from the elastic deflection and expansion of forging dies or material overflow which is difficult to calculate. Complex rotationally symmetrical 2D geometries are calculated easily using CAD and according to Guldin's law. Extremely complex 3D geometries are handled using the volumetric calculation modules of 3D CAD systems.

■ 6.4.2 Process plan

The formed part and the relevant volumetric calculation represent the starting point for developing the process plan. A creative process is initiated in which a wide range of boundary conditions – such as starting diameter, material, prescribed strength levels, lot size, heat treatment, number of stations, distance between stations, press force, off-center load capacity of the press and internal operating circumstances – must be considered simultaneously.

On the basis of the calculated volume, the billet dimensions (preferred diameter and length) are defined. For many parts, particularly rod-shaped parts, the billet diameter is selected from a diameter which already exists in the part to be formed. This must take into account the clearance necessary to transport and locate the part from one station to the next, and the preferred diameter. For all flat parts which are produced by upsetting operations, the upset ratio and the length/diameter ratio of the billet must be considered, as these influence the possibility of buckling.

Starting with the geometry of the formed part, development work proceeds to determine the previous stages, working backwards towards the starting billet. Initially, rough process plan alternatives are sufficient here. These preliminary process plans are used as a basis for defining the forming processes. Volumetric adjustment is not yet necessary at this stage, as only the degree of true strain is calculated and compared with the corresponding process limits. If the existing conditions are critical, it may be possible to alleviate these by changing billet dimensions and the sequence of operating stations. If this is not possible, intermediate heat treatment is unavoidable. If one, two or more intermediate heat treatments are necessary, warm forming can offer another more economical alternative.

Once a promising process plan alternative has been found as a result of this intuitive procedure, volume calculations are carried out. Furthermore, the feasibility is verified based on force and energy calculations and on the computation of the die load.

If the use of process combinations means that there are several processes being performed in a single station, reliable prediction of the material flow is difficult even for experienced forming engineers. In this case, metal flow simulation with the aid of a suitable finite element program can serve to illustrate the metal flow in the relevant forming process. This allows to predict and eliminate problems such as crack formation, underfilling and excessive die loads prior to actual die try-out *(Fig. 6.4.1).*

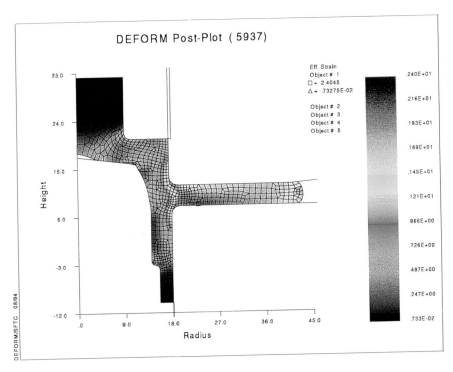

▲ **Fig. 6.4.1** FEM study of material flow for a flange upsetting process with relief borehole

6 Solid forming (Forging)

■ 6.5 Force and work requirement

Determination of the force and work requirement is used as a basis for selection of the forming machine and the structural design of the forming dies. In practice, calculation of these characteristics simultaneously produces values of process-related parameters such as the overall degree of true strain, flow stress, forming pressure, compressive stresses on the inside wall of the extrusion container, etc. (cf. Sect. 2.2). The process limits and the respective case of application of the forming method must also be taken into consideration here.

The formulae provided below are based on only one method of calculation and make no claim of completeness. The flow stress levels are average values obtained from the flow curves of VDI 3200 Sheets 2 and 3. To calculate the process variables concerning machine selection, the upper limit of a flow curve should be selected. However, in making calculations regarding die configuration, the lower limit of a flow curve should be selected. Table 6.5.1 provides an overall view of the process limits for the main extrusion processes.

■ 6.5.1 Forward rod extrusion

The forward rod extrusion method transforms a billet into a solid part with a reduced cross section (cf. Fig. 6.1.1). The die opening which determines the shape of the extruded parts is formed solely by the press container. As a result of the wall friction which occurs – also during the ejection process – the ratio of height to diameter of the initial billet h_0/d_0 [–] should be no greater than 5 to 10 (Fig. 6.5.1). If the required

Table 6.5.1: Process limits of the most important extrusion methods taking into account economical die life

Process	Forward rod extrusion			Forward tube extrusion			Backward cup extrusion		
material / limiting value	$\varepsilon_{A\,max}$	φ_{max}	$(h_0/d_0)_{max}$	$\varepsilon_{A\,max}$	φ_{max}	$(h_0/d_0)_{max}$	$\varepsilon_{A\,min}$	$\varepsilon_{A\,max}$	$(h_2/d_1)_{max}$
NF-materials aluminium (e.g. Al99,5), lead, zinc	0.98	4		0.98	4		0.10	0.98	6
copper (E-Cu)	0.85	1.9		0.85	1.9		0.12	0.80	4
brass (CuZn37 - CuZn28)	0.75	1.4		0.75	1.4		0.15	0.75	3
steels easy to form (e.g. QSt32-3, Cq15)	0.75	1.4	10	0.75	1.4	15	0.15	0.75	3
difficult to form (e.g. Cq35, 15MnCr5)	0.67	1.1	6	0.67	1.1	12	0.25	0.65	2
more difficult to form (e.g. Cq45, 42CrMo4)	0.60	0.9	4	0.60	0.9	8	0.35	0.60	1.5

dimensions are greater than these values, free or open die extrusion must be used as a supplementary extrusion process. In addition, the ratio h_0/d_0 in the remaining (non-extruded) portion of the billet must be no smaller than 0.5. Otherwise, the so-called "suck-in" defects may occur.

The maximum true strain achievable with the forward rod extrusion of steel is around $\varphi = 1.6$ depending on the material. In the case of materials with low formability, this value is as low as 0.5 to 0.9. The die opening angles $2\,\alpha$ [°] must be within a range of approx. 45 and 130°, which results in tapers on the formed part with angles between 22.5 and 65°.

The part volume is calculated from the individual volumetric values as illustrated in *Fig. 6.5.1:*

$$V_0 = V_1 + V_2 + V_1',$$

whereby V_0, V_1, V_1' and V_2 are calculated by means of

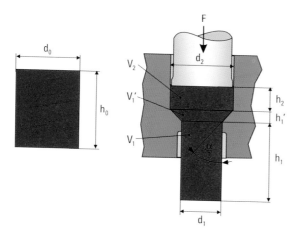

$$V_0 = d_0^2 \cdot \frac{\pi}{4} \cdot h_0 \qquad\qquad \left[mm^3\right]$$

$$V_1 = d_1^2 \cdot \frac{\pi}{4} \cdot h_1 \qquad\qquad \left[mm^3\right]$$

$$V_1' = \frac{\pi \cdot h_1'}{12} \cdot \left(d_2^2 + d_1^2 + d_2 \cdot d_1\right) \qquad \left[mm^3\right]$$

$$V_2 = d_2^2 \cdot \frac{\pi}{4} \cdot h_2 \qquad\qquad \left[mm^3\right]$$

In this calculation, it is possible to use $d_2 \approx d_0$ as an approximation; the height of the truncated cone h_1' and the forming stroke h are determined by means of:

$$h_1' = \frac{d_2 - d_1}{2} \cot \alpha \qquad\qquad h = h_0 - h_2 \quad [mm]$$

The amount of deformation (strains) can be determined from the degree of true strain φ and the specific cross section change ε_A. The latter can be obtained quickly from the relevant squares of the diameters instead of using the surfaces:

$$\varphi = \left(\ln \frac{d_2^2}{d_1^2}\right)$$

$$\varepsilon_A = \frac{A_0 - A_1}{A_0} \cdot 100\%$$

$$\varepsilon_A = \frac{d_2^2 - d_1^2}{d_2^2} \cdot 100\%$$

The ratio between billet height h_0 and billet diameter d_0 and the die opening angle are taken into account with the correction coefficients k_h and $k_{2\alpha}$ [–] when estimating the specific punch pressure and thus also when calculating the forming force.

$$k_h = 0.2 \cdot \frac{h_0}{d_0} + 0.8 \qquad k_{2\alpha} = \frac{1}{200} \cdot (2\alpha) + 0.7$$

The flow stress at the start of the forming process k_{f0} [N/mm²] and towards the end of the forming process k_{f1} can be derived from the sheets of VDI Guideline 3200 for the material used in each case. Both values are used to estimate the specific and ideal forming work w_{id}:

$$w_{id} \to \approx \frac{k_{fo} + k_{f1}}{2} \cdot \varphi \qquad \left[\frac{Nmm}{mm^3}\right]$$

As the actual forming force deviates from the theoretically ideal forming force, it is corrected by using the effiency factor η_F [–].

$$\eta_F = \sqrt{0.703125 \cdot \frac{\varepsilon_A(\%)}{100}}$$

The average forming pressure p_{St} [N/mm²] and the forming force F_U [kN] are thus estimated by using:

$$p_{St} = \frac{w_{id}}{\eta} \cdot k_h \cdot k_{2\alpha} \qquad \left[\frac{N}{mm^2}\right]$$

$$F_U = p_{St} \cdot d_2^2 \cdot \frac{\pi}{4} \cdot 10^{-3} \qquad [kN]$$

The forming work W is the product of forming force and forming stroke h

$$W = F_U \cdot h \qquad [J\ resp.\ Nm]$$

For die design, the pressure on the container wall p_i [N/mm²] is determined:

$$p_i = \frac{1}{2}\left(p_{St} + k_{f0} + k_{f1}\right) \qquad \left[\frac{N}{mm^2}\right]$$

Example:
The sample calculation provided here is based on the workpiece material 1.0303 (QST 32-3). The starting values are specified on the left, the characteristic values subsequent to the pressing operation on the right.

$d_0 = 75$ mm $\quad d_1 = 45$ mm $\quad a = 45°$
$h_0 = 110$ mm $\quad h_2 = 65$ mm
$\quad\quad\quad\quad d_2 = 75.3$ mm
$\quad\quad\quad\quad => h = 45$ mm

The resulting deformation or ideal effective strain is $\varphi = 1.03$, which corresponds to a relative cross section change of $\varepsilon_A \approx 64$ %, and the correction coefficients are $k_h \approx 1.09$ and $k_{2\alpha} \approx 1.15$:

$$\varphi = \ln\left(\frac{75.3^2}{45^2}\right) = 1.03 \qquad\qquad \varepsilon = \frac{\left(75.3^2 - 45^2\right)}{75.3^2}\cdot 100\,\% = 64\,\%$$

$$k_h = 0.2\cdot\left(\frac{110}{75}\right) + 0.8 \approx 1.09 \qquad k_{2\alpha} = \frac{1}{200}\cdot 2\cdot 45 + 0.7 \approx 1.15$$

If the amount of deformation is 1.03, the resulting value for k_{f1} is approx. 740 N/mm², (k_{f0} 250 N/mm²), meaning that the specific forming work is approx. 471 Nmm/mm³. With an efficiency of

$$\eta_F = \sqrt{0.703125\cdot 64/100} = 0.67$$

the specific die loads, force and work result in:

$$p_{St} = 510/0.67\cdot 1.09\cdot 1.15 \approx 957\,\left[N/mm^2\right]$$

$$F_U = 957\cdot 75.3^2\cdot \pi/4\cdot 10^{-3} \approx 4,261\,[kN]$$

$$W = 4,261\cdot 0.045 \approx 192\,[kJ\ resp.\ kNm]$$

$$p_i = 1/2\,(957 + 250 + 740) \approx 974\,\left[N/mm^2\right]$$

■ 6.5.2 Forward tube extrusion

Forward tube extrusion involves the production of a tubular part with a reduced wall thickness starting *(cf. Fig. 6.1.1)* from a cup or cylinder. The die opening, which determines the shape of the produced part, is formed by the press container and the punch. If the mandrel is perma-nently mounted in the punch, it is called a moving mandrel: As a result of material flow, additional frictional forces occur which can lead to critical levels of tensile stress in the mandrel. This can be avoided by using a moving mandrel, that is moved, within certain limits, by the deforming material. Friction-related tensile stresses occur during for-ward extrusion with the moving mandrel and during stripping of the workpiece in both forward tube extrusion methods. Therefore, geomet-rical conditions must be set such that mandrel load does not exceed 1,800 N/mm². This is generally the case when the ratio of height to diameter in the starting material h_0/d_0 does not exceed 10 to 15. The achievable levels of deformation depend on the material and corre-spond to those achievable in forward rod extrusion. Thus, comparable (shoulder) geometries can also be formed (Table 6.5.1).

■ 6.5.3 Backward cup extrusion and centering

A (thin-walled) tubular part is produced from a solid billet. The die opening which determines the shape of the part is formed by the punch and container, or the die *(cf. Fig. 6.1.1)*. In the backward cup extrusion of steel, a certain limiting ratio of penetration depth to punch diameter $h_2/d_1 = 3$ to 3.5 cannot be exceeded. Depending on the mate-rial, the maximum achievable degree of true strain is $\varphi = 0.9$ to 1.6 ($\varepsilon_A = 60$ to 80%). However, economical tool life is only achieved with somewhat lower levels of true strain (Table 6.5.1).

Due to a maximum admissible punch stress of $p_{St\,max} \leq 2,300$ N/mm² when processing steel, a minimum level of true strain of at least $\varphi = 0.16$ to 0.43 ($\varepsilon_A = 15$ to 35%) must be ensured, i.e. it is not possible to back-ward extrude a cup with any optional internal diameter relative to the outside diameter. It may be possible to produce small bores with addi-tional secondary geometries (e.g. a protrusion for extrusion piercing), provided these relieve the material flow into the cup wall.

The minimum achievable wall thicknesses for steel are around 1 mm, for aluminium around 0.1 mm. Extremely thin walls can be produced by a subsequent ironing process following backward cup extrusion. For steel, the minimum cup bottom thickness is 1 to 2 mm, for NF metals 0.1 to 0.3 mm. The bottom thickness should always be greater than the wall thickness (underfilling). In the case of thick-walled cups which are additionally expected to satisfy high concentricity requirements, it is advisable to carry out a centering operation prior to cup extrusion. Centering is a process very similar to backward cup extrusion, although with only a small deformation stroke. Centering is generally performed using guided punches.

6.5.4 Reducing (open die forward extrusion)

The principle of reducing is similar to that used for forward rod extrusion as far as the die design is concerned (cf. Fig. 6.1.2). However, in contrast to forward rod extrusion, the billet material is not pressed against the container wall. Thus, only a limited degree of deformation is possible without buckling. During the forming process, the material must neither be deformed (upset) in the container nor buckle. In the case of starting billets with $h_0/d_0 > 10$, reductions in cross section can only be achieved using this process. Larger amounts of deformation require several successive reducing operations. The achievable level of deformation depends on the material, the preliminary state of strain hardening and the die opening angle. An overall level of deformation (calculated from the first to the nth reducing operation) of 50 % should not be exceeded due to the formation of central bursts (chevrons). Reducing or open die forward rod extrusion is a process which is not used in practical warm forming operations, as the achievable degree of deformation is too low due to significantly reduced flow stress in warm forming.

With a die opening angle of approx. 20°, it is possible to achieve a maximum level of deformation of $\varphi = 0.3$. If the material used has already been strain hardened, for example as a result of a previous reducing operation, a somewhat higher degree of deformation is possible. However, with an increasing die opening angle, the achievable level of deformation drops substantially.

Die angles range between 10 and 45° and generate tapers with shoulders between α = 5 and 22.5° on the pressed part *(cf. Fig. 6.1.2)*. The punch load is of minor importance, and depends on the billet material. It should not upset the billet plastically, and cause buckling stresses in slender billets. Neither of these stress values should be exceeded. The process is highly sensitive to fluctuations in the strength of the starting materials and to the quality of lubrication, i.e. phosphating and soaping.

■ 6.5.5 Ironing

Ironing is a process whereby a tubular part with a bottom (e.g. backward extruded or drawn cup) is drawn to reduce the wall thickness *(cf. Fig. 6.1.2)*. This is a process in which the force is introduced mainly through the bottom of the workpiece and partially by the mandrel friction force. As a result, in addition to compressive stress at the die shoulder inlet, tensile stress also occurs after exit from the die in the wall of the workpiece (drawing process). This is in direct contrast to reducing, in which the force and thus also the compressive stress occurs in the workpiece cross section before the die shoulder inlet (press through process). The part shape is generated by the drawing mandrel and the ironing (die) ring. It is also possible to arrange several ironing rings one after the other.

The maximum degree of true strain achievable when ironing steel is around φ = 0.5 (ε_A = 40%) depending on the material characteristics. For difficult to form materials this value drops to φ = 0.35 (ε_A = 30%). Die opening angles may range up to around 45°, which results in tapers up to 22.5° in the finished part *(cf. Fig. 6.1.2)* Customary die opening angles for ironing range between 5 and 36°. A minimum punch force is achieved at an ironing angle of around 12°, when the probability of wall fracture is minimum. Ironing is a process in which an extremely good surface quality and excellent tolerances are achieved, and it is, therefore, used frequently as a sizing operation.

■ 6.5.6 Upsetting

Upsetting is defined as "free forming", a process in which the part height is reduced, generally between flat parallel die surfaces (upsetting dies) *(cf. Fig. 6.1.2)*. If upsetting is carried out in a closed die without flash formation, for example in order to press cylindrical or lens-shaped pre-forms, this is referred to as sizing.

(Free) upsetting *(cf. Fig. 2.2.1)* is characterized by three process limits: Depending on the material, the maximum true strain of $\varphi = 1.6$ can be achieved. The upsetting ratio h_0/d_0 (height to diameter of the starting workpiece) may not exceed 2.3 for single pressing, 4.5 for dual pressing and 8.0 for triple pressing, as otherwise buckling may occur and the fiber flow is interrupted *(cf. Fig. 6.2.1)*. The maximum die pressure is 2,300 N/mm^2.

In contrast to upsetting in which the exterior contour and dimension are free formed, in sizing the workpiece is fully enclosed (trapped die). The same process limiting values exist as for free upsetting. However, the same magnitude of the upsetting punch load is reached at lower levels of deformation, as the material leans against the die wall.

■ 6.5.7 Lateral extrusion

Lateral extrusion generates shapes with flanges, collars or other geometric features (teeth, protrusions), whereby the die opening which gives the workpiece its shape remains unchanged during the entire forming process *(cf. Fig. 6.1.1)*. This is achieved by closing the dies with a high force before the slide reaches the bottom dead center. After the dies are closed, the material is pressed into the impression by the penetrating punch. It is possible to produce for example flanges with a diameter ratio $D_1/d_0 \leq 2.5$ in a single forming operation. The ratio between flange thickness and starting s/d_0 should not exceed 1.4 because the material may separate at the neutral surface.

The required forces are substantially lower than for upsetting, in particular when pressing secondary geometric features. To date, the actual force requirement has been estimated only through experimental tests. Because of the state of compressive stress, acting on all sides, local levels of true strain of up to $\varphi = 5$ have been achieved using this non-stationary process.

6 Solid forming (Forging)

■ 6.6 Part transfer

Cold and warm forming are mass production processes whose high output levels and economy are achieved in particular by automation. Automation in this context includes all devices which, in addition to loading the wire coil or filling the containers with sheared billets, permit the transfer of the billet, the preformed part between forming, and where applicable any scrap (piercing, trimming operations) within the press.

   Installations that use wire coil have the advantage that once the wire end has been fed in, the billet material always assumes a defined position during the straightening (cf. Sect. 4.8.3), drawing and shearing (cf. Sect. 6.3). The material is transported within the press by means of a mechanical transfer device, and the parts leave the press following the last forming station by sliding down through a chute. In installations that use sheared billets, the billets must be fed in, oriented, possibly weighed and conveyed to the pick-up point of the transfer device. The main focus of investigation part handling is the transportation of workpieces from one station to the next within the press. The objective is to utilize kinematic factors and die layout so to:

- guarantee absolutely reliable part transfer – this entails preventing the part from assuming an undefined, i.e. free position at any time –,
- ensure that no collision is possible either between the die elements or between die elements and grippers during idle stroking of the press (e.g. on start-up or when running the press empty), and
- minimize contact times, in particular in the case of warm forming.

There is a broad range of possible solutions for part transfer, including transport devices in the broader sense, such as tilting mechanisms for billet containers, steep conveyors (pusher or chain conveyors), vibrating conveyors, separating devices and scales. Therefore, these will not be dealt with here. This chapter will concentrate instead on the loading station and the transfer study for a mechanical cold forming press with 2D transfer (cf. Sect. 4.4.6).

■ 6.6.1 Loading station

The wide diversity of parts that must be fed into a forming press calls for a number of different loading station concepts. Where cylindrical billets with a length/diameter (L/D) ratio greater than 0.8 are being processed, they are loaded from above into the feeding station by means of a feed channel *(Fig. 6.6.1, top)*. The bottom-most parts are held in a holder, generally cylindrical in shape, that is provided with recesses on the side to allow for the closing movement of the gripper. Following the removal of the first part, the weight of the column causes it to drop onto the base plate at the level of the transport plane. Where billets with different dimensions are fed, if necessary the feed channel and the exchangeable parts or the complete feed unit can be exchanged with the aid of quick-acting clamps. The grippers are equipped for cylindrical parts with prismatic active gripper elements whose height should be approx. 50% of the billet length to ensure a reliable gripping action.

If the L/D ratio is below around 0.8 to 1, or if formed parts with relatively flat geometry are being processed *(Fig. 6.6.1, bottom)*, then the parts are conveyed into the loading station side by side in specially designed feed channels. The parts are positioned initially below the transport plane, and are raised by a pusher to transport level in synchronization with the press cycle. After the pusher returns to its starting position, the next part slips into the loading station to be raised to the transport plane. The starting height and feed angle of the feed channels are configured in such a way that the billets continue to slide under their own weight. In the case of formed parts, the grippers are designed for shaped fit if a prismatic gripper exhibits an unfavorable L/D ratio and there is a danger that billets could rotate while being held by the gripper.

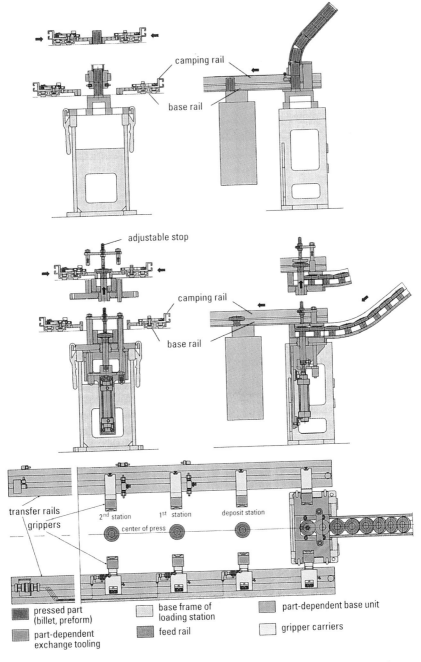

▲ **Fig. 6.6.1** Loading station concepts for cylindrical *(top),* disc-shaped or flat shaped parts *(bottom)*

For cold forming processes, the gripper units are mounted on clamping rails which are fixed in turn on a base rail *(Fig. 6.6.1)*. When resetting for producing a different part, all that has to be exchanged are the clamping rails including the pre-set grippers. The front grippers are equipped with proximity switches to monitor part transfer.

Parts with a higher L/D ratio (approx. 3 to 8) must be transferred between stations by a 3D transfer. This involves the use of an additional horizontal pusher which moves the part together with its retainer into a position from where it can be raised vertically. Here, too, the level of the base plate on which the parts are positioned is located below the transport plane. Shaft-shaped parts with a billet L/D ratio of > 6 can also be moved into the uplift position using special devices such as rotating sleeve retainers.

In warm forming, the loading station has the additional function of segregating parts which either have not been fed in correct synchronization with the press, or which have an incorrect temperature. To allow these parts to be ejected, an opening is released in the base plate of the loading station through which the parts drop down into a chute for removal.

■ 6.6.2 Transfer study

Before conducting a transfer study, data on the kinematics of the slide movement must be available in the form of a time-displacement diagram. The stroke is defined based on the range of parts, the required forming process and the die layout. The slide curve is given by the stroke height and press kinematics (eccentric drive, knuckle-joint drive, modified knuckle-joint drive) *(cf. Fig. 3.2.3)*. The slide curve, whose bottom dead center is at 180°, can only be displaced vertically for transfer study.

The ejector stroke is determined by the range of parts being processed and the respective position of the forming stations in the die. The largest necessary ejector stroke is equal to the sum of the inlay depth of the part in the die and the part length. The ejector stroke can be achieved by means of a mechanical and an additional pneumatic displacement. The stroke covered mechanically must eject the part, which is stuck in the tool as a result of elastic deflection of the container. In addition, the pneumatically generated displacement is able to raise the part. Our present example has a mechanical ejector curve with a stroke of 200 mm which is travelled through at 80° crank angle: In this case, no pneumatic ejector is required. The starting point for the ejector curve can be optionally selected, allowing the curve in the diagram *(Fig. 6.6.2)* to be displaced horizontally. The slope of the ejector curve can be configured in such a way that it corresponds approximately to the linear portion of the slide curve.

Starting with the transverse movement of the connecting rod, the slide ejector curve is generated by the corresponding kinematic *(Fig. 6.6.2, dashed line)*. For the transfer study, this curve can be displaced only vertically.

The curve of the pneumatic ejector, whose actual distance from the slide curve must be determined with the transfer study, runs parallel to the slide curve *(Fig. 6.6.2, broken line)*. The slide ejector acts mechanically briefly after the bottom dead center and can be operated pneumatically after that.

In addition to these main press motions, in universal transfer devices the curves for opening and shutting the grippers are also significant. The opening and closing times can normally be adjusted within certain limits, e. g. within a 30° range, making the curves horizontally displaceable in the transfer study. The opening stroke is determined by the required diameter of the upper die (Fig. 6.6.2, opening stroke 100 mm). Where mostly slender punches are used, this stroke is smaller than for female dies used on the side of the slide, for example when producing long rod-shaped parts or parts in closing devices. Depending on the opening stroke, these movements require a smaller (approx. 30° for around 40 mm) or a larger (approx. 60° for around 100 mm) crank angle range. Through the forward and reverse movement, lateral transportation takes place. Forward motion can be initiated as soon as the grippers engage the part, and must have been completed before the grippers open, i. e. the upper die elements make contact with the transported part. The forward and return motions require a crank angle of approx. 80 to 100°. The return motion takes place during the infeed and pressing cycles.

Transport studies of 3D transfer devices involve a lift-up motion in addition to the movements previously described. This motion starts briefly after the ejector movement and after the closing of the grippers. It ends with the lowering motion which should have been completed

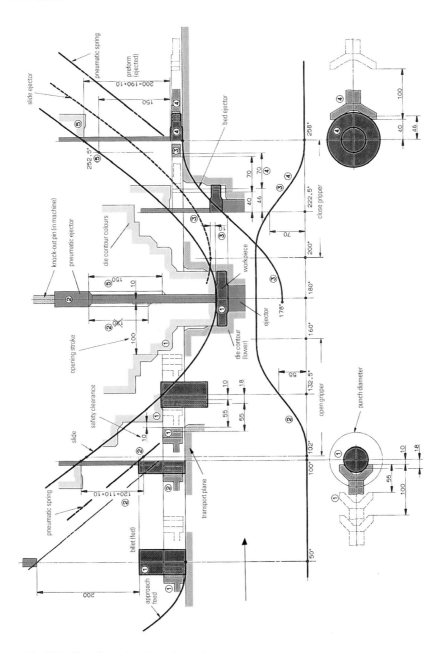

▲ **Fig. 6.6.2** Two-dimensional transfer study:
1 define opening starting point; 2 determine pneumatic spring travel for fed part;
3 define shortening of ejector bolt in the machine and starting point of
ejector in bed; 4 define closing end point;
5 determine pneumatic spring travel for ejected part

just before the opening of the grippers. Here, too, depending on the lift-up motion, a greater or smaller crank angle range may be required. A lift-up motion of around 100 mm can require a crank angle of around 60°. All the transport motions of the press and transfer device are limited by maximum allowable acceleration levels, and must be separately designed for each individual case under consideration.

■ 6.7 Die design

In industrial practice, extrusion or cold forging dies are frequently sub-ject to parametric designs and adaptations, making the configuration of the punches and dies of central importance. Less frequently, for exam-ple where a new press is used, questions related to die holder design and spare and active die sets must be addressed. Modern cold and warm forg-ing dies have a multiple-station configuration. Most of the main forging operations are completed within one to three stations. However, using four or even five-station die sets it is possible to execute other sizing, piercing and trimming operations in one and the same processing step. Thus, it is possible to approach the objective of net shape or near net shape production, and to reduce the transport and processing costs involved in an additional pass through the press.

Figure 6.7.1 illustrates the basic schematic layout of a modern multi-ple-station die for a universal press. An intermediate plate is mounted on the press bed and distributes the press force in the form of a pressure cone which increases in size as it acts on the press bed. The intermediate plate is equipped with recesses for connecting the shafts of the transfer device clamping boxes, and is able to accommodate sensors for press force measurement or roller brackets for die set changeover. Hydraulic die clamps for complete die change operations can also be located on the intermediate plate.

The die base plate houses the die holding sleeves. As demonstrated in the example, this plate can also hold hydraulic closing devices and scrap chutes. However, where these are used, the overall height of the plate must be configured to be higher. The dies themselves can be mounted in a retainer block which reaches over the entire width of the

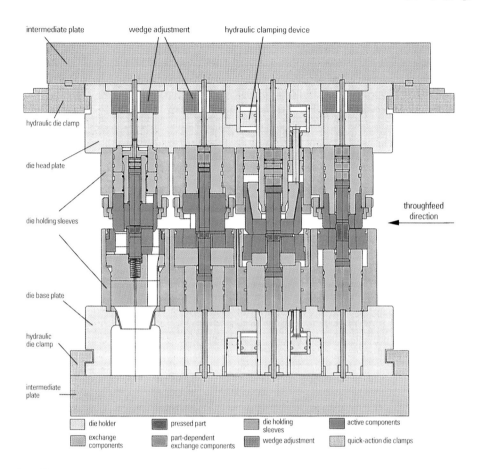

intermediate plate   wedge adjustment   hydraulic clamping device

hydraulic die clamp

die head plate

throughfeed
direction

die holding sleeves

die base plate

hydraulic
die clamp

intermediate
plate

| die holder | pressed part | die holding sleeves | active components |
| exchange components | part-dependent exchange components | wedge adjustment | quick-action die clamps |

▲ **Fig. 6.7.1** Four-station die holder with hydraulic die clamping and closing system
in the second station

die. However, for warm forming and in some cases also for cold forming, the use of single die holding sleeves is preferred. Thus, the thermal
expansion is absorbed in each individual die holder, independent of
other die stations. Using a suitable cooling device, it is possible to prevent the motion of die block, as a result of thermal expansion, relative
to the die halves mounted on the slide.

In principle, the layout of the upper die is similar. However, additional wedge adjustment devices *(Fig. 6.7.2)* are located in the head plate.
These are used for individual height adjustment of the dies. The wedge
adjustment devices have an adjustment height of about 3 to 5 mm.

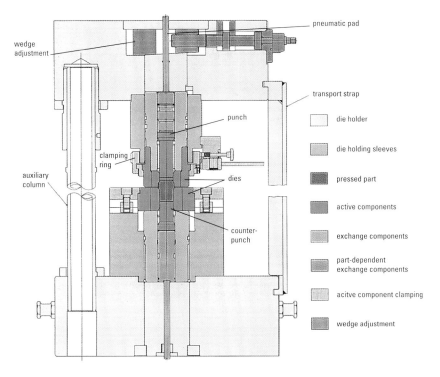

wedge
adjustment

pneumatic pad

punch

clamping
ring

auxiliary
column

dies

counter-
punch

transport strap

☐ die holder

☐ die holding sleeves

☐ pressed part

☐ active components

☐ exchange components

☐ part-dependent
exchange components

☐ acitve component clamping

☐ wedge adjustment

▲ **Fig. 6.7.2** Section through a die holder with wedge adjustment
(auxiliary columns and transport straps are built in for die holder change)

If no slide height adjustment is available in the press, the wedges must
then have a longer adjustment height of approx. 8 to 12 mm in the individual stations. Before actuating a wedge adjustment device, the clamping ring is released, thus allowing the downward motion of the dies. The
upper dies are relatively high, as they are designed to accommodate
pneumatically supported pads for workpiece transfer *(Fig. 6.7.3)*. As a
result, reliable part transfer is ensured. Thus, higher effective stroking
rates and increased output can be obtained. The stroke of the pneumatic ejectors is determined by means of a transfer study (cf. Sect. 6.6.2).

Active and exchange die components are located in the die holders.
The active die components contact the workpiece contour and are subject to wear. These must be reworked or replaced periodically. The
active die components include the containers, the die inserts, punches
and counterpunches, as well as for example stripper sleeves. The part-
dependent tooling must be exchanged together with the active die

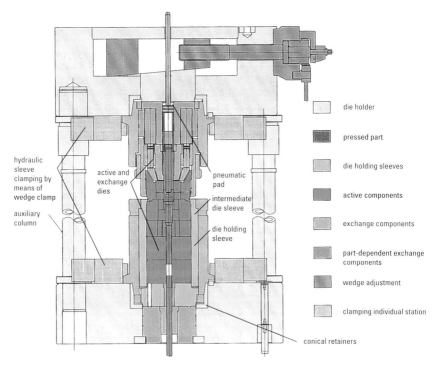

hydraulic sleeve clamping by means of wedge clamp

auxiliary column

active and exchange dies

pneumatic pad

intermediate die sleeve

die holding sleeve

conical retainers

die holder

pressed part

die holding sleeves

active components

exchange components

part-dependent exchange components

wedge adjustment

clamping individual station

▲ **Fig. 6.7.3** Section through a die holder with wedge adjustment and hydraulic sleeve clamping

components when setting up the press for a new product. Other universally used exchange tooling components remain in the holding sleeves. The exchange tooling include mainly pressure plates, pressure pins, spring elements and guide sleeves. For the structural design of active and exchange components of die sets, please refer to the VDI Guidelines 3176, 3186 Sheets 1 to 3.

■ 6.7.1 Die holders

The die holders hold both the active and exchange tooling components. Their function is to ensure that the lower and upper die operate as on-center as possible in relation to each other. The number of stations, the distance between stations and the holder sizes are generally grouped together in standardized series. Table 6.7.1 illustrates a modu-

Table 6.7.1: Modular system for possible die holder sizes (widths) (**bold** = frequently used sizes) for number of and distance between stations

Number of stations	Die holder widths with distances between stations in mm				
	200	*250*	*300*	*350*	*400*
2	400	500	**600/750**	700	800
3	600	750	**900/1,000**	1,050	1,200
4	800	**1,000**	**1,200/1,250**	1,400/1,500	1,600
5	**1,000**	**1,250**	**1,500**	1,850	**2,000**
6	**1,200/1,250**	**1,500**	–	–	–

lar system of this type for a universal press design frequently used for forming parts of approx. 0.1 to 15 kg. Of course, other component and die sizes can also be implemented.

The die change system is closely linked with the die holder design. The range of parts to be produced and the individual batch sizes are the main criteria that influence the die holder design and the layout of the die changing system.

The complete die holder change, required when starting work on a completely new part, involves moving the entire die holder out of the machine to be replaced by a ready-prepared holder located in a prescribed waiting position. The die head and base plate of the die holders are hydraulically clamped *(Fig. 6.7.1)*. For moving in and out of the press, the upper and lower part are moved together by means of auxiliary columns and locked together with safety straps for transport *(Fig. 6.7.2)*. A third die holder can be assembled and set-up in the die shop during the pressing or exchanging the die holders. For this method of die holder exchange, parts of the gripper rails have to be uncoupled, fixed on the die holder and removed together from the press. Depending on the size of the holder, the exchange can take anywhere between 20 and 40 min. These times include resetting of the feeding station and, in the case of warm pressing, of the induction furnace as well.

In exchanging the sleeves, the base and the head plate are left in the press and the sleeves in the table and slide are replaced individually. For this purpose, the base and head plate are fastened at the bed and slide. They are equipped with conical centering rings *(Fig. 6.7.3 and 6.7.4)*.

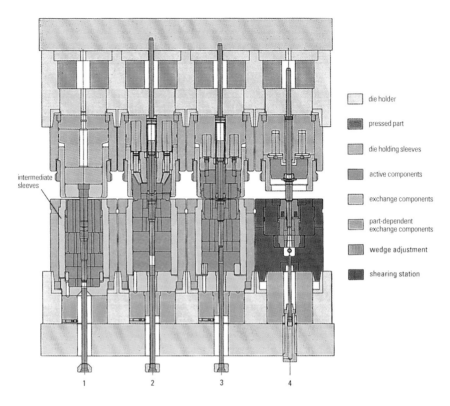

▲ **Fig. 6.7.4** Four-station die holder with hydraulic sleeve clamping and shearing station in the fourth station

Each die holding sleeve is fastened by means of two to four hydraulic wedge-type clamps *(Fig. 6.7.3).* This design is used for large batch sizes when the tools wear at very different rates in the individual stations and have to be individually replaced. In this case, ready prepared die holding sleeves are located in a prescribed loading position and can be mounted into the die station by a die changing arm *(cf. Fig. 3.4.5).* It takes between 4 and 8 min to replace one sleeve. *Figures 6.7.3* and *6.7.4* illustrate the use of intermediate sleeves which can be assembled with axial pre-stressing. Multiple-station forging permits subsequent operations to be executed in one processing step. As illustrated in *Fig. 6.7.4* and *6.7.5,* in the case of this specific die holder, the end of the extruded shaft is sheared to the required length by a wedge-action pusher die in the fourth die station.

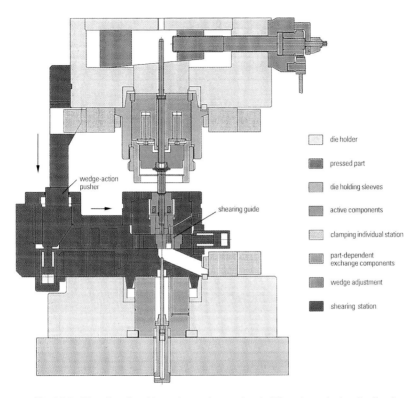

die holder

pressed part

die holding sleeves

active components

clamping individual station

part-dependent
exchange components

wedge adjustment

shearing station

wedge-action
pusher

shearing guide

▲ **Fig. 6.7.5** Shearing die with wedge-action pusher, holding sleeve hydraulically clamped

■ 6.7.2 Die and punch design

Dies are subjected to high degrees of internal pressure which can quickly lead to die failure using single ring extrusion containers manufactured with die steels available today. The extremely high stresses, present on the internal die wall of the extrusion container (cf. Sect. 6.5.1)
can be reduced by pre-stressing (shrink-fit) the container with one or
more shrink rings. Punches are subjected initially to pressure and then
additionally, depending on the process, to bending stress as a result of
off-center loads. Less frequently, the possibility of punch buckling must
be taken into consideration. The calculation methods outlined here for
dies and punches are described in detail in VDI Guidelines 3176, VDI
3186 and ICFG Doc. No. 5/82.

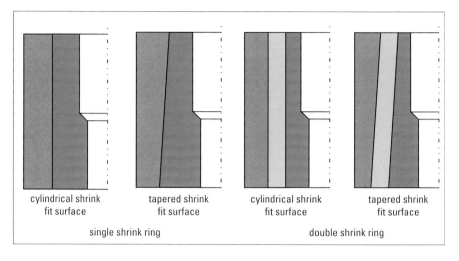

cylindrical shrink fit surface	tapered shrink fit surface	cylindrical shrink fit surface	tapered shrink fit surface
single shrink ring		double shrink ring	

▲ **Fig. 6.7.6** Shrink fits with single and double rings – each with cylindrical and tapered fits

Die design

In *Fig. 6.7.6*, various shrink ring designs are illustrated. Cylindrical shrink fits are simpler to produce, but can only be used up to interferences which can be achieved by warm shrink fitting, in order to avoid cold welds during assembly.

Tapered shrink fits permit simpler, scratch-free disassembly. A drawback of this method are the high production costs. The tapered shrink fit exhibits different levels of radial pre-stress, which can lead in case of very long containers to dimensional differences on the forged part. Usual taper angles are 0.5 and 1°; if the height to diameter ratio of the container is below 0.8, taper angles of 2 to 3° are used in order to avoid disassembly of the shrink rings during part ejection. Analytic approximate operations are based on the method by LAMÉ and were published by Adler/Walter. These assume an ideal loading situation where an infinitely long thick-walled cylinder is subjected to a constant hydrostatic pressure over its entire length. Using Tresca's yield condition, which can be applied in this case, the stress distribution can be calculated. It is found that the maximum values are at the internal wall of the container and correspond approximately to the value of the internal pressure p_i, for radial stress σ_r and tangential stress σ_t *(Fig. 6.7.7, left)*. The internal pressure can therefore, due to

$$\sigma_v = \sigma_r + \sigma_t = 2\,p_i$$

amount to no more than $\dfrac{R_{p0.2}}{2}$, for the effective stress $\sigma_v = R_{p0.2}$. Thus, for material values of $R_{p0.2} = 2{,}000\ \text{N/mm}^2$, maximum pressure levels of $1{,}000\ \text{N/mm}^2$ are permissible. If, in the case of thick-walled pipes subjected to internal pressure, the effective stress exceeds the yield strength of a material with sufficient toughness, then plastic flow occurs at the inside wall of the container. If the ultimate strength is reached in containers or inserts from brittle materials, cracks occur.

As a result of adding compressive stress, the tangential stress and thus also the effective stress status at the inside wall of the extrusion container (or insert) are reduced *(Fig. 6.7.7 center, right)*.

The radial pre-stress is generated by means of shrink rings. A shrink ring has an internal diameter that is smaller than the outer diameter of the corresponding inner ring by a selected dimension (interference). By maintaining the outer diameter unchanged, the permissible internal pressure can be increased by up to 100 % as a result of the shrink ring compared to a container without a shrink ring. For a given permissible internal pressure, the outside diameter of the container, using shrink

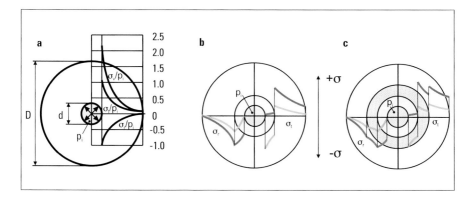

▲ **Fig. 6.7.7** Theoretical stress distribution in a thick-walled single-piece hollow cylinder *(left),* a two-part *(center)* and three-part *(right)* shrink fit design (red: without internal pressure, blue: with internal pressure)

Table 6.7.2: Increase of the allowable internal pressure $p_{i\,zul.}$ or reduction of the outside diameter d_a through single and double shrink rings

Characteristic value	Press container without shrink ring	with single shrink ring	with double shrink ring
$p_{i\,zul.}$ [N/mm²]	690	1,160	1,395
d_a [mm] with d_i = 20 mm	100	40	36.4

rings, can be reduced by about 60% (Table 6.7.2). In multiple-station forming machines this results in a substantial reduction of die diameters, the distances between stations, the dimensions of the die holder and thus also the machine bed size.

Carbide cannot be subject to tensile stresses. Therefore, when using dies or inserts from carbide, the extrusion container must be pre-stressed in such a way that no tangential tensile stress occurs under internal pressure. Accordingly, the interference dimensions are considerably higher than those used for tool steels. Guidelines also have been established for the calculation of axial pre-stress levels; in the thin, ring-shaped lateral split, surface pressures of 700 to 1,000 N/mm² are assumed as guideline values.

In extrusion dies, the actual stress conditions differ from the ideal assumptions made here, however. The extrusion containers have a finite length, a limited and possibly eccentric pressure area, no uniform internal pressure exists, and the containers have often off-center die openings. Realistic conditions in the configuration of shrink fit assemblies can be determined by using discrete approximation techniques, for example the finite element method (FEM). For practice-oriented application, various calculations have been conducted by making a parametric study using FEM. These results have been made accessible for practical application in the form of Nomograms (VDI 3176, ICFG Doc. 5/82) and corresponding computer programs.

Punch design
Extrusion punches are subjected to an axial force with an average compressive stress. They are calculated as follows:

$$p_{St} = \frac{F_{St}}{A_{St}}$$

The off-center applied force causes additional bending stresses. The overall stress in the punch then amounts to:

$$\sigma_{St\,ges} = p_{St}(1 + 8 \cdot e / d)$$

whereby d [mm] is the punch diameter and e [mm] the off-center position of force application. Values for critical buckling stress are provided in *Fig. 6.7.8*. With backward cup extrusion, the eccentricity generally amounts to e/d = 0.01 to 0.02. The calculated punch stress is compared to the compressive yield strength of the punch material in its hardened state.

The counterpunch is subjected to an average compressive stress of

$$p_G = \frac{F_{St}}{A_0}$$

during backward cup extrusion.

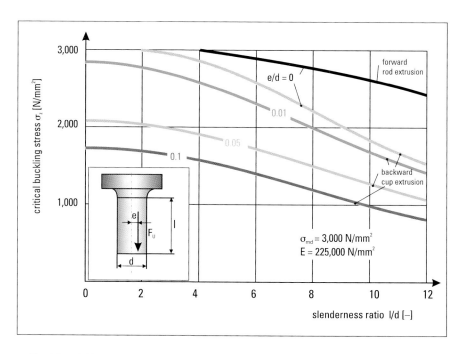

▲ **Fig. 6.7.8** Critical buckling stress in function of the slenderness ratio (l/d) of a punch made of high-speed steel: 1 forward rod extrusion; 2 backward cup extrusion; d punch diameter; e off-center position of force application; l (free) punch length

■ 6.7.3 Die and punch materials

Materials used for cold and warm forming are required to have high strength, high toughness and high hardness. For an economical die life, extremely good wear and tempering resistance are necessary. Die materials should be chosen, where possible, to ensure good machinability, and as a result, low manufacturing costs (cf. Sect. 4.1.3).

All these requirements cannot be fulfilled simultaneously by each type of tool steel. The characteristics wear resistance and toughness, for example, show opposing tendencies. The materials 55NiCrMoV6 (1.2713), 57NiCrMoV7 (1.2714) and X3NiCoMoTi1895 (1.2709) are extremely tough and are used for shrink rings and inserts which are subjected to high levels of elongation. Their wear characteristics, in

Table 6.7.3: Survey of the most frequently used cold, warm and hot forging tool steels

Die materials I / Cold forging tool steels / Active components

Material No. to DIN		DIN Germany	ANSI USA	JIS Japan	Composition in %										
					C	Si	Mn	P	S	Co	Cr	Mo	Ni	V	W	Ti
1.2363		X100CrMoV51	A2	SKD11	1.00	0.30	0.55	0.03	0.03	–	5.00	1.10	–	0.20	–	–
1.2369		81MoCrV4216			0.81	0.25	0.35	–	–	–	4.00	4.20	–	1.00	–	–
1.2379		X155CrVMo12 1	D2	SKD11	1.55	0.30	0.35	0.03	0.03	–	12	0.70	–	1.00	–	–
1.2709		X3NiCoMoTi1895			0.03	0.10	0.15	0.01	0.01	9.25	0.25	5.00	18	–	–	–
1.2713	1. NiCrMo	55NiCrMoV6	6F2	SKT4	0.55	0.30	0.60	0.03	0.03	–	0.70	0.30	1.70	0,10	–	–
1.2714		57NiCrMoV7		SKT4	0.58	0.30	0.70	0.03	0.03	–	1.00	0.50	1.70	0.10	–	-
1.2767		X45NiCrMo4	6F7		0.45	0.25	0.30	0.03	0.03	–	1.35	0.25	4.00	–	–	–
1.3207	HSS	S10-4-3-10	T42	SKH57												
1.3343		S-6-5-2	M2	SKH51	0.90	0.45	0.40	0.03	0.03	–	4.15	5.00	–	1.85	6.35	–
1.3344		S-6-10-2	M3/2		1.20	0.45	0.40	0.03	0.03	–	4.15	5.00	–	3.00	6.35	–

Die materials I / Hot-warm forging tool steels / Active components

Material No. to DIN		DIN Germany	ANSI USA	JIS Japan	Composition in %										
					C	Si	Mn	P	S	Co	Cr	Mo	Ni	V	W	Ti
1.2713	1. NiCrMo	55NiCrMoV6	6F2	SKT4	0.55	0.30	0.60	0.03	0.03	–	0.70	0.30	1.70	0.10	–	–
1.2714		57NiCrMoV7		SKT4	0.58	0.30	0.70	0.03	0.03	–	1.00	0.50	1.70	0.10	–	–
1.2343	2. CrNiMoV	X32CrMoV51	H11	SKD6	0.38	1.00	0.40	0.03	0.03	–	5.30	1.10	–	0.40	–	–
1.2344		X40CrMoV51	H13	SKD61	0.40	1.00	0.40	0.03	0.03	–	5.30	1.40	–	1.00	–	–
1.2365		X32CrMoV33	H10		0.32	0.30	0.30	0.03	0.03	–	0.30	2.80	–	–	–	–
1.2367		X40CrMoV53		SKD7	0.30	0.20	0.30	0.03	0.03	–	2.40	–	–	–	4.30	–
1.2606	3. WCrV	X40CrMoW51		SKD62												
1.2622		X60WCrCoV93		SKH51												

Table 6.7.4: *Heat treatment of cold, warm and hot forging tool steels*

Die material II / Cold forging tool steels / Active components

No.	Heat treatment (°C) Annealing	Hardening	Quenching medium	Tempering	Hardness (customary)	Cooling Water	Application/hardness (HRC/N/mm²)
1.2363	800-840	930-970	Ö,WB 400	180-400	60 +/–1		blanking/punching dies
1.2369	800-840	1,070-1,100	Ö,WB 450-550	550	61 +/–1		punches, blanking/punching dies
1.2379	840-860	1,040-1,080	Ö,L,WB 400	180-250	60 +/–1		punches, dies
1.2709	840	480	L	–	55		shrink ring
1.2713	650-700	830-870	Ö	300-650	45 +/–1	P,Ö,(W)	shrink/intermediate ring and
1.2714	650-700	860-900	L	300-650	45 +/–1	P,Ö,(W)	pressure pin (53+/-1 HRC, 1,150 N/mm²)
1.2767	610-630	840-870	W,Ö,L	160-250	54 +/–1		punch, mandrel, one piece press container
1.3343	1,100-900	790-820	Ö,L,WB 550	540-560	62 +/–1		punch, die (insert), press container and
1.3344	1,100-900	770-820	Ö,L,WB 550	550-570	62 +/–1		counterpunch, mandrel

Die material II / Hot, warm forging tool steels / Active components

No.	Heat treatment (°C) Annealing	Hardening	Quenching medium	Tempering	Hardness (customary) HRC	Cooling	Application/hardness
1.2713	650-700	830-870	Ö	300-650	42 +2	P,Ö,(W)	shrink/intermediate ring and
1.2714	650-700	860-900	L	300-650	42 +2	P,Ö,(W)	pressure pin (52+2 HRC, 1,150 N/mm²)
1.2343	750-800	1,000-1,040	L,Ö,WB 500-550	550-650	50 +2	W,P,Ö	die (insert), shrink ring (45+/–1 HRC)
1.2344	750-800	1,020-1,060	L,Ö,WB 500-550	550-650	50 +2	P,Ö	mandrel and counterpunch
1.2365	750-800	1,020-1,060	Ö,WB 500-550	500-670	50 +2	(W),L,P,Ö	die (insert), punch and
1.2367	750-800	1,060-1,100	L,Ö,WB 500-550	600-700	54 +2	(W),L,P,Ö	counterpunch
1.2606	750-790	1,020-1,050	Ö,L,WB 500-550	550-650		W,L,P,Ö	
1.2622	760-800	1,150-1,200	Ö,WB 500-550	500-650	56 +2	L,Ö	blanking die

W = water   WB = water bath   L = air   P = compressed air   Ö = oil

contrast, are not quite as favorable. The high-speed steel S10-4-3-10 (1.3207) offers outstanding wear characteristics, but tends to be more brittle. The materials X155CrVMo121 (1.2379) and S-6-5-2/S-6-10-2 (1.3343/44) represent a compromise concerning wear and toughness characteristics: Their tempering resistance can also be considered to be very good at 550 °C. The most important criterion to be considered when selecting a tool material is the type and extent of load, followed by the layout and geometry of the die.

Table 6.7.3 provides a summary of the most commonly used tool materials for cold and warm forming. The table also specifies the designations of comparable steels from the USA or Japan and their compositions. Table 6.7.4 describes the heat treatment, i.e. the annealing temperature, the hardening temperature, the quenching medium and the tempering temperatures. The levels of hardness generally required for dies used in cold forming applications are also given. Table 6.7.5 describes the characteristics of wear, toughness, machinability and grinding using a 10-point scale, whereby a high number indicates especially good capability. The tables make no claim to completeness, but present a substantially reduced number of actually required tool steels in following the principles of optimised and reduced inventory.

The most frequent tool material used for cold forging is high-speed steel, while for warm forging temperature-resistant tool steel is used.

Table 6.7.5: Properties of cold, warm and hot forging tool steels

Die materials III / Cold forging tool steel / Active components

No.	Characteristics				Remark
	V	Z	B	S	
1.2363	7	6	8	7	
1.2369	7/8	5			
1.2379	8	4	4	4	12 % Cr-steel
1.2709	5	9	5	7	special steel
1.2713	2	10	6	8	
1.2714					
1.2767	4	10	6	8	
	9	2	4	5	
1.3343	9	4	4	5	
1.3344	10	3	3	4	

Die materials III / Hot, warm forging tool steel / Active components

No.	Characteristics				Remark
	V	Z	B	S	
1.2713	2	10	6	8	44 HRC 1,400 -1,480 N/mm^2 52 HRC 1,800 - 1,900 N/mm^2
1.2714					
1.2343	5	8			50 HRC 1,700 - 1,800 N/mm^2
1.2344	5	8	8	8	surface welding Capilla 521 for erosive machining
					surface welding Capilla 5200 for machining
1.2365	4	8	8	8	50 HRC 1,700 - 1,800 N/mm^2
1.2367	5	7	8	8	54 HRC 1,925 - 2,050 N/mm^2
1.2606					
1.2622					56 HRC 2,050 - 2,200 N/mm^2

V = wear resistance   Z = toughness   B = machinability   S = grindability

Powder metal-manufactured high-speed steels demonstrate very uniform carbide distribution and are segregation-free. This results in improved toughness properties and extremely high compressive strength. If a larger wear resistance is required, for example for forward rod extrusion or ironing dies, for large-production series carbides are also used. The carbide types used comprise tungsten carbide as a phase in a cobalt matrix. The cobalt content which determines the characteristics of the carbide, lies between 15 and 30%. With a rising tungsten carbide content, hardness, compressive strength and wear resistance all increase. However, at the same time toughness, notch impact strength, bending and buckling resistance are reduced. The grain size exerts an influence here: fine-grained carbides are unsuitable as a result of poor toughness characteristics. Table 6.7.6 provides a summary of the mechanical characteristics of carbides. The strength values apply to static loads and must be reduced by 50% for stress cycles greater than 10^6.

Tool steels for warm forming
The materials X40CrMoV53 (1.2367), X38CrMoV51 (1.2343), X40-CrMoV51 (1.2344) and X60WCrCoV93 (1.2622) are currently used for warm forming and in conjunction with the water cooling method generally applied today; In the case of shrink rings, 55NiCrMoV6 (1.2713)/57NiCrMoV7 (1.2714) or X38CrMoV51 (1.2343)/X40CrMoV-51 (1.2344) can be used. Due to their low thermal shock resistance, carbides, high-speed steels and cold forming tool steel 81MoCrV4216 (1.2369) are only suitable when using water-free cooling methods which were common in the past. On closer examination, it becomes clear that materials for warm forging are generally those used in the classical field of hot forging. Special materials for warm forging, which are capable of withstanding relatively high pressure levels and relatively high temperatures simultaneously, have not yet been developed for implementation in series production. Developments of die materials for warm forming are moving in the following directions:

Compared to hot forging tool steels, high-speed steels already exhibit good wear and heat resistance properties as well as improved tempering resistance. By a gradual reduction of the carbon content, attempts have been made to achieve a further improvement in thermal shock resistance of high-speed steels. With simultaneous deterioration of the wear and heat resistance, however, the thermal shock resistance level

Table 6.7.6: Mechanical properties of carbides

Carbide types			E-modulus	Ultimate strength	
Co content	Density	Hardness		Compressive strength	Bending strength
% by weight	g/cm³	HV	N/mm²	N/mm²	N/mm²
15	14.0	1,150	530,000	4,000	2,400
20	13.5	1,050	480,000	3,800	2,600
25	13.1	900	450,000	3,300	2,600
30	12.7	800	420,000	3,100	2,700

was far below that of hot forming tool steels. In view of the successful implementation of environmentally friendly water-soluble lubricants, the factor thermal shock resistance is expected to continue to be a subject of particular interest.

Thus, investigations are concentrated on tool steels which are derived from hot forming tool steels, i.e. those which have a relatively low carbon content. These tool steels demonstrate superior wear and heat resistance properties as well as very good thermal shock resistance. A number of different die materials are being developed concurrently, for example using nickel-based alloys (Inconell 718), which have already produced good results in bar extrusion (temperatures 1,150 °C, pressures of 900 to 1,100 N/mm²) and isotherm forging applications.

Die coatings (VDI 3198)
The cost-effectiveness of cold forging processes depends largely on the tool costs and accordingly on the achievable service life of dies. Die wear influences the dimensional stability and surface quality of the produced workpieces. There are two main types of wear resistant coatings considered for use in forging technology. Coatings, for example using adding-on processes such as CVD or PVD processes, and coatings obtained through reaction layers such as nitriding and ion implantation. Processes which have proven successful for cold forming dies include the CVD and PVD techniques, which permit service life to be more than doubled (cf. Sect. 4.1.3).

An essential prerequisite for a successful coating is good adhesion between the coating material and the substrate, whose hardness and compressive strength must provide adequate support. High-speed steels, ledeburitic 12% chromium steels (1.2379) and carbides fully comply with these requirements.

Coatings are applied mainly by specialized companies. This means that particular attention must be paid to the preparation of dies by the die manufacturer. Prior to coating, the work surfaces must be free of grooves and polished to a surface roughness of $R_z < 1$ mm. The limit for optimum coating of internal holes is around l (length) = d (diameter) when using the PVD technique. With the CVD technique, there are no known limitations. The limits for the ratio between the diameter D and thickness H of coated panels in cases where a "deflection" of < 0,02 mm is required after treatment, are as follows when using the CVD technique:

$D < 7 \cdot H$ for steels and

$D < 15 \cdot H$ for carbides

and when using the PVD technique

$D < 20 \cdot H$

It is only possible to coat pre-stressed dies in cases where the insert is pressed out, coated and pressed in again after coating. In order to prevent the formation of beads, sharp edges must be broken beforehand. In cases where a mean deviation of the mandrel length L and diameter D after coating of < 0.01 mm is required, the CVD process should be used in preference for the following dimensions

$L < 10 \cdot D$ in the case of steel

$L < 15 \cdot D$ in the case of carbide

and the PVD process for the length

$L < 20 \cdot D$

Reliable improvement in die life, as a result of coating warm forming dies with the CVD or PVD techniques, has not been demonstrated yet, classical nitriding and welding-on techniques are generally used to enhance performance. The problem faced here is that the base materials have lower degrees of hardness compared to the requirements of cold forging, and the necessary supporting effect of the substrate material is not achieved. Once a continuous increase in the hardness of the substrate material through to the actual coating has been successfully achieved, more success can be expected. It may be possible to achieve this result using plasma nitriding and subsequent coating with CrN or TiAlN, or through the development of special sandwich coatings.

■ 6.7.4 Die closing systems (multiple-action dies)

Recent developments in the field of die closing systems have substantially increased the range of parts that can be produced using the cold and warm forging methods. Flashless forging of spiders, spherical hubs, tie rod ends, T-fittings for radiator construction, rotors and parts with different gearing has only become possible with the development of die closing systems.

With an additional closing force initiated in the die set, female dies are closed before the bottom dead center of the slide is reached. Somewhat later, the material is then formed, flashless, in the closed die by penetration of the punch. Flashless pressing is only possible if the pre-form can be completely located in the dies when these are closed. The closing force itself can be initiated by means of elastomers, hydraulic systems, or mechanically by the kinematics of the press.

Elastomers offer the advantage that their design is simple and their spring characteristic can be adjusted within certain limits by altering the material composition. However, they only permit limited travel. The elastomers also have to be exchanged periodically. Hydraulic closing systems generate high closing forces within a minimum of space, and can thus be positioned adjacent to each other in multiple-station dies. By adjusting the pressure, a greater closing force range can be covered. However, their design is more complex, and they call for a number of safety precautions to protect both the operator as well as the dies.

In modern presses, mechanical initiation of closing motions can be achieved over a large crank angle range. However, the investment here is substantially higher than that needed for elastomer or hydraulic systems.

Figure 6.7.9 illustrates the basic layout of a hydraulic closing system integrated fully into the die. In the provided example, there is a closing system integrated into both the upper and the lower die. This type is used predominately in cases where there is only one forging step or when only one station of a multiple-station die is required to work with a closing system (cf. Sect. 6.7.1). Typical workpieces produced in this die include gearing elements which are being formed from pre-formed blanks (rings, lenses). Closing systems can also be accommodated in separate neutral plates *(Fig. 6.7.10)* which always remain in the press irrespective of the actual die in use. Depending on the die set, mounted in the press, the closing systems which are currently required can be activated and connected by means of pressure pins to the female die. A typical application for this system is the production of tie rod heads and T-fittings which require up to five closing systems. Additional parts include inner races of constant velocity joints where an additional piercing process is required and the grooves, formed in the previous station, must maintain this shape accurately during the piercing operation.

The closing force can be individually adjusted according with the (annular) piston surfaces. The press forces which are introduced down-

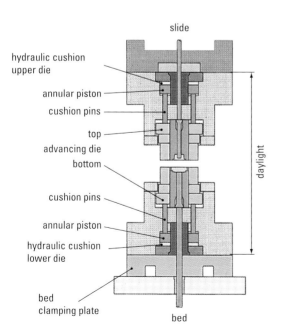

Fig. 6.7.9
Single-station die set with hydraulic closing system at the top and bottom

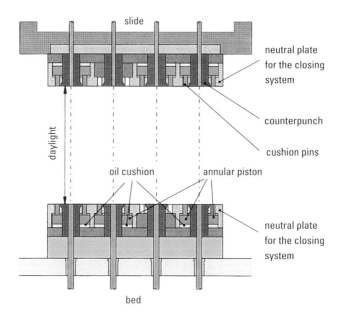

neutral plate
for the closing
system

counterpunch

cushion pins

daylight

oil cushion   annular piston

neutral plate
for the closing
system

bed

▲ **Fig. 6.7.10**   Multiple-station die set with hydraulic closing systems
at the top and bottom of each station

wards into the press bed via a counterpunch are absorbed centrally.
This determines the minimum inside diameter of the annular piston.
As a guideline, the closing force can be estimated to be 0.6 to 0.9 times
the punch load. The required closing force depends on whether a completely closed (or trapped) die set is being used or forging takes place
through extrusion shoulders into a free space. In the case of completely closed (trapped) die sets, the closing system also serves to apply an
axial pre-stress on the parting line of the dies, that are subjected to high
loads.

6 Solid forming (Forging)

■ 6.8 Presses used for solid forming

■ 6.8.1 Choice of press

The most suitable press for solid forming is selected by taking into account the following factors:

– part geometry and material,
– batch sizes and
– the user's existing equipment (e. g. annealing and phosphating lines).

On the basis of this information, process engineers can select the most economical production method and recommend a suitable press.

As a general rule, *mechanical presses* are recommended for large production lots and batch sizes. The maximum possible part length depends on the slide stroke and the available ejector stroke. Another determining factor is the time-motion curve of the press and the automatic feed system. In the case of shaft-type parts, the maximum transport length applies instead of the maximum part length. For the preliminary stages, this dimension is often greater than that of the finished part. For parts with an internal contour, the penetration depth of the punch must be taken into consideration. As a rough guideline, the maximum possible part or transport length suitable for continuous automatic operation of mechanical presses can be calculated using the following formula:

for parts with internal contour: 0.35 to 0.45 times slide stroke

for shaft-type parts: 0.45 to 0.60 times slide stroke

Using special devices such as separately driven feed systems with lifting stroke, it is possible to manufacture lengths up to 0.8 times the slide stroke. In this case, presses operate in the automatic single-stroke mode at a production output of approx. 15 to 18 parts per minute. In general, a transfer study is required in setting up automated mechanical presses in order to ensure a reliable sequence of motions between the slide and feeding system (cf. Sect. 6.6.2).

In comparison to sheet metal forming, solid forming requires substantially longer forming strokes in the partial and nominal load range, and accordingly a greater press energy capacity (cf. Sect. 3.2.1). In addition, powerful ejector systems are generally used both in the press bed and the slide *(cf. Fig. 3.2.12)*. When using multiple-station dies, a two-point slide drive is beneficial to enhance the off-center loading capability of the press (cf. Fig. 3.1.3).

Regarding production lot sizes, *hydraulic presses* operate at substantially lower speeds than mechanical presses. However, they have longer slide and ejector strokes as well as high press forces and they are considerably simpler to implement with a lower investment volume *(cf. Fig. 4.4.2)*. For these reasons, operation in the single stroke mode is also economical. The application of hydraulic presses varies considerably. Concerning the operating sequence and necessary press force, the differences between mechanical and hydraulic systems are minimal. The working speed of the slide is a particularly important factor in determining the economical application of hydraulic presses. High approach and return speeds of approx. 400 to 500 mm/s substantially increase the press cycles *(cf. Fig. 3.3.1)*. Even more important is the highest possible velocity under load (approx. 40 to 60 mm/s). These values call for the use of pumps and motors with adequate capacities *(cf. Fig. 3.3.3)*.

Equally fundamental to economical forming are reliable control systems using proportional valves for precise control of the slide motion. Positive stops that can be automatically adjusted are used for sizing and coining operations.

Figure 6.8.1 gives rough guideline values for the required press forces depending on the formed part, the material and the production process. Material-dependent differences for the required forming force are considerably lower for warm and hot forming than for cold forming *(cf. Fig. 6.1.5)*.

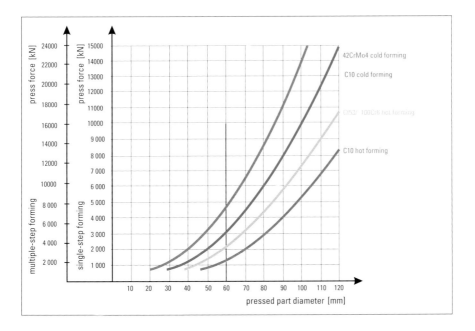

▲ **Fig. 6.8.1** Guideline values for press force requirement

A number of different presses used for solid forming are presented in the following. For a description of the fundamental principles of press design, please refer to Sects. 3.1 to 3.3.

■ 6.8.2 Mechanical presses

Mechanical presses are used primarily in conjunction with automatic feed systems for large production series of over 1.5 million parts per year and batch sizes of more than 10,000 parts. Both vertical and horizontal press designs are used. Hourly production rates, achieved using vertical systems, range between 1,800 and 3,600, while the corresponding figure for horizontal presses lies between 3,600 and 9,000. The systems that are most widely used are discussed below.

Knuckle-joint presses with bottom drive
The principle layout of this type of press with nominal press forces from 1,600 kN is described in section 3.2.2 *(cf. Fig. 3.2.5). Figure 6.8.2* illus-

◀ **Fig. 6.8.2**
Knuckle-joint press with bottom drive
(nominal press force 6,300 kN, four
stations with transfer feed)

trates a typical press. Here, the press frame moves the upper die up and down. In the case of single-station dies, part transfer takes place from the back towards the front (or vice-versa). For multiple-stage dies, the press frame incorporates lateral openings which provides space for the rails of the transfer system and for part feed.

One of the benefits offered by this design principle is the low center of gravity, and the minimal machine weight. These factors result in lower capital investment compared to presses, offering similar capabilities, but with the top drive. Slide strokes are in the range of 150 to 250 mm (in special designs up to 450 mm) and they are substantially larger than those available in coining or blanking presses.

Multiple-station presses with modified top knuckle-joint drive system
The main characteristic of this type of press with rated press forces from 4,000 kN is its modified knuckle-joint drive (cf. sect. 3.2.2). In addition to the advantages mentioned above, this design principle offers the benefit of a process-adjusted low slide velocity in the work range *(Fig. 6.8.3*

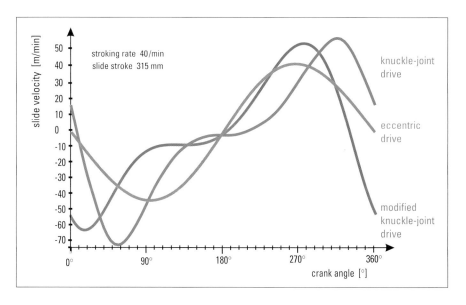

slide velocity [m/min]

stroking rate 40/min
slide stroke 315 mm

knuckle-joint drive

eccentric drive

modified knuckle-joint drive

crank angle [°]

▲ **Fig. 6.8.3** Slide velocity curves for mechanical cold forging presses

and *cf. Fig. 3.2.3)*. This is supplemented by acceleration of the slide downward and upward to achieve an overall higher stroking rate. As a result, the velocity of the slide when impacting the workpiece material is reduced and remains approximately constant during forming. As a result, the dynamic loads on the press and die are reduced. For extrusion processes particularly, this helps to extend the service life of dies and reduce the stress on the drive system. Another positive side-effect is a reduction of noise development by about 6 to 8 dB(A).

A further benefit in comparison to conventional drive systems is a three to four-fold increase in the deformation stroke, with a drive torque value comparable to that of conventional mechanical presses. This makes production of press parts, requiring a high level of deformation energy and a large nominal working stroke (> 20 mm) very economical (cf. Sect. 3.2.1). The eight-track slide gib operates according to the "fixed-loose bearing" principle, thus compensating for thermal expansion in the slide. This makes this press ideally suited for warm and hot forming *(cf. Fig. 3.1.7)*. The slide stroke ranges between 250 and 400 mm (special designs up to 520 mm) *(Fig. 6.8.4)*.

▲ **Fig. 6.8.4** Multiple-station press with modified knuckle-joint top drive
(nominal press force 10,000 kN, four stations, loading system and transfer feed)

Multiple-station presses with eccentric drive
The main field of application for these presses with nominal press forces from 6,300 kN lies in warm forming (680 to 820 °C) and automated flashless forging *(Fig. 6.8.5)*. Due to their large slide stroke of 630 to 800 mm, these presses are also ideally suited for cold forming of long, shaft-type pressed parts. The slide is guided in the same gib system as illustrated in *Fig. 3.1.7*. This guidance concept also permits narrow gib clearances in warm and hot forming applications, meaning that it is no longer necessary to provide for an additional clearance to compensate for the unavoidable thermal expansion of the slide.

▲ **Fig. 6.8.5** Warm forming line comprising a multiple-station press with eccentric drive (nominal press force 20,000 kN, induction furnace, transfer feed and die change support)

Consequently, extremely precise pressed parts can be produced. Central offset, in particular, is considerably lower than when using conventional gib systems. As a result, additional measures such as lateral punch adjustments or column gibs can be eliminated.

In warm and hot forming applications, this concept is supplemented by highly efficient cooling systems which substantially reduce the overall heat generation in the dies.

Horizontal multiple-station presses for manufacture of pressed parts made of wire

The major benefit of horizontal multiple-station presses is that wire, suitably pre-treated for cold forming, can be used. Thus, the manufacture of (largely cold pressed) contoured parts is possible without additional chemical or thermal treatment. The wire is drawn and straightened by a wire feed device. An integrated shearing system with closed shearing dies is used to produce volumetrically precise slugs. The special gripper transfer system transfers the parts from one station to the next (cf. Sect. 6.6).

The main design feature of the press line illustrated in *Fig. 6.8.6* is the laterally positioned die mounting area which is accessible from the operator's side. This concept permits an ergonomically favorable tool setting and convenient tool change. Other important design features include the highly robust slide gib against off-center loading, the cast monobloc frame structure *(cf. Fig. 3.1.1)* and individually adjustable gripper and ejection systems for each press station.

The production stroking rates for this type of press are substantially higher than that found in comparable vertical multiple-station presses. Thus, the use of these presses is highly economical for large-scale pro-

▲ **Fig. 6.8.6** Horizontal multi-station press suitable for fasteners and form parts manufactured from wire (nominal press force 1,300 kN, five stations, quick tool change system, wire diameter max 18 mm)

duction despite the high initial investment costs involved. By using semi-automatic and automatic die resetting and die change systems, set-up times can be substantially reduced, meaning that even medium batch sizes of 20,000 to 40,000 parts can be economically manufactured *(cf. Fig. 3.4.5)*. The presses are used at nominal press forces of between 1,000 and 6,300 kN for wire diameters ranging from 10 to 36 mm.

Horizontal presses – single-station with knuckle-joint drive
These press systems are used mainly for the production of extruded parts made of aluminium and aluminium alloys *(Fig. 6.8.7)*: These machines are also ideally suited for pre-forming steel slugs. The presses are fully automated and make use of simple, cam-controlled pushers for part feed. The horizontal design, simple part feed and discharge and the superior forming characteristics of aluminium permit high stroking rages of up to 300/min. Compared with vertical presses, substantially higher stroking rates also result when forming steel. Nominal press forces range between 1,500 and 12,000 kN.

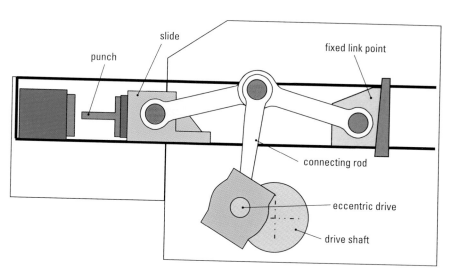

▲ **Fig. 6.8.7** Single-station horizontal press with knuckle-joint drive (nominal press forces 1,500 to 12,000 kN)

Horizontal presses with modified knuckle-joint
If longer forming strokes are necessary, a modified knuckle-joint drive system can be used as a special variation of the standard knuckle-joint power unit. The force-stroke curve and the slide velocity correspond to the vertical press shown in *Fig. 6.8.3*.

■ 6.8.3 Hydraulic presses

Hydraulic presses are also designed in vertical and horizontal layouts. They enjoy extremely universal application in solid forming both on a single-station and multiple-station basis, either with manual operation for small batch sizes or fully automatically for production rates of up to 1.8 million pressed parts per year. They are frequently used as presses for long shafts and shaft-type parts, or for large parts with weights of up to approx. 15 kg.

As in mechanical presses, the selection of the necessary press size is made on the basis of part geometry, workpiece material and of the required manufacturing process. The basic principles are described in Sects. 3.3 and 6.5.

Vertical design
The hydraulic components and oil tanks are generally accommodated in the press crown *(Fig. 6.8.8)*. A service platform offers convenient maintenance of the easily accessible units. The uprights are generally configured in the form of a welded monobloc or panel structure *(cf. Fig. 3.1.1)*.

In special cases (for example in the event of space problems or to reduce noise), the pumps and motors can also be mounted in the foundation or in a separate room. Vertical hydraulic presses are generally used for cold forming. However, it is not unusual for them to also be used for hot forming under certain circumstances (cf. Sect. 6.8.5). Particularly where large parts are involved (weights above 5 kg), hydraulic presses are frequently used for hot forging. The usual press forces lie between 1,600 and 36,000 kN, and presses with slide strokes of up to 1.5 m have already been manufactured.

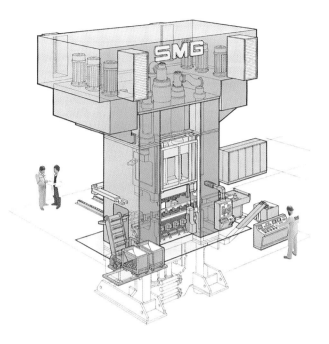

▲ **Fig. 6.8.8** Hydraulic multiple-station press (nominal press force 20,000 kN, four stations, loading system, transfer feed and die change support)

Horizontal design

Long tubular parts such as shock absorber sleeves are drawn from flat blanks or backward cup extrusions made from circular blanks *(cf. Fig. 6.1.1)*. A subsequent ironing process over several stages then reduces the wall thickness *(cf. Fig. 6.1.2)*. The workpiece is held by a punch, which draws it through one or more ironing rings. The bottom thickness remains unchanged. For this process, extremely long forming strokes are required. Therefore, in such applications hydraulic presses are most frequently used.

A horizontal press layout has a number of benefits to offer here:

- set-up on floor level,
- simple gravity feed,
- simple discharge by means of synchronized conveyor belts,
- simple interlinking, for example with automatic edge trimming devices.

Figure 6.8.9 illustrates the layout of a particularly economical press system for ironing. The slide is driven on two sides by plunger pistons. The die sets are located on both sides with loading chambers positioned in front. The right and left-hand punches are guided immediately in front of the workpiece loading chamber using an additional punch support. With each stroke, alternatively forming and stripping of the formed part take place on either side. Each new workpiece pushes the previously stripped part through the workpiece holder out of the press. With the same forming process taking place in the left and right-hand die sets, it is possible to almost double the output compared to a conventional vertical press design. Alternatively, the part produced in one of the die sets can be further processed in the other die set following intermediate treatment. In the case of thick-walled blanks, an upstream deep-drawing process can be integrated as an initial operation. Precise slide guidance on the press bed and precisely adjustable punch guidance permit to obtain extremely narrow tolerances concerning the concentricity of dimensions.

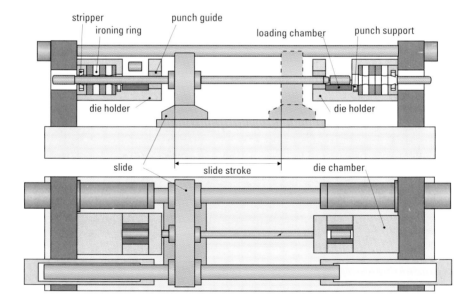

▲ **Fig. 6.8.9** Horizontal hydraulic press with one slide and two press beds

■ 6.8.4 Supplementary equipment

A large number of special measures and equipment are available to address ever more stringent demands for maximum possible production outputs, low set-up times and the minimum achievable manufacturing costs. The most important elements here are systems for the automatic loading and transfer of parts through various die stations.

Loading systems for blanks and pre-formed workpieces
Where parts weighing less than 150 g are produced, presses are generally loaded using vibration feeding devices which are filled using inclined conveyors. Heavier parts are generally fed using inclined conveyors with rotating steel belts or – in particular where pre-formed parts are being fed – by means of pushing conveyors. The outfeed area of these systems is equipped with devices for part orientation to ensure that the parts are correctly aligned when they enter the in-feed channel and the loading station *(Fig. 6.8.10).*

▲ **Fig. 6.8.10** Loading device for cold extrusion press lines

Feed systems

A number of different feed systems can be used to transfer the parts through the press and into the dies. The so-called transfer feed system is the dominating method for vertical multiple-step presses. Feed systems come in a dual-axis version, or with an additional lifting stroke for long parts, in the form of a tri-axis transfer. The control cams are driven in mechanical presses by direct coupling with the press drive system. In the case of hydraulic presses, the various units are actuated by servo drive systems *(Fig. 6.8.11)*.

Hydraulically driven swivel-mounted grippers are used to transport heavy pressed parts with weights over 5 kg *(Fig. 6.8.12)*. The controlled grippers are configured for a substantially higher clamping force than is possible using transfer feed systems. The swivel-mounted grippers generally execute a lifting movement. A loading arm (swivel or linear movement) is used to transfer the parts from the loading station into the first pressing station, and one gripper arm is allocated to each die station.

◀ **Fig. 6.8.11**
Two-axis transfer feed

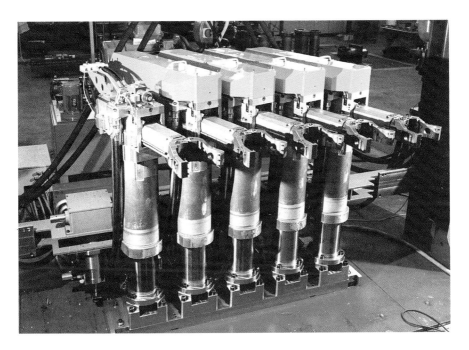

▲ **Fig. 6.8.12** Swivel arm feed for pressed parts with piece weights up to 15 kg

In addition to these systems, a number of other feed devices is used including the freely programmable handling robot. In horizontal multiple-station presses, gripper arms powered individually by cams or servo drive systems are used.

Weight control
When billets are prepared, weight differences are unavoidable. These are the result not only of length tolerances occurring during the billet separation, but also (sometimes to a far greater extent) of diameter tolerances in the starting material. Particularly when using closed (trapped) die systems, exceeding the admissible blank weight may cause die failure.

When transporting the container filled with billets to the intermediate treatment processes and to the press, it is also possible for parts to be accidentally placed in the wrong containers. This can naturally lead to die damage when pressing. In order to eliminate the risk of feeding incorrect parts, automated press lines are frequently equipped with a device which checks part weight. The devices used here are electronically controlled automatic scales, which are mounted between the loading system and the feeding station. Parts which are too heavy or too light are automatically ejected. The weighing accuracy is below 0.2% of workpiece weight. These devices are able to weigh automatically up to 80 parts per minute.

Die change
Fast die change can be a decisive factor when it comes to the economic viability of a press, in particular when working with small batch sizes. Both die set and individual die changing systems are used *(cf. Fig. 3.4.5)*. The basic die change principle used in presses for sheet metal forming and solid forming are very similar. A more detailed description is provided in Sect. 3.4. However, die change is only one aspect which determines short set-up times in automatic press lines. Fast resetting of the loading and feed systems is equally important here.

■ 6.8.5 Special features of hot and warm forming lines

The special supplementary equipment described in Sect. 6.8.4 is also used for hot forming. An additional and essential requirement of hot forming systems is an effective cooling and lubrication system. Depending on the degree of forming difficulty, basically two different systems are used:

Flooding
Dies are cooled and lubricated by continuous flooding using special lubricants, some of which are mixed with water at a ratio of between 1 : 2 and 1 : 4. The lubricant and water mixture is fed towards the die through boreholes and spray rings, and drains away via generously dimensioned channels. The lubricant is supplied from a central reservoir with a capacity of between 400 and 1,000 l by means of medium pressure pumps (working at a pressure level of 10 to 20 bar) (cf. Sect. 3.2.11). Used lubricating emulsion is collected, purified and returned to the reservoir, which is equipped with a re-cooling system. Filters in the return and pressure circuits take care of the necessary purification. The lubricant is mixed with bactericides in order to prevent bacterial infes-

tation. As this would render the lubricant unusable, bactericides are generally added to the lubricants by most suppliers. The lubricant quantities are set using simple manual control devices. Automatic devices for controlling the lubricant quantity are available. Where lubricants contain graphite, an agitating device in the reservoir is essential in order to prevent solid graphite particles from settling down in the reservoir.

Spray systems with compressed air as carrier
Mist coolant application systems provide an extremely fine surface film of coolant and lubricant for optimum cooling and lubrication. Coolants and lubricants are fed using a joint pipe. Water is used for cooling, while depending on the process involved, different lubricating agents can be used. Graphite-free emulsions containing water-mixable oils are generally used for simple forming operations. Where more complex forming processes have to be performed (such as constant velocity joint components used in vehicle drive systems), emulsions made of oil and graphite are used. These emulsions comprise a 1 : 1.3 to 1 : 4 mixture with water. The water and lubricant quantities are controlled alternatively by electro-valves. Compressed air is used as a carrier medium for the lubricant. Compressed air and water or lubricant are mixed to create a mist in a mixing valve and fed to the die through a joint pipe. The electro-valves are controlled individually for each station.

Cooling systems for forging
During automatic forging, high-pressure spray systems (80–120 bar) are used for cooling purposes. There is generally no need for additional lubrication. In order to achieve the necessary degree of heat dissipation, considerable quantities of water are required (between 40 and 60 m^3/h). The cooling water is cleaned, recooled and then re-used. The scale collected in the water must be removed using screens and filters.

Billet heating
Billets are heated by induction. For the temperature range of 760 to 800 °C, inductor outputs of approx. 0.24 kW are required per kg of steel. Depending on the billet diameter, the frequency range lies between 2 and 4 kHz. For forging temperatures from 950 to 1,100 °C, around 0.40 to 0.45 kW/kg of steel are required at frequencies of 1 to 2 kHz. In gen-

eral, aluminium and aluminium alloys are also heated by induction to temperatures between 420 and 480 °C, depending on the alloy. The required inductor output is approx. 0.35 to 0.4 kW/kg with a frequency of 0.5 to 1 kHz.

Thyristor-controlled systems are used as converters. Considerable quantities of cooling water are required to cool the inductors. Recooling is performed in the plant's own recooling systems or in separate installations.

Billet preparation
Warm forming (680 to 800 °C) of shaft-shaped parts made of steel frequently calls for coating (pre-graphitizing) of blanks. This measure serves to increase die service life and is particularly beneficial for first pressing operations (generally forward extrusion or reducing operations). The graphite can either be tumbled on prior to heating, or sprayed on during the heating process. This entails inductive pre-heating of the blanks (approx. 180 °C), followed by spray application of the graphite-water mixture. The water evaporates over a relatively short travel, during transport to the main induction heater. Thus, the billets are heated with a dried-on coating of graphite. Although a proportion of the applied graphite burns, a relatively high percentage (approx. 60 %) still permits lubrication and, together with a (very thin) scale layer, creates the desired lubricating and separating effect in the die.

■ 6.8.6 Sizing and coining presses

Sizing of forged and sintered parts
The dimensions, tolerances and surface quality of forged or hot formed parts frequently fail to meet the standards expected from finished products. Consequently, often costly finish processing operations are necessary. By introducing a sizing operation, substantially narrower tolerances can be achieved and subsequent machining processes can be reduced.

Many forged parts are finish machined on machining centers. These call for precise clamping and contact surfaces in order to assure a reproducible and well defined workpiece clamping. It is frequently not possible to achieve, in forged parts, a surface quality that is required for adequate clamping. In this case, cold sizing of the appropriate surfaces is an essential pre-requisite for successful further processing *(Fig. 6.8.13)*. Sintered parts are also frequently subjected to a subsequent compacting or sizing process in order to compensate for warping caused by heat treatment *(Fig. 6.8.14)*. All these operations require a large forming force, coupled with minimal machine and die deflection. Due to its favorable force distribution, the knuckle-joint bottom drive press has been shown to offer the most suitable forming system for coining applications. The basic principle of this drive type is described in Sect. 3.2.2 and 6.8.2 *(cf. Fig. 3.2.5)*.

Compared to presses for solid forming, the slide stroke and work length can be substantially shorter. Sizing is generally performed in a single station, so allowing the die mounting area or the press bed to be smaller. Ejector systems in the slide and press bed are only necessary for coining work.

Cutlery embossing presses
In addition to possible applications already outlined here, this type of sizing press is also used with a good degree of success for the forming and embossing of cutlery. This operation allows even the finest chiselling effects to be precisely embossed.

◀ **Fig. 6.8.13**
Coined forgings

▲ **Fig. 6.8.14** Coined sintered parts

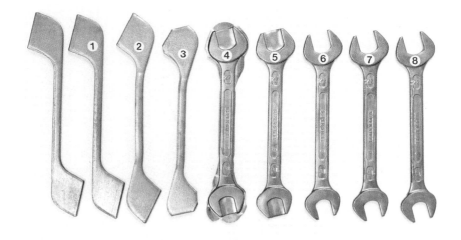

▲ **Fig. 6.8.15** Operating sequence for the manufacture of open wrenches

Embossing of hand tools

Another field of application for calibrating presses is the embossing of hand tools. *Figure 6.8.15* illustrates the sequence of operations involved in the manufacture of an open wrench. Using slugs blanked from plate, the parts are manufactured in eight operations on a knuckle-joint press with bottom drive, nominal press force 16,000 kN *(cf. Fig. 3.2.5).* The production stroking rate is 35 parts per minute. The plant operates on a fully automatic basis. The slugs are stored, and a two-axis transfer is used as a feed system *(Fig. 6.8.11).* In order to assure fast die change, die change sets are used. These can be moved in and out, controlled by pushbutton, by a die change cart *(Fig. 6.8.16).* The die sets are fastened by means of hydraulic clamps *(cf. Fig. 6.7.1).*

▲ **Fig. 6.8.16** Line for the manufacture of open wrenches
(nominal press force 16,000 kN, transfer feed and die change cart)

■ 6.8.7 Minting and coin blanking lines

Rimming and edge lettering machines

Coin blanks are manufactured on high-speed blanking presses from an indexed sheet metal strip *(cf. Fig. 4.6.6 and 4.6.8)*. Depending on the material characteristics and the condition of the die, a rough cut surface and/or blanking burr are formed at the outside surface of the coin blank *(cf. Fig. 4.7.8)*. To achieve a high level of coin quality, the outside edge of the coin blank must be smoothed. This is achieved by a radial upsetting process in rimming and edge lettering machines.

When smoothing the edge, the coin blank is sized within extremely narrow tolerance limits (+/– 0.02 mm). Reducing the diameter by 0.2 to 0.5 mm causes material to accumulate at the periphery of the coin blank. Depending on the intended purpose, different upsetting shapes can be selected *(Fig. 6.8.17)*. The rimming or upsetting process helps to accumulate the material necessary for forming the rim area of a coin. Thus, this material does not have to be forced outwards from the center of the coin blank. As a result, the coining process, conducted after upsetting, requires less force. Reduction of the coining force leads in turn to an increase in the service life of the coining dies.

Rimming machines are used for a diameter range from 14 to 50 mm both with vertically and horizontally arranged upsetting shafts. In rimming machines with a vertical upsetting shaft, the blanks are fed towards the upset die, which comprises the upset ring and upset seg-

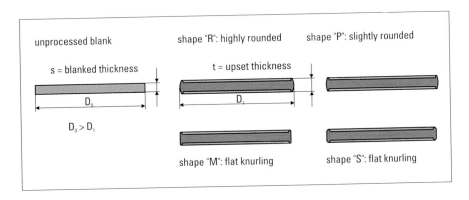

▲ **Fig. 6.8.17** Upsetting shapes on coin blanks

ment, by means of two diametrically arranged mechanical hoppers and channels *(Fig. 6.8.18)*. During the rolling process between the upset ring and upset segment, diameter reduction takes place. Rimming machines with horizontally arranged upsetting shaft can be used both for rimming or upsetting, and also for embossing deep-set edge lettering and decoration of coin blanks. Edge lettering is carried out around the periphery of the coin blank in a separate work cycle in which the previously upset blank is rolled and deep-set between the upset ring and the lettering jaws.

Depending on the machine type and the diameter of the coin blank, rimming machines achieve an output of 130,000 and 600,000 pcs/hour.

Minting presses
For minting applications (change coins with a diameter of between 14 and 40 mm) preferably presses with knuckle-joint drive or with modified knuckle-joint drive system are used (cf. Sect. 3.2) with nominal press forces from 1,000 to 2,000 kN and stroking rates between 650 and 850/min *(Fig. 6.8.19)*.

◀ **Fig. 6.8.18**
High-performance rimming (upsetting) machine

Minting presses are designed in a vertical and also horizontal version. Horizontally working presses are used to produce round coins. This machine design is characterized by simple operation and a low number of coin-specific exchange parts *(Fig. 6.8.20)*. Vertically operating presses can be universally applied for the production of round, multiple-edged and bimetal coins *(Fig. 6.8.21)*; however, a greater number of coin-specific exchange parts is required here.

The monobloc press frame made of nodular cast iron offers good damping properties *(cf. Fig. 3.1.1)*. The press is driven by means of a controllable electric motor, flywheel, eccentric shaft and knuckle-joint. The bearing points in the knuckle-joint (friction bearing) are configured for a maximum specific compression stress of 110 N/mm^2. The slide is mounted without backlash in pre-tensioned roller gibs. The indexed dial feed plate for infeed and outfeed transport of the coin blanks and coins is driven by means of a synchronous timing belt from the eccentric shaft. The coin blanks are fed by means of mechanical hoppers and channels towards the indexed dial feed plate. The ejector system is directly linked to the slide and safeguarded against overloading. The coining dies can be exchanged (coining punch and minting

▲ **Fig. 6.8.19** Press shop of a modern mint

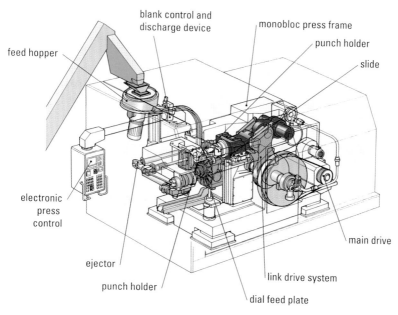

blank control and
discharge device

monobloc press frame

punch holder

feed hopper

slide

electronic
press
control

main drive

ejector

link drive system

punch holder

dial feed plate

▲ **Fig. 6.8.20** Horizontal minting press

collar) within 2 to 3 min. The machine rests on vibration absorbing elements. As a result, no more than 10 % of the static load is transmitted into the foundation as dynamic loading. This means that the presses can be set up on upper floors. These presses are generally equipped with sound enclosures in order to minimize noise generation.

The coin blanks are fed towards a mechanical hopper via an inclined conveyer with storage space or other metering systems *(Fig. 6.8.22)*. The hopper operates according to the centrifugal principle, which ensures particularly gentle handling of the blanks. The possibility of scouring is minimal, and noise development is also accordingly low.

The hoppers have a slight incline. The feed plate is directly driven by an adjustable gear motor. The outfeed channel is tangentially located. Depending on the diameter of the blanks, this arrangement permits feed rates of up to 3,000 blanks/min.

When resetting for the production of different coin sizes, only the blank outfeed channel and thickness need to be adjusted. No exchange parts are required. Depending on the press design, the hopper and feed

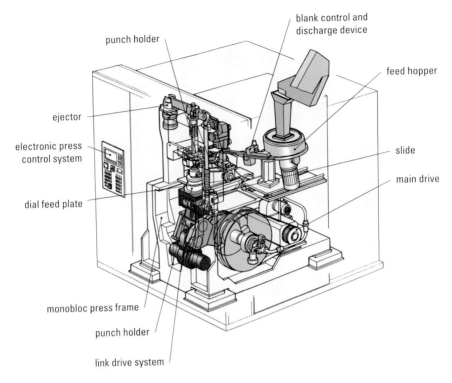

▲ **Fig. 6.8.21**   Vertical minting press

channel can be swivelled out or moved completely out of the press for die change.

In the following channel, the blanks are checked for diameter and thickness related defects, and ejected if found to be outside the permissible tolerance *(Fig. 6.8.23)*. The standstill periods caused by off-size blanks would seriously impair the efficiency of high-speed machines. Therefore, coin blanks with defects are automatically segregated in a control station while the machine is running. If a defective blank reaches the control station, the flow of blanks is interrupted. The interruption triggers a sensor, which disables further blank feed from the hopper by means of a stop cylinder. The gauge jaws are opened by means of a lift cylinder and the off-size blank is ejected into a collecting tray. The gauge jaws return to their home position, the stop cylinder opens and blank feed continues. The described cycle takes only fractions of a second, ensuring a continuous supply of blanks and preventing production disruptions.

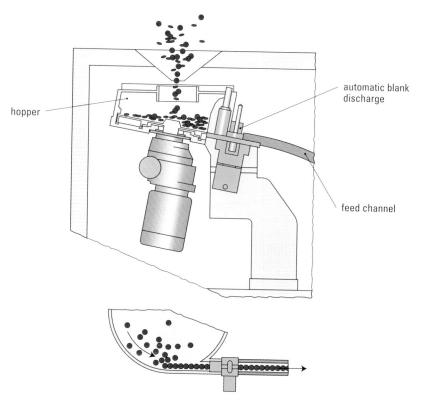

▲ **Fig. 6.8.22** Schematic of the rotary feed system

In *horizontal operating* machines, the blanks drop under the force of gravity into the indexed dial feed plate, where they are fed towards the coining station. To carry out the coining operation, the indexed plate stops. After coining, the coin is ejected from the minting collar by the ejector, which is connected to the press slide, and discharged. The short stroke of the link drive system permits the ejector movement to be derived directly from the slide. The ejector travel is reduced again in proportion to the slide stroke by means of a system of levers. Consequently, temporal offset between the two movements, as a result of tolerances and clearances in the drive elements, is eliminated. The system is not subject to wear, as neither cams nor rollers are used.

The elastic deflection of the press frame which takes place during coining is measured by a quartz crystal transducer and fed via a charge

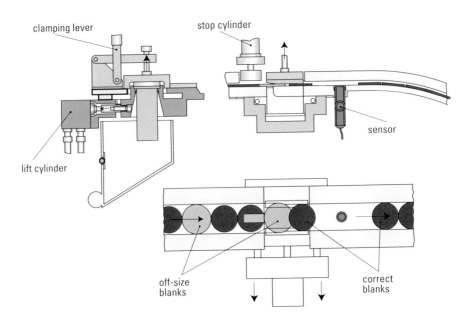

clamping lever

stop cylinder

sensor

lift cylinder

off-size
blanks

correct
blanks

▲ **Fig. 6.8.23** Automatic circular blank control and discharge device

amplifier with peak value memory to the programmable logic controller for evaluation. The pulses are counted after the coining force exceeds a certain minimum value. This ensures that only coining processes which have really been carried out can be picked up and counted. At the same time, the coining force is displayed and monitored relative to a preset limiting value. In case of a force value that is above or below the prescribed tolerance limits, the coining press is automatically switched off.

In the case of *vertically operating* presses, the blanks are fed via a channel into a filling pipe. From here, they are pushed into the indexed plate by means of a cam-controlled feeding device *(Fig. 6.8.24)*. The drive is equipped with a backlash-free mechanical indexing gear with extremely high indexing accuracy and optimum acceleration and deceleration characteristics. The indexing gear is enclosed and runs in an oil bath.

To protect the indexing drive against overloads which are not process-related, all coining presses are equipped with a synchronous

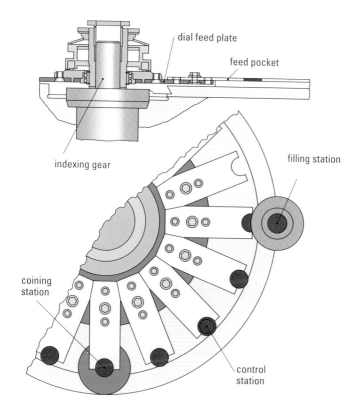

dial feed plate

feed pocket

indexing gear

filling station

coining station

control station

▲ **Fig. 6.8.24** Feed dial plate with overload safeguard

overload clutch. This overload clutch is attached without backlash to the output drive shaft of the indexing drive. The connection between the drive shaft and indexed plate is formed by means of several spring-loaded rollers located in appropriate recesses. In case of an overload, the rollers disengage and the part of the clutch which is attached to the drive shaft can still rotate in ball bearings with respect to the blocked indexed plate. It is only possible to re-engage the clutch in a position ensuring synchronous movement with the press.

The crank angle of approx. 120°, required for the indexed plate to switch around, is achieved in coining presses with a normal knuckle-joint drive system *(cf. Fig. 3.2.5)* by means of a sufficiently dimensioned idle stroke, i. e. when there is no connection between the slide die and slide. In presses with modified knuckle-joint drive *(cf. Fig. 3.2.6)*, the standstill of the slide is reached at the bottom or rear dead center. The idle phase continues over one third of the total cycle time *(Fig. 6.8.3)*. During this comparatively long standstill phase, the produced coin can be reliably discharged and simultaneously the next coin blank moved precisely into the coining position. In addition, a firm connection is provided between the slide and die. The benefit of this firm connection is that it prevents the impact of the slide on the back of the punch (hammer effect) and resulting unwanted vibrations.

Coining dies

To enhance the service life of coining dies, the coining punches are hard chromium plated. The achievable service life ranges between approx. 400,000 to 800,000 coining strokes. This figure increases three or four-fold when using titanium nitride-coated punches *(Fig. 6.8.25)*. Generally rings with a carbide insert are used as minting collars. They have a service life of between 70 and 100 million coining processes. The admissible specific compressive strain for the coining punch ranges between 1,500 and 1,800 N/mm².

Universal minting press for coins and medals

In cases where both change and collectors' coins with improved surface quality have to be produced, a vertical universal minting press with a nominal press force of 3,000 kN is used. The design of this press is the same as that used for coining presses for the production of change coins. However, this press also offers the facility to preselect several coining blows in order to improve surface quality. Up to four coining blows can be pre-selected at the control desk. In contrast to standard minting presses, the dial feed plate is driven in this press type by a servo motor in conjunction with an indexing gear. The dial feed plate comes to a standstill for execution of the coining impacts. The same applies to the ejector. To manufacture coins with a particularly high standard of surface quality, the press is equipped with a manual feeding device. The press also provides a suitable facility for punch cleaning. The machine produces within a diameter range of 14 to 50 mm, and reaches stroking rates of 400/min when a single coining blow is used, and between 75 and 150/min in multiple blow operation.

◀ **Fig. 6.8.25**
Titanium nitride-coated coining punches

Universal press for joining, minting and punching

Based on the underlying concept of the vertical 1,500 kN minting press, in addition to the minting of standard, multi-sided and bimetal coins, two further application possibilities are offered:

- The press can be used to separate the rings and centers of used bimetal coins which are being withdrawn from circulation. Separation of the different materials offers clear advantages for recycling.
- Designed for piercing, the press permits the manufacture of outer rings from upset coin blanks.

Normally, *rings for bimetal coins* are produced at high stroking rates in multiple dies on a high-speed blanking press *(cf. Fig. 4.6.6 and 4.6.8)*. Due to the minimal stability of the rings, however, subsequent upsetting is only possible to a limited degree.

Experience has shown that the reduction in diameter which takes place here lies around 0.1 to a maximum of 0.2 mm. Due to the low web width, it may be necessary to upset the edge in several passes in order to achieve the specified diameter. In practice, a cylindrical upsetting shape with a 45° chamfer on each side was found to be successful *(Fig. 6.8.17)*. In the case of rings produced on universal presses in mints, however, any optional upsetting shapes and edge decoration can be used if the coin blanks are rimmed or edge lettering is carried out prior to piercing.

Figure 6.8.26 indicates the individual phases required for the piercing process:

- The upset coin blank is centered in the piercing station by the feed insert. The slide is positioned in the bottom dead center.
- The slide moves upwards, the coin blank is held between the female piercing die and the spring-mounted stripper.
- The coin blank is inserted in the holding collar and pressed against the fixed piercing punch.
- At the top dead center, the piercing process is completed, the scrap web is pushed through the piercing punch into the scrap hole of the female piercing die.
- On the return stroke of the slide, the spring-mounted stripper pushes the pierced coin blank out of the holding collar back into the feed insert. The pierced coin blank is discharged by the indexed dial feed plate, while a new upset unpierced coin blank is fed for the next cycle.

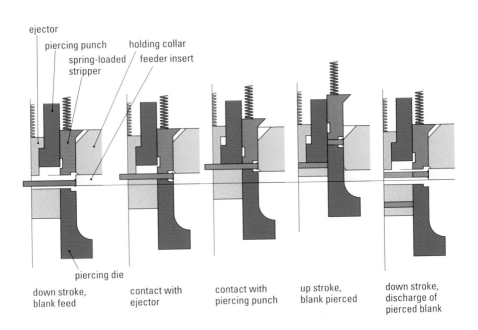

▲ **Fig. 6.8.26** Piercing sequence

Medal minting presses

Presses with a nominal force of 2,000, 3,600 and 10,000 kN are used to produce medals and commemorative coins. Depending on the press type, the maximum medal diameter is between 38 and 80 mm. The stroking rates range from 45 to 120/min. The resulting output depends on the number of coining strokes:

Single stroke:	25 – 40 medals/min
Double stroke:	16 – 30 medals/min
Triple stroke:	12 – 24 medals/min

Complex designs and even difficult medal contours can thus be manufactured in a single stroke. For high-grade medals, knuckle-joint presses are configured for multiple stroke operation, as it is only with repeated coining that the required standard of surface quality can be achieved.

The press frame serves as a slide which is moved by the knuckle-joint, whereby a rigid connection exists between the press bed and the press housing *(cf. Fig. 3.2.5)*. This results in practically shock-free closure of the dies.

Automatic medal minting machines are equipped with mechanical, cam-driven ejectors. The drive cam disk sits directly on the crankshaft and ensures synchronous operation with the press work cycle. For the medal minting process, the ejector is controlled to wait until the number of blows, preselected in the multiple stroke device, has been completed.

For the manufacture of top quality medals, ultra-clean, dust-free operation must be guaranteed. The machines are equipped with ventilating and filtration devices which permit only filtered air to enter the die area. The die mounting area is also enclosed and is slightly pressurized. The highest degree of quality is achieved in medal minting presses using manual feed methods *(Fig. 6.8.27)*.

In order to ensure the quality of the medals, the punches must be cleaned at short intervals. By pressing a button, the ejector moves to the cleaning position, the bottom die is raised around 2 mm out of the minting collar and dirt particles or spilled cleaning materials can be wiped away.

Medals with enhanced surface quality for collectors are often produced in high piece numbers. A cassette feed system is ideally suited for this type of application *(Fig. 6.8.28)*. In view of the stringent quality requirements, an automatic cassette feed system must ensure particularly gentle transport of the medal blanks. Only one blank is removed from the cassette magazine at a time and lowered onto the transfer level. Plastic grippers lift the blank into an intermediate deposit station and transport it to the minting station.

The subsequent blank remains clamped in the medal magazine until the transport area is clear. After the minting process, the grippers lift the minted medal from the bottom die, which is located in the ejection position, and place it on a conveyor belt which transports it out of the minting area. A turning station, connected to the conveyor belt, permits the operator to inspect both sides of the medal during running production.

◀ **Fig. 6.8.27**
Medal minting press with
manual feed

CNC-controlled hydraulic medal minting presses

These presses permit path or pressure-dependent control of each stroke of the slide, as well as synchronous operation of the slide and ejector by two CNC axes. In case of multiple stroke operation, the die remains closed without opening, i. e. in contact with the medal. Only the coining stress between the upper and lower die is released. As a result, penetration of abraded matter between the dies and the coined piece is prevented.

The nominal press force of the machine series ranges between 4,000 and 12,000 kN. In addition to minting of medals, coins and jewellery, the machine can also be used for the cold hobbing of tools, for example in the manufacture of coining punches, die sinks, jewellery moulds, dies for fasteners, dies for cutlery, etc. The die sinking velocity required for this technique ranges between 0.01 and 4 mm/s *(Fig. 6.8.29)*.

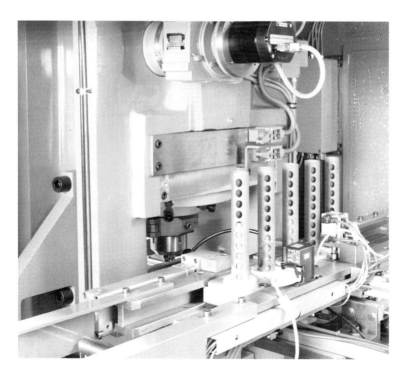

▲ **Fig. 6.8.28** Cassette feed for medal production

To comply with the very highest quality requirements and when producing difficult and complex reliefs, it is possible to use an oscillating technique with varying press forces *(Fig. 6.8.30)*.

The hydraulic double-action press comprises a press frame with a closed frame structure, an integrated drive cylinder and separate power drive. The press slide (coining cylinder) is CNC-controlled and has a multiple-surface design.

Based on the principle of differential pressure surfaces, optimum levels of output are achieved with low energy consumption. The ejector cylinder, also controllable by CNC, is integrated underneath in the press frame. The cheeks of the press yoke are positioned directly against the die system and are slightly larger than the piston rod, so guaranteeing optimum rigidity. The traversing motions are controlled by means of servo control valves with incremental setpoint inputs and mechanical actual value feedback.

▲ **Fig. 6.8.29** Hydraulic medal minting press

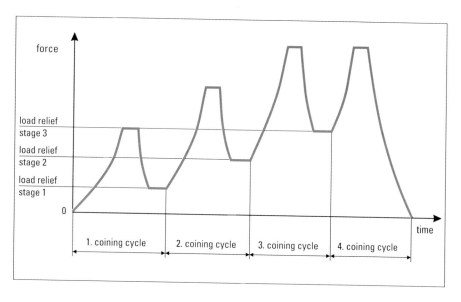

▲ **Fig. 6.8.30** Coining cycles when using multiple stroke with programmable forces and times

Bibliography

Beisel, W.: Aufwand reduzieren, feine Gravuren durch Prägen, Industrieanzeiger (1995) 10.

Beisel, W. , Stahl, K.: Kaltfließpressen als Alternative, Werkstatt u. Betrieb 128 (1995) 3.

DIN 1543 : Steel flat products; hot rolled plate 3 to 150 mm thick; permissible deviations of dimenison, weight and form, Beuth Verlag, Berlin (10.81).

DIN 1544: Flat steel products; cold rolled steel strip, dimensions, permissible variations on dimensions and form, Beuth Verlag, Berlin (8.75).

DIN 17 006 : Stahlguß, Grauguß, Hartguß, Temperguß, Beuth Verlag, Berlin (10.49).

DIN 17 100 : Steels for general structural purposes; Quality standard, Beuth Verlag, Berlin (9.66).

DIN 17 200 : Steels for quenching and tempering, technical delivery conditions, Beuth Verlag, Berlin (12.69).

DIN 17 210 : Case hardening steels, technical delivery conditions, Beuth Verlag, Berlin (12.69).

DIN 17 240 : Heat resisting and highly heat resisting materials for bolts and nuts; Quality specifications, Beuth Verlag, Berlin (1.59).

DIN 17 440 : Stainless steels; technical delivery conditions for plate and sheet, hot rolled strip, wire rod, drawn wire, steel bars, forgings and semi-finished products, Beuth Verlag, Berlin (12.72).

Geiger, R.: Stand der Kaltmassivumformung - Europa im internationalen Vergleich, Neuere Entwicklungen in der Massivumformung, DGM Verlag, Oberursel (1993).

Hoffmann, H., Rahn, O., Waller, E., Blaschko, A.: Prozeßüberwachung an Maschinen der Kaltmassivumformung, Industrie Anzeiger (1988) 33.

Körner, E., Knödler, R.: Möglichkeiten des Halbwarmfließpressens in Kombination mit dem Kaltfließpressen, Umformtechnik 26 (1992) 6.

Körner, E., Schöck, J. A.: Rationalisierungspotentiale im Rahmen der Werkzeug- und Presseninbetriebnahme, Umformtechnik 26 (1992) 6.

Körner, E., u.a.: Mechanische Fließpreßanlagen für die Kalt- und Halbwarmumformung, Neuere Entwicklungen in der Massivumformung, DGM Verlag, Oberursel (1995).

Kudo, H., Takahashi, A.: Extrusion technology in the Japanese automotive industry, VDI Berichte Nr. 810, VDI-Verlag, Düsseldorf (1990).

Lange, K.: Umformtechnik, Volume 2: Massivumformung, Springer-Verlag, Heidelberg (1988).

Schwebel, R.: Prägen mit minimalen Toleranzen, Bleche Rohre Profile (1995) 4.

Sheljaskow, Sh., Dübler, A.: Entwicklungsstand und Einsatzbereich der Halbwarmumformung, Verein deutscher Werkzeugmaschinenfabriken e. V. (VDW) Study (1991).

Stahlschlüssel-Taschenbuch, Verlag Stahlschlüssel, Wegst GmbH, Marbach/N.

VDI 3138 Bl. 1 : Kaltfließpressen von Stählen und NE-Metallen, Arbeitsbeispiele, Wirtschaftlichkeit, Beuth Verlag, Berlin (10.70).

VDI 3138 Bl. 2 : Kaltfließpressen von Stählen und NE-Metallen, Anwendung, Beuth Verlag, Berlin (10.70).

VDI 3138 Bl. 1 : Kaltfließpressen von Stählen und NE-Metallen, Grundlagen, Beuth Verlag, Berlin (10.70).

VDI 3176 : Vorgespannte Preßwerkzeuge für das Kaltumformen, Beuth Verlag, Berlin (10.86).

VDI 3186 Bl. 1 : Werkzeuge für das Kaltfließpressen von Stahl, Aufbau, Werkstoffe (ICFG Data Sheet 4/70), Beuth Verlag, Berlin (12.71).

VDI 3186 Bl. 2 : Werkzeuge für das Kaltfließpressen von Stahl, Gestaltung, Herstellung, Instandhaltung von Stempeln und Dornen (ICFG Data Sheet 5/70), Beuth Verlag, Berlin (12.71).

VDI 3186 Bl. 3 (6.74): Werkzeuge für das Kaltfließpressen von Stahl, Gestaltung, Herstellung, Instandhaltung, Berechnung von Preßbüchsen und Schrumpfverbänden (ICFG Data Sheet 6/72), Beuth Verlag, Berlin und Cologne.

VDI 3187 Bl. 3 : Scherschneiden von Stäben, Schneidfehler (ICFG Data Sheet 5/71), Beuth Verlag, Berlin (8.73).

VDI 3198 : CVD-PVD-Verfahren, Beuth Verlag, Berlin (9.91).

VDI 3200 Bl. 1 : Fließkurven metallischer Werkstoffe, Grundlagen, Beuth Verlag, Berlin (10.78).

VDI 3200 Bl. 1 : Fließkurven metallischer Werkstoffe, NE-Metalle, Beuth Verlag, Berlin (10.78).

VDI 3200 Bl. 2 : Fließkurven metallischer Werkstoffe, Stähle, Beuth Verlag, Berlin (4.82).

Verson, M. D.: Impact machining, Verson Allsteel Press Company, Chicago (1969).

Zebel, H.: Die Trennflächen von abgescherten Stababschnitten, Werkstattstechnik (1964) 54.

Index and information about the SCHULER Group

Schuler GmbH (Holding)
Postfach 1222
73012 Goeppingen/Germany
Phone +49-71 61-66-0
Fax +49-71 61-66-850

A global leader in machine tool production

With 4.000 employees and production facilities in Europe, America and Asia, the Schuler Group is a major player among the world's machine tool manufacturers. Founded in 1839, the production of dies and sheet metal working machines has been the heart of the company's business since the mid-19th century. This early

specialization in metal forming has ensured the company's growth and success up to the present day.

With its core competence in the fields of sheet metal forming, hydroforming and solid forming, the Schuler Group provides tailor-made, turn-key solutions for the whole metal-working industry. The Group's global presence and network of international manufacturing facilities guarantee the successful realization of cost-effective engineering projects throughout the world.

**Schuler Pressen
GmbH & Co**
Postfach 929
73009 Goeppingen/Germany
Phone +49-71 61-66-0
Fax +49-71 61-66-233

Expertise in press shop planning and equipping

To meet the demands of an internationally-oriented automotive industry with varying production structures and conditions, Schuler Presses offers in its Large Mechanical Presses product line a full range of leading-edge metal forming equipment. The range consists of cut-to-length blanking

presses for the production of blanks, large-panel transfer presses with crossbar or tri-axis transfer, automated press lines for auto body panels and in addition, transfer presses for ready-to-install vehicle component production as well as equipment for presetting of tooling and transfer simulation. Technical know-how and a wealth of experience guarantee cost-effective solutions.

SCHULER PRESSES

**Schuler Pressen
GmbH & Co**
Postfach 929
73009 Goeppingen/Germany
Phone +49-7161-66-0
Fax +49-7161-66-233

Equipment for stamping, blanking, solid forming and minting

From high-speed blanking lines to flexible notching machines, the Standard Machines product line of Schuler Presses offers a wide range of tailor-made plant concepts for both large-scale

production and individual component manufacturing.

A further key component of the range are the production lines, dies and process technologies for solid metal forming. In this sector Schuler is also able to draw on the expertise of the Liebergeld GmbH & Co, Nuremberg.

Schuler is the leading supplier of minting equipment in the world.

THE SCHULER GROUP

Pfleghar GmbH & Co
Heinrich-Hertz-Str. 6
88250 Weingarten/Germany
Phone +49-751-4006-0
Fax +49-751-4006-190

Schuler Werkzeuge GmbH & Co
Postfach 1129
73011 Goeppingen/Germany
Phone +49-7161-66-0
Fax +49-7161-66-622

Complete die technology – from engineering to full production launch

With Pfleghar Engineering and Schuler Tool & Die, the Schuler Group covers the process chain from planning of auto body subassemblies to production dies. From component design, computer-

The main focus of Schuler Tool & Die is the production of dies with continuous data transfer from planning through design, manufacturing, testing and prototype part production to full production launch.

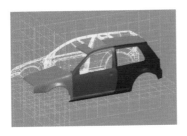

aided process planning, simulation, prototyping and 2D or 3D ExactSolid design to complex foundry patterns, Pfleghar offers a complete range of services from a single source.

SCHULER CASTINGS

Schuler Guss GmbH & Co
Postfach 1309
73013 Goeppingen/Germany
Phone +49-71 61-66-0
Fax +49-71 61-66-616

Gray and nodular cast iron for high-quality requirements

With the aid of modern casting facilities and large-scale capacities, Schuler's foundry is able to produce more high-grade cast iron than is required by the Schuler Group and can therefore offer its products on the open market.

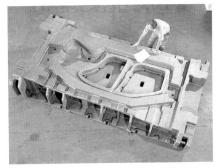

SMG PRESSES

**SMG Süddeutsche
Maschinenbau
GmbH & Co**
Louis-Schuler-Str. 1
68753 Waghaeusel/Germany
Phone +49-72 54-988-0
Fax +49-72 54-988-214

The leader in hydraulic press technology

SMG offers hydraulic press systems as cost-effective engineering solutions for international customers in the automotive and household appliance industries and their suppliers. The company's presses are ideally suited for forming operations requiring variable parameters of press force and slide speeds. In the field of sheet metal forming this includes presses for blanking, deep

For solid forming processes, hydraulic presses are especially suited for the manufacture of long shafts, axles and hollow cylinders. In addition, SMG presses are used in the processing of reinforced plastics and for the precision fine blanking technology of Feintool AG, Switzerland.

drawing, transfer as well as hemming presses and press centers for die try-out.

**SMG Engineering
für Industrieanlagen
GmbH & Co**
Louis-Schuler-Str. 1
68753 Waghaeusel/Germany
Phone +49-7254-988-411
Fax +49-7254-988-409

Partner for complete manufacturing systems and turnkey production facilities

As a subsidiary of SMG Presses, SMG Engineering provides everything from planning to realization of complete manufacturing systems and turnkey production facilities.

SMG Engineering services include technical planning of the manufacturing process, specification of the production equipment, preparation of the plant layout, design and engineering of equipment for metal forming, coordination with all sub-contractors as well as the on-site start-up of all equipment. Expertise and efficient project management methods ensure that deadlines are met and budgets not exceeded.

SMG Engineering is founded on lean production principles and characterized by cost-efficiency and productivity.

**Hydrap Pressen
GmbH & Co**
Postfach 249
73652 Pluederhausen/Germany
Phone +49-71 81-80 06-0
Fax +49-71 81-80 06-36

Hydraulic presses for special applications

Hydrap Presses, a subsidiary of SMG Presses, is specialized in the manufacture of hydraulic presses between 630 and 20.000 kN. The company provides production lines for exhaust emission technology, the household appliance and electrical industry as well as presses for the manufacture of sintered parts, seals and radiators.

**Schäfer Hydroforming
GmbH & Co**
Auf der Landeskrone 2
57234 Wilnsdorf/Germany
Phone +49-27 39-808-0
Fax +49-27 39-808-110

Hydroforming
opens new markets

One SMG Presses subsidiary, Schäfer Hydroforming, specializes in the development of processes and production systems which utilize high-pressure hydroforming technology. Hydroforming enables the forming of complex-shaped hollow components by inflating, compressing and expanding metal tubes.

Schäfer supplies machines and equipment for the manufacture of fittings in all sizes and shapes for use in pipe applications.

Schäfer technology is used in the automotive industry to produce parts such as exhaust components, integrate structural components, dashboard panel reinforcements as well as components for tubular frame construction.

From prototyping through die manufacture, production equipment to volume production Schäfer Hydroforming, together with SMG Presses, is able to provide a full range of cost-effective and efficient services.

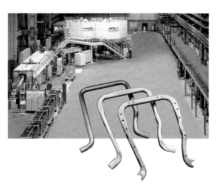

**GMG Automation
GmbH & Co**
Louis-Schuler-Str. 9
75050 Gemmingen/Germany
Phone +49-7267-809-0
Fax +49-7267-809-180

Press shop automation, hemming and welding technology, machine tool automation

GMG Automation produces automation components for the automotive industry and its suppliers, for machine tool manufacturers and the household appliance

industry. For the automotive industry, GMG supplies press feeder and robot automation, blank destackers/loaders, blank washers, panel racking systems and blank stack turnover devices. For the automation of cutting-type machine tools, the company's

product line includes linear storage units for flexible manufacturing systems as well as circular pallet storage devices and milling head changers. In the field of hemming

and welding technology, GMG supplies laser applications and machinery for the manufacture of tailored blanks.

**Schleicher Automation
GmbH & Co**
Louis-Schuler-Str. 1
91093 Hessdorf/Germany
Phone +49-91 35-715-0
Fax +49-91 35-715-103

Competent partner
for coil feeding, slitting,
stacking and roll forming

Schleicher Automation is a leading producer in the field of coil processing and roll forming equipment.

Conventional and compact coil lines, as well as high-performance

cut-to-length lines with washing machine, oscillating shear and stacking units, made by Schleicher

Automation are used in the automotive industry, the household appliance industry and by steel service centers.

The company's high-performance roll forming lines are supplied

throughout the world for the production of such diverse products as lamp housings and garage doors.

GRÄBENER PRESSES

Gräbener Pressensysteme
GmbH & Co. KG
Wetzlarer Str. 1
57250 Netphen/Germany
Phone +49-27 37-962-0
Fax +49-27 37-962-100

Extrusion and precision embossing systems

Gräbener Presses specializes in
the cold forming of high-quality,
flat workpieces.
At the heart of the product line
for sheet and solid metal forming
are the company's precision
embossing, sizing and solid form-

ing presses. Using tried and trusted
knuckle-joint technology, these
machines guarantee a particularly
high standard of precision.
In addition, the company supplies
presses for the production of
radiator panels.

THE SCHULER GROUP

Spiertz Presses S. A.
BP 26
67023 Strasbourg
Cedex 1/France
Phone +33-3-88-65-70-80
Fax +33-3-88-40-34-90

Schuler Incorporated
2222, South Third Street
Columbus,
Ohio 43204-2494, USA
Phone +1-614-443-9481
Fax +1-614-443-9486

Production facilities in France, the USA, Brazil and China

Spiertz Presses, a member of the Schuler Group, is the leading supplier of press shop equipment for the automotive industry in France.

Located in Columbus/Ohio and Dearborn/Michigan, Schuler Incorporated provides the Schuler Group with an effective organization at the heart of the American automotive industry.

Prensas Schuler S. A.
Caixa Postal 136
09901-970 Diadema-SP, Brazil
Phone +55-11-7458444
Fax +55-11-7452766

Shanghai
Schuler Presses Co. Ltd.
No. 8, Lane 750
East Luo Chuan Road
Shanghai 200 072, PR China
Phone +86-21-56650081
Fax +86-21-56036930

Tianjin SMG Presses Co. Ltd.
11, Bohai Road, Teda
Tianjin 300457 PR China
Phone +86-22-25326933
Fax +86-22-25322960

Prensas Schuler, Brazil, is by far the most significant producer of mechanical and hydraulic presses in Latin America.

A key member in Schuler's inter-national manufactoring network, Prensas Schuler is equipped with leading-edge technology and capable of supplying even the largest press lines and transfer presses for the automotive industry.

The Schuler Group is represented in the fast-growing Chinese market by its two joint-venture companies – Shanghai Schuler Presses and Tianjin SMG Presses.

THE SCHULER GROUP

Schuler GmbH (Holding)
Postfach 1222, 73012 Goeppingen/Germany
Phone +49-71 61-66-0, Fax +49-71 61-66-850

Schuler Pressen GmbH & Co
Postfach 929, 73009 Goeppingen/Germany
Phone +49-71 61-66-0, Fax +49-71 61-66-233

Schuler Werkzeuge GmbH & Co
Postfach 1129, 73011 Goeppingen/Germany
Phone +49-71 61-66-0, Fax +49-71 61-66-622

Pfleghar GmbH & Co
Heinrich-Hertz-Str. 6, 88250 Weingarten/Germany
Phone +49-751-40 06-0, Fax +49-751-40 06-190

Schuler Guss GmbH & Co
Postfach 1309, 73013 Goeppingen/Germany
Phone +49-71 61-66-0, Fax +49-71 61-66-616

SMG Süddeutsche Maschinenbau GmbH & Co
Louis-Schuler-Str. 1, 68753 Waghaeusel/Germany
Phone +49-72 54-988-0, Fax +49-72 54-988-214

SMG Engineering für Industrieanlagen GmbH & Co
Louis-Schuler-Str. 1, 68753 Waghaeusel/Germany
Phone +49-72 54-988-411, Fax +49-72 54-988-409

Schäfer Hydroforming GmbH & Co
Auf der Landeskrone 2, 57234 Wilnsdorf/Germany
Phone +49-27 39-808-0, Fax +49-27 39-808-110

Hydrap Pressen GmbH & Co
Postfach 249, 73652 Pluederhausen/Germany
Phone +49-71 81-8006-0, Fax +49-71 81-8006-36

GMG Automation GmbH & Co
Louis-Schuler-Str. 9, 75050 Gemmingen/Germany
Phone +49-72 67-809-0, Fax +49-72 67-809-180

Schleicher Automation GmbH & Co
Louis-Schuler-Str. 1, 91093 Hessdorf/Germany
Phone +49-91 35-715-0, Fax +49-91 35-715-103

Liebergeld GmbH & Co
Edisonstr. 43, 90431 Nuremberg/Germany
Phone +49-911-961-77-0, Fax +49-911-961-77-30

Gräbener Pressensysteme GmbH & Co KG
Wetzlarer Str. 1, 57250 Netphen/Germany
Phone +49-27 37-962-0, Fax +49-27 37-962-100

Spiertz Presses S. A.
BP 26, 67023 Strasbourg Cedex 1/France
Phone +33-3-88-65-70-80, Fax +33-3-88-40-34-90

Schuler Incorporated
2222 South Third Street
Columbus, Ohio 43207-2494/USA
Phone +1-614-443-9481, Fax +1-614-443-9486

Detroit Office
15300 Commerce Drive North, Suite 104
Dearborn, Michigan 48120-1222/USA
Phone +1-313-441-0084, Fax +1-313-441-2498

Prensas Schuler S. A.
Caixa Postal 136
09901-970 Diadema-SP/Brazil
Phone +55-11-7458444, Fax +55-11-7452766

Shanghai Schuler Presses Co. Ltd.
No. 8, Lane 750, East Luo Chuan Road
Shanghai 200 072/PR China
Phone +86-21-56650081, Fax +86-21-56036930

Tianjin SMG Presses Co. Ltd.
11, Bohai Road, Teda, Tianjin 300457/PR China
Phone +86-22-25326933, Fax +86-22-25322960

Schuler Iberica S. A.
Avda. de Roma, 2 - 4, 08014 Barcelona/Spain
Phone +34-93-4231421, Fax +34-93-4243408

Schuler India Pvt. Ltd.
201, Madhava, Bandra-Kurla Complex
Bandra (East), 400 051 Mumbai (Bombay)/India
Phone +91-22-6450552, Fax +91-22-6450549

SMG Engineering (Malaysia) Sdn. Bhd.
Suite 8.01, 8th Floor, IGB Plaza
Jln. Kampar, off Jln. Tun Razak
50400 Kuala Lumpur/Malaysia
Phone +60-3-441-4441, Fax +60-3-441-1171

SMG Engineering (Thailand) Co. Ltd.
703 Ratchada Suite Building, 7th Floor
Wongsawang Rd. Bang Sue
Bangkok 10800/Thailand
Phone +66-2-9107076, Fax +66-2-9108408

On the Internet:
http://www.schulergroup.com
e-mail: info@schulergroup.com